DATE DUE FOR RETURN

This book may be recalled before the above date.

Plants that Hyperaccumulate Heavy Metals

Plants that Hyperaccumulate Heavy Metals

their Role in Phytoremediation, Microbiology, Archaeology, Mineral Exploration and Phytomining

Edited by

Robert R. Brooks
Emeritus Professor of Geochemistry
Department of Soil Science
Massey University
Palmerston North, New Zealand
(E-mail: r.brooks@massey.ac.nz)

CAB INTERNATIONAL

CABI *Publishing* is a division of CAB *International*

CABI Publishing
CAB International
Wallingford
Oxon OX10 8DE
UK

Tel: +44 (0)1491 832111
Fax: +44 (0)1491 833508
Email: cabi@cabi.org
Web site: http://www.cabi.org

CABI Publishing
10 E. 40th Street
Suite 3203
New York, NY 10016
USA

Tel: +1 212 481 7018
Fax: +1 212 686 7993
Email: cabi-nao@cabi.org

A catalogue record for this book is available from the British Library, London, UK.
A catalogue record for this book is available from the Library of Congress, Washington DC, USA.

ISBN 0 85199 236 6

First printed 1998
Reprinted 2000

Printed and bound in the UK at the University Press, Cambridge, from copy supplied by the editor.

Contents

Contributors

Frances A. Bennett, BSc
Postgraduate Student
Department of Soil Science
Massey University
Palmerston North
New Zealand

Thomas J. Beveridge, PhD
Professor of Microbiology
Department of Microbiology
College of Biological Science
University of Guelph
Guelph, Ontario
Canada N1G 2W1

Robert S. Boyd, PhD
Professor of Botany
Department of Botany and
Microbiology
Auburn University
Auburn
AL 36849-5407
USA

Robert R. Brooks, DSc
Emeritus Professor
Department of Soil Science
Massey University
Palmerston North
New Zealand

Michael F. Chambers
Retired Researcher
PO Box 13511
Reno
NV 89507
USA

Alessandro Chiarucci, PhD
Herbarium Director
Dipartimento di Biologia Ambientale
Università degli Studi di Siena
Siena
Italy

Colin E. Dunn, PhD
Head of Geochemical Research
Geological Survey of Canada
601 Booth St
Ottawa
Canada K1A OE8

Paul E.H. Gregg, PhD
Associate Professor
Department of Soil Science
Massey University
Palmerston North
New Zealand

Tanguy Jaffré, DSc
Director of Research
ORSTOM
BP A5
Nouméa
New Caledonia

Steven P. McGrath, PhD
IACR-Rothamsted
Harpenden
Herts AL5 2JQ
UK

Larry J. Nicks, MS
Retired Researcher
PO Box 650
Fernley
NV 89408
USA

Brett H. Robinson, MSc
Postgraduate Student
Department of Soil Science
Massey University
Palmerston North
New Zealand

Robert B. Stewart, PhD
Senior Lecturer
Department of Soil Science
Massey University
Palmerston North
New Zealand

Emily K. Tyler, BSc
Department of Soil Science
Massey University
Palmerston North
New Zealand

Acknowledgements

Acknowledgement is gratefully made to the following sources for kind permission to reproduce material

American Chemical Society for Figures 9.4, 9.5, 9.6, 15.2, 15.3 and 15.4.
American Nuclear Society and the Argonne National Laboratory for Figure 9.8.
Elsevier Science, Kidlington for Figures 7.2 and 7.14.
Elsevier Scientific Publishing Co., Amsterdam for Figures 4.4, 4.5, 13.8 and 15.1.
Gustav Fischer Verlag, Stuttgart for Figure 7.7.
Intercept Ltd., Andover for Fig.4.1.
Journal of Environmental Quality for Figure 13.9.
Journal of Plant Physiology for Table 2.14 and Figure 13.5.
Kluwer Academic Publishers, Dordrecht for Figures 13.6 and 13.7 and Table 4.3.
Marcel Dekker, New York for Figure 13.10 and Tables 13.4 and 13.5.
National Academy of Sciences, Washington for Figure 13.1.
Nature Publishing Co., New York for Figure 9.7.
New Phytologist for Table 12.10.
Oikos for Fig.3.3.
Prentice Hall Europe for Tables 2.3 and 3.4.
Royal Society of London for Figure 3.17.
Società Toscana di Scienze Naturale, Pisa for Table 2.4.
Soil Science Society of America for Figure 13.4.

Preface

The past twenty years have witnessed the extraordinary increase of interest in plants that hyperaccumulate heavy metals. My colleagues and I first introduced the term "hyperaccumulation" in a 1977 paper and it must be admitted that we did not then realise the full potential of these plants. Since then, these unusual species have found a ready application in such diverse fields as phytochemistry, archaeology, mineral exploration, ecology, phytoremediation and phytomining.

This is the first book to appear in which the various uses and disciplines have been summarised within a single volume. Of particular interest is the field of *phytoextraction* with its subdivisions of phytoremediation and phytomining. The latter is not proven at this stage, but the former is now well established and there are companies that are now offering a commercial service as an alternative to the older more expensive classical techniques such as soil removal or acid leaching of pollutants.

This work was only made possible by the important contributions of my students and colleagues as well as by other authors of the chapters contained therein. I am grateful to them all.

The chapter authors represent a wide range of disciplines extending from chemistry to microbiology and each contributor has been able to imprint a personal viewpoint of the way in which hyperaccumulator plants have served to enhance or initiate a particular research field.

I do not consider that either myself or my colleagues and former students can be regarded as the catalysts for the emergence of the manifold disciplines derived from the study of hyperaccumulators. The due acknowledgement is to three pioneering scientists. The first of these was A. Baumann, who in 1885 analysed specimens of *Viola calaminaria* and *Thlaspi calaminare* growing over the calamine deposits of Aachen, Germany. He found over 1% (dry weight) zinc in these plants and recorded the first of the hyperaccumulators of any heavy metal. In the 1930s O.A. Beath and his coworkers discovered hyperaccumulation of selenium in *Astragalus* plants from the western USA.

Finally, credit must be given to the late O. Vergnano Gambi from the University of Florence, who, with C. Minguzzi in 1948 discovered the unusual hyperaccumulation of nickel by the Tuscan serpentine plant *Alyssum bertolonii*.

None of the above discoveries excited much interest at the time, and even when we introduced the term hyperaccumulator in 1977, another decade was to pass before much attention was paid to the possible application of hyperaccumulators in diverse fields. This renewed interest was stimulated by the "green" movements of the past 20 years that have caused nations and commercial companies to examine the possibility of using plants to remediate our polluted environment.

Although about half of this book has been devoted to the use of hyper-accumulator plants for phytoextraction, some attention has also been paid to the "mirror image" of *hypervolatilisation* of such elements as mercury, boron, arsenic and particularly selenium. This is a field that has a potential for much future progress, albeit limited to locations that have an excess of the above elements.

It is to be hoped that this book will be a forerunner of many more dealing with the scientific and commercial uses of plants that hyperaccumulate heavy metals and if this turns out to be the case, then my colleagues and I will be more than satisfied.

Robert R. Brooks

Palmerston North
August 1997

Chapter one:

General Introduction

R.R. Brooks
Department of Soil Science, Massey University, Palmerston North, New Zealand

Historical Introduction

A small perennial shrub in Tuscany, Italy was destined to lead the way to a whole range of new technologies and discoveries in a wide gamut of seemingly unrelated disciplines such as archaeology, mineral exploration, entomology, phyto-chemistry, land restoration and mining. It all began when a 16th century Florentine botanist named Andrea Cesalpino (Fig.1.1) reported that he had noticed the ubiquitous presence of an *"alyson"* growing over *"black stones"* (ultramafic rocks) in the Upper Tiber Valley in Tuscany (Cesalpino, 1583).

The plant was later described by Desvaux (1814) who named it *Alyssum bertolonii*. Its place in the context of the wider serpentine vegetation of Tuscany was established 30 years later by Amidei (1841). The definitive work on the serpentine flora of Tuscany is by Pichi Sermolli (1948) who summarised his own work and that of other workers in an exhaustive treatise on the subject.

Before proceeding further, it may be appropriate to mention an important matter of terminology. *Serpentine* [$Mg_3Si_2(OH)_5$] is in fact a common rock-forming mineral found predominantly in ultramafic (high magnesium and iron) rocks (formerly known as ultrabasic rocks). The distinctive stunted vegetation growing over soils derived from these rocks was (and still is) known by botanists as *serpentine vegetation*. This term still persists despite modern use of the more accurate term *ultramafic vegetation*, and will be used throughout this book.

The year 1948 was significant not only for the publication of the Pichi Sermolli book, but also for the appearance of an initially unrecognised paper by another Florentine couple, C. Minguzzi and O. Vergnano. This paper (Minguzzi and Vergnano, 1948) described the unusual accumulation of nickel by *Alyssum bertolonii* from the Impruneta region near Florence. They found up to 0.79 % (7900 μg/g) nickel in dried leaves of plants growing in soils containing only 0.42 % of this element. On an ash-weight basis the leaves contained 9.21 % nickel.

The original paper by Minguzzi and Vergnano excited little interest and even the discovery of a second "nickel plant", *Alyssum murale* by a Russian scientist (Doksopulo, 1961) excited no ripple in the scientific community. Another decade passed before Severne and Brooks (1972) and Cole (1973) reported the unusual accumulation of nickel by the West Australian species *Hybanthus floribundus*. The plant was found to contain up to 1.38% nickel in soils containing typically only 0.07-0.10% of this element.

Fig.1.1. The 16th century Florentine botanist A. Cesalpino.

Three years later, Jaffré and Schmid (1974) reported an almost unbelievable maximum of 4.75% nickel in dry leaves of *Psychotria douarrei* from New Caledonia. This was followed a year later by a paper by Kelly *et al.* (1975) in which they reported unusual accumulation of nickel by two more species from New Caledonia, *Hybanthus caledonicus* and *H.austrocaledonicus*.

One year later, Jaffré *et al.* (1976) reported an unusual tree from New Caledonia which contained a blue sap (*sève bleue*) that was later found to contain about 10% nickel in the fresh material and 20% in the dried sap. This tree is illustrated in Fig.3.6 in Chapter 3. This was the most unusual discovery so far made in the field of "nickel plants". Dr Jaffré later showed the tree to myself and a colleague, Dr R.D.Reeves. We commented somewhat wryly among ourselves "... if only it had been gold..." It was left to others to see the real potential of our discoveries.

In 1977, only seven "strong accumulators" of nickel had been reported and then in 1977, we analysed some 2000 specimens of *Hybanthus* and *Homalium* supplied by herbaria throughout the world (see Chapter 4). This work (Brooks *et al.*,1977) resulted in the rediscovery of all of the then known plants containing >1000 μg/g nickel (dry weight) as well as the discovery of five more. In this paper we coined the term *hyperaccumulator* to describe plants containing >1000 μg/g (0.1%) nickel in dry material. This term is not completely arbitrary as it represents a concentration about 100 times greater that the highest values to be expected in non-accumulating plants growing over serpentine.

Fig.1.2. Ornella Vergano Gambi seated among serpentine vegetation at Monteferrato near Florence, Italy. The plant near her right knee is *Alyssum bertolonii*, identified by her as containing about 1% nickel in its dry mass. Photo by R.R.Brooks.

The late Dr Vergnano Gambi (Fig.1.2) had not been able to analyse significant quantities of herbarium material because the analytical techniques of the time required virtually the whole of the specimen. The availability of modern sensitive analytical techniques such as atomic absorption spectrometry (see Chapter 2) now permits the use of very small samples (milligram quantities) that can result in the determination of a wide range of elements at the μg/g (ppm) or even ng/g (ppb) level. This enabled the original work of Brooks *et al.* (1977) in which 2000 very small herbarium samples were analysed.

Following upon the identification of two *Alyssum* species that were hyperaccumulators of nickel, Brooks *et al.* (1979) then analysed all except one of the 168 recognised species of *Alyssum* (Dudley, 1964). This work resulted in the identification of 45 hyperaccumulators of nickel of which 31 were identified for the first time. Since the latter study, the number of known species of hyperaccumulators of nickel has climbed to nearly 290 including a recently discovered 20 in Cuba (R.D.Reeves - pers. comm. 1997), perhaps the "last frontier" of discovery. The total may well reach 400 at some time in the future but the rate of discovery will certainly slow down. The reader is referred to Chapter 3 for a more detailed discussion of hyperaccumulators of nickel and other elements such as copper, cobalt, selenium and zinc.

Disciplines Initiated or Arising from Discovery of Hyperaccumulator Plants

Numbers of known hyperaccumulators

It may first be appropriate to summarise the number of known hyperaccumulators for each element (Table 1.1) and the families in which they are found predominantly. See Table 3.2 (Chapter 3) for a listing of threshold concentrations that delineate hyperaccumulation.

Table 1.1. Numbers of known plant hyperaccumulators for eight heavy metals and the families in which they are most often found.

Element	No.	Families
Cadmium	1	Brassicaceae
Cobalt	26	Lamiaceae, Scrophulariaceae
Copper	24	Cyperaceae, Lamiacaeae, Poaceae, Scrophulariaceae
Manganese	11	Apocynaceae, Cunoniaceae, Proteaceae
Nickel	290	Brassicaceae, Cunoniaceae, Euphorbiaceae, Flacourtiaceae, Violaceae
Selenium	19	Fabaceae
Thallium*	1	Brassicaceae
Zinc	16	Brassicaceae, Violaceae

*Leblanc *et al.* (1997).

A word of caution must be sounded in respect of Table 1.1. The number of known hyperaccumulators may not present a picture of the total that exist for any one element. The table largely reflects the amount of effort that has been expended in identifying plants that hyperaccumulate a given element. This is particularly true for nickel. The number of "manganese plants" may well exceed that for nickel, but there has been little interest in the former.

When the number of known hyperaccumnulators started to increase in the 1970s and 1980s, few would have thought that they were anything but a scientific curiosity until scientists in a wide range of disciplines perceived their potential use and set in train an extraordinary range of different studies in a variety of disciplines. Some of these studies have enormous potential for remediation of the environment. The rest of this chapter will be devoted to a brief introduction to the disciplines that have been stimulated by the discovery of hyperaccumulator plants.

Phytochemistry and studies of plant tolerance to heavy metals (Chapters 2 and 12)

Until the past two decades, phytochemical studies were largely restricted to radiochemical investigations in which trace elements were added to the plants as radioactive isotopes and then counted in selected tissues and organs. This was because the amounts extracted by the plant were quite small and often below the limits of detection of the analytical methods employed at that time. Even if the target elements could be determined analytically, it was seldom possible to isolate sufficient of the compounds with which the metals were usually complexed.

When scientists began to study hyperaccumulators they were, for the first time, able to isolate milligram quantities of the complexes and in elucidating their composition and structure, were able to understand more fully the processes involved in metal accumulation and the tolerance of plants to these elements.

Tolerance studies have also been very important because it follows that a plant that has an inordinately high metal content, must by definition, also be very tolerant to the element concerned. Such plants should have an important role to play in the revegetation of terrain polluted by heavy metals.

Evolutionary aspects of hyperaccumulation of elements by plants is a subject of great interest to scientists because therein lies the answer to possible ways in which hyperaccumulating genes might be introduced into plants of greater biomass, or selection be made of wild plants to produce strains with greater biomass and metal-accumulation capacity.

Mineral exploration (Chapters 3 and 4)

Hyperaccumulators of heavy metals are usually endemic to a given type of geological substrate and their actual presence is a sure indication that a specific type of rock or mineralisation is present. There are several nickel hyperaccumulators that are confined to ultramafic rocks (such as nearly 50 *Alyssum* species). These however only indicate rock type and not mineralization within it.

The well-known colonisers of base metal (copper-lead-zinc) deposits such as *Viola calaminaria*, *Thlaspi calaminare,* and *T. caerulescens* are faithful indicators of mineralisation and have been as such for over 100 years.

Some plants, such as the copper indicator *Becium homblei* from Central Africa were used in the 1950s to delineate copper deposits in the Zambian Copper Belt.

Perhaps one of the most interesting examples of ore indicators has been shown by species of *Astragalus*. Although these plants are indicators and hyperaccumulators of selenium, this element is associated geochemically with uranium in the mineral *carnotite*. Scientists at the U.S. Geological Survey plotted the distribution of *Astragalus* in the Colorado Plateau and thereby indicated indirectly the presence of uranium mineralisation (Cannon, 1960, 1964).

Prokaryotic microorganisms and bioremediation (Chapters 6 and 13)

Microorganisms have been used for remediation of the environment for many years. As defined by Bollag and Bollag (1995), *bioremediation* is the use of microorganisms or plants to detoxify a degraded or polluted environment. Strategies include:

    1 - stimulation of the activity of indigenous microorganisms by transforming or degrading many organic pollutants;

    2 - inoculating the site with specific microorganisms;

    3 - application of immobilised enzymes;

    4 - use of plants to remove, contain or transform pollutants.

    Care must be taken in the nomenclature of environmental remediation. By the term *bioremediation* we usually understand that microorganisms will be involved. If plants are to be used, it is more appropriate to use the term *phytoremediation*. This is illustrated in the recent publication by Skipper and Turco (1995) that has the title "Bioremediation Science and Applications". Of the 18 Chapters, only one (Cunningham and Lee, 1995), is devoted to plant-based systems.

    There are some fields in which it is not clear whether microorganisms and/or plants are involved. This is particularly true of volatilisation phenomena in which elements such as arsenic and selenium can be removed from soils by use of plants or the microflora associated with them.

    The term *biomining* is often used erroneously. In its true sense it applies to the use of microorganisms to solubilise sulphide minerals, particularly those of copper, in which organisms such as *Thiobacillus ferrooxidans* oxidise insoluble metal sulphides to their soluble sulphates that can then be leached from a heap of low-grade sulphide ore (Brierley, 1982). It is sometimes applied to the new idea of *phytomining* (Chapters 11,14 and 15) in which a plant is grown to harvest its metal content for economic return.

Phytoarchaeology and hyperaccumulators (Chapter 7)

An unusual use of hyperaccumulators has emerged in recent years in which plants can be used to detect ancient artefacts and trade routes. Brooks and Johannes (1990) have coined the term *phytoarchaeology* to describe this new discipline.

    In Central Africa (Zaïre), ancient African artesans smelted copper in furnaces built over abandoned termite mounds. They brought ore to the sites and after a few years abandoned them. Later the mounds were weathered to ground level but the soil remained poisoned and was colonised by copper hyperaccumulators such as *Haumaniastrum katangense*. Open areas showing a carpet of these plants were later examined by archaeologists who found various native artefacts beneath the surface including copper crosses used as currency at that time.

    Ancient trade routes between Anatolia and Corsica (at the town of Bastia) were identified by a colony of *Alyssum corsicum* over ultramafics at Bastia. The seeds had been brought along with cargoes of wheat and showed the existence of these ancient Venetian trade routes.

A *raison d'être* for hyperaccumulation of metals by plants (Chapter 8)

Boyd and Martens (1992) have examined the question as to why plants hyperaccumulate heavy metals that are usually phytotoxic. The following were some of the hypotheses that they examined:

1 - tolerance to, or disposal of the element from the plant;
2 - a drought resistance strategy;
3 - a means of avoiding competition from less metal-tolerant plants;
4 - inadvertent uptake of heavy metals;
5 - defence against herbivores or pathogens.

The above authors concluded that the last hypothesis was the most likely. They found that hyperaccumulators when grown in non-metalliferous soil were more susceptible to soil fungi and pathogens. In experiments with the "nickel plant" *Streptanthus polygaloides* they found that larvae of the generalist Brassicaceae herbivore *Pieris rapae*, died when fed a diet of field-collected leaves of this plant.

Hyperaccumulation in the aqueous environment (Chapters 5 and 9)

Some of the most extraordinary hyperaccumulations of heavy metals are to be found in the aqueous environment where vascular plants, and algae such as seaweeds, can easily accumulate heavy metals to a level 10,000 times higher than their concentrations in the surrounding water. There are three associated fields of research that are of interest to scientists:

1 - uptake of trace elements by seaweeds can be used to concentrate economically valuable elements such as iodine or to act as indicators of nearby mineralisation (Chapter 5);

2 - floating or rooted fresh-water vascular plants (FVPs) can be used to remove heavy metals from polluted waters (Outridge and Noller, 1991). The water hyacinth is one of the most effective of these because it combines a high degree of extraction together with an appreciably high biomass (Chapter 9);

3 - rhizofiltration (Chapter 9) is a technique whereby terrestrial plants can be grown with their roots in polluted waters and are able to concentrate quite high amounts of selected elements because of their high biomass (Dushenkov *et al.*,1995). Plants such as sunflower (*Helianthus annuus*) and Indian mustard (*Brassica juncea*) seem to have some potential for this purpose.

Phytoremediation (Chapter 12)

It is appropriate that more detailed attention should be paid to the potential use of hyperaccumulators for the purpose of phytoremediation (McGrath *et al.*, 1993) because this subject is one of the most important to be discussed in this book. Cunningham *et al.* (1995) have traced the development of this methodology from its inception in the early 1980s.

Remediation of the environment can be carried out by a number of strategies that either remove the contaminants or stabilise them within the soil. Procedures can include, acid leaching of contaminants, excavation and storage of the soil itself, physical separation of the pollutants, and electrochemical processes (Acar and Alshawabkeh, 1993).

The overriding advantage of phytoremediation is that the procedure is carried out *in situ* and can be very much less expensive than physical methods such as removal of soil (Cunningham *et al.*, 1995). Classical remediation methods can cost around $100,000-$1,000,000 per hectare for *in situ* remediation of water-soluble pollutants (other procedures are much more expensive). Cunningham *et al.* (1995) give an estimate of $200-$10,000 per hectare for phytoremediation techniques.

Phytoremediation is not an entirely new concept and it is difficult to decide who first thought of the idea. One of the first proposals was however, made by Yamada *et al.* (1975) and I can recall a letter to me from Professor K.W.Brown of Texas A & M University who wrote to me in the late 1970s requesting seed of the cobalt hyperaccumulator *Haumaniastrum katangense*. His experiments would have involved using this plant to extract radioactive cobalt from a site in Texas. A few years later, Utsunomiya (1980) took out a Japanese patent on the use of plants to phytoextract cadmium, though hyperaccumulators were not involved.

The possibility of using hyperaccumulators to phytoremediate contaminated soils was examined by Chaney (1983) and later reviewed by Baker and Brooks (1989) who were the first to suggest the possibility of *phytomining* (see Chapters 11, 14 and 15).

Phytoremediation does suffer from a number of limitations that have been summarised by Cunningham *et al.* (1995) as follows:

1 - hyperaccumulators often only accumulate one specific element and have not been found for all elements of interest. However there are several that can hyperaccumulate a pair of elements such as zinc and cadmium by *Thlaspi caerulescens* and copper and cobalt by *Haumaniastrum katangense;*

2 - many hyperaccumulators grow slowly and have a low biomass. There are of course exceptions such as the fast-growing South African *Berkheya coddii* that can have a biomass as high as 20 t/ha;

3 - little is known about the agronomic characteristics of these plants such as their fertiliser requirements and susceptibility to disease and insect attacks.

The ultimate solution to the above problems will lie in crossing these plants with high-yielding high-biomass relatives, or screening wild plants for mutants that could be raised individually to produce progeny with the same enhanced characteristics.

Genetic engineering is another technique that might be applied advantageously to the search for better phytoremediators and has already achieved limited success.

Among the target metals involved in the need for soil remediation are elements such as lead which is usually extremely immobile and not readily accumulated, let alone hyperaccumulated by any plant species except perhaps by *Thlaspi rotundifolium* subsp. *cepaeifolium* (Reeves and Brooks, 1983). A novel solution to this problem has been reported by Huang and Cunningham (1996). The technique is based on growing non-accumulatory plants such as *Zea mays* in lead-contaminated soils and then adding a chelating agent to the soil just as the plant reaches its maximum biomass. The accumulated lead is of course toxic to the plant which is harvested before it dies. In experiments with corn the above authors were able increase the lead content of the plant from 40 to 10,600 $\mu g/g$ (0.004 to 1.06%) in the dry matter. The chelating agent was HEDTA (ethylene-diaminetetraacetic acid as opposed to its various salts such as Na_2EDTA). The initial soil content was 2500 $\mu g/g$ and it was calculated that the site could be cleaned up in 7-8 years using the above procedure. The big question that remains is what will happen to the residual HEDTA?

Phytomining (Chapters 11,14 and 15)

Baker and Brooks (1989) appear to have been the first workers to suggest that it might be possible to grow a crop of a metal such as nickel, but it was left to Nicks and Chambers (1995) to carry out the first practical study in this newly developing field (see Chapter 14). They showed that an economic crop of nickel might be obtained if the energy of combustion of the dried matter could be utilised. The experiments were carried out in California using the nickel hyperaccumulator *Streptanthus polygaloides* that is endemic to ultramafic soils of this state. Later, Robinson *et al.* (1997) carried out an analogous experiment in Italy using *Alyssum bertolonii*, the first plant to have been assigned to hyper-accumulator status.

The principles of phytomining are very similar to those of phytoremediation, but with one important difference. Economics dominate the parameters of the mining technique. Whereas with phytoremediation, the value of the recovered metal is of no real importance, though it can be a bonus if it turns out to be valuable, in phytomining the value of the metal is paramount. This does of course impose restrictions on the potential use of this technology. Robinson *et al.* (1997) have calculated the required biomass for a hyperaccumulator with a metal content of 1% (dry weight) that would return $500/ha at current world prices (see Chapter 15). The calculated biomasses range from 0.0037 t/ha for gold to 61.2 t/ha for lead.

Assuming that the upper possible limit for any annual crop is 30 t/ha (i.e. the same as *Zea mays*), only cobalt, nickel, tin, cadmium, copper, manganese and the noble metals would be amenable to this new procedure.

If we apply the further restriction of only considering metals known in nature to exceed 1 % dry weight in plants, the final list is reduced to only cobalt, nickel, copper and manganese. Of these, cobalt is by far the most promising because the price of the metal is currently $48,000/t and there are several plants (all from Zaïre) known to accumulate up to 1 % cobalt in their dry tissue. A disadvantage of phytomining for cobalt is that unlike the case of nickel, there are not vast areas of subeconomic grade ores or lateritic soils available for phytomining operations.

The restrictions on phytomining discussed above are not so limiting if the biomass and/or metal content of selected plants can be increased by adding fertilisers or sequestering agents to the soils. It is here that future significant advances will be made and could move phytomining from the realm of uncertainty to that of economic reality.

As a corollary, I would like to quote a few words from Harry V. Warren the "father" of biogeochemical prospecting for minerals.

... we have raised the credibility of biogeochemistry from general disbelief through benevolent scepticism to general acceptance...

What was true for biogeochemical prospecting in the 1950s and 1960s is certainly as likely to be true for phytoextraction. Its associated field of phytoremediation is now approaching general acceptance and phytomining has perhaps moved from the *general disbelief* to the *benevolent scepticism* category.

In Conclusion

This introductory chapter has focused on the manifold and seemingly unrelated uses to which hyperaccumulators have been applied during the past decade. Common ground is not obvious between phytochemistry, seaweeds, mineral exploration, phytoarchaeology, insectivory, plant evolution, and of course phytoextraction with its twin disciplines of phytoremediation and phytomining. The common thread is only the highly unusual capacity of a few rare plants to hyperaccumulate a number of heavy metals.

If any of the above disciplines have captured the imagination of the public and scientific community, it is surely phytoextraction. There is a good reason for this. There is a great deal of money involved. This is a lure that may attract some workers in the field who might be intent on harvesting the financial support as well as the plants. Others, no doubt the majority, will be attracted by the possibility of finding a "green" solution to a remediation of the environment that is expected to cost upwards of 300 billion dollars in the United States alone, just to clean up in the 1235 polluted ("Superfund") sites identified by the United States Environmental Protection Agency (USEPA, 1993 - see Fig.1.3).

The majority of the Superfund sites are contaminated with organic and organometallic compounds , but nevertheless, the remainder, polluted by heavy metals, will still require an enormous investment in capital, and will demand the expertise of a large number of scientists well into the 21st century.

Final site
Proposed site

Fig.1.3. Location of 1235 ("Superfund") sites earmarked by the United States
Environmental Protection Agency for remediation. Source: USEPA (1993).

Because of the overriding importance of phytoremediation among the
multitude of different disciplines enumerated above, it may be appropriate to
bring this chapter to a close by discussing future research needs and opportunities
that will arise in this discipline during the next decade.

As pointed out by Bollag and Bollag (1995), we know very little about the
actual processes involved in metal uptake by plants. We are particularly deficient
in knowledge of the agronomic characteristics of the new crops that will be
needed to perform phytoremediation of polluted sites. If we have to add
sequestering agents to the soil to enhance metal uptake, what effect will this have
on the soil for future crops? What are the microbial breakdown mechanisms, if
any, for relatively stable complexing agents such as EDTA? By adding this
compound to the soil, are we merely substituting one pollutant by another? These
and many other questions will have to be answered before we will be able to
employ phytoremediation on a large scale with complete confidence.

Other barriers that will have to be surmounted include the problems of dealing
with local and state governments in order to obtain permissions for remediation.
This is where a certain degree of education of the local authorities and the general
public will be required. Before this can be done, the workers in the field of
phytoremediation will have to show that they have mastered all the intricacies and
problems associated with this new discipline. When this has been achieved by

cooperative research by workers in many diverse fields such as agronomy, soil science, botany, chemistry, genetics, microbiology and others, then the future will surely be bright for phytoremediation.

References

Acar,Y.B. and Alshawabkeh,A.N.(1993) Principles of electrokinetic remediation. *Environmental Science and Technology* 27, 2638-2647.

Amidei,G.(1841) Specie di piante osservate nei terreni serpentinosi. *Atti 3ᵃ Riunione degli Scienzie Italiana,* 523.

Baker,A.J.M. and Brooks,R.R.(1989) Terrestrial higher plants which hyper-accumulate metal elements - A review of their distribution, ecology and phytochemistry. *Biorecovery* 1, 81-126.

Bollag,J.-M. and Bollag,W.B.(1995) Soil contamination and the feasibility of biological remediation. In: Skipper,H.D. and Turco,R.F.(eds) *Bioremediation: Science and Applications.* Soil Science Society of America and others, Madison, pp.1-12.

Boyd,R.S. and Martens,S.N.(1992) The *raison d'être* of metal hyperaccumulation by plants. In:Baker,A.J.M.,Proctor,J. and Reeves,R.D.(eds) *The Vegetation of Ultramafic (Serpentine) Soils.* Intercept, Andover, pp.279-289.

Brierley,C.L.(1982) Microbial mining. *Scientific American* September 1982, 247, 42-50.

Brooks,R.R. and Johannes,D.(1990) *Phytoarchaeology.* Dioscorides Press, Portland, 224 pp.

Brooks,R.R.,Lee,J.,Reeves,R.D. and Jaffré,T.(1977) Detection of nickeliferous rocks by analysis of herbarium specimens of indicator plants. *Journal of Geochemical Exploration* 7, 49-57.

Brooks,R.R.,Morrison,R.S.,Reeves,R.D.,Dudley,T.R. and Akman,Y.(1979) Hyperaccumulation of nickel by *Alyssum* Linnaeus (Cruciferae). *Proceedings of the Royal Society (London) Section B* 203, 387-403.

Cannon,H.L.(1960) The development of botanical methods of prospecting for uranium on the Colorado Plateau. *United States Geological Survey Bulletin* 1085-A, 1-50.

Cannon,H.L.(1964) Geochemistry of rocks and related soils and vegetation in the Yellow Cat area, Grand County, Utah. *United States Geological Survey Bulletin* 1176, 1-127.

Cesalpino,A.(1583) *De Plantis Libri v.16.* Florentiae, 369.

Chaney,R.L.(1983) Plant uptake of inorganic waste constitutes. In: Parr,J.F.,Marsh,P.B. and Kla,J.M.(eds) *Land Treatment of Hazardous Wastes.* Noyes Data Corp., Park Ridge, pp.50-76.

Cole,M.M.(1973) Geobotanical and biogeochemical investigations in the sclero-phyllous woodland and shrub associations of the Eastern Goldfields area of Western Australia with particular reference to the role of *Hybanthus floribundus* (Lindl.) F.Muell. as a nickel indicator and accumulator plant. *Applied Ecology* 10, 269-320.

Cunningham,S.D.,Berti,W.R. and Huang,J.W.(1995) Phytoremediation of contaminated soils. *Trends in Biotechnology* 13, 393-397.

Cunningham,S.D. and Lee,C.R.(1995) Phytoremediation plant-based remediation of contaminated soils and sediments. In: Skipper,H.D. and Turco,R.F.(eds) *Bioremediation: Science and Applications*. Soil Science Society of America *et al.*, Madison, pp.145-156.

Desvaux,N.A.(1814) Coup d'oeil sur la famille des plantes Crucifères. *Journal de Botanie* 3, 145-187.

Doksopulo,E.P.(1961) Nickel in Rocks, Soils, Waters and Plants Adjacent to the Chorchanskaya Group (in Russian). University of Tbilisi, Georgia.

Dudley,T.R.(1964) Synopsis of the genus *Alyssum*. *Journal of the Arnold Arboretum* 45, 358-373.

Dushenkov,V.,Kumar,N.P.B.A.,Motto,H. and Raskin,I.(1995) Rhizofiltration: the use of plants to remove heavy metals from aqueous streams. *Environmental Science and Technology* 29, 1239-1245.

Huang,J.W. and Cunningham,S.D.(1996) Lead phytoextraction: species variation in lead uptake and translocation. *New Phytologist* 134, 75-84.

Jaffré,T.,Brooks,R.R.,Lee,J. and Reeves,R.D.(1976) *Sebertia acuminata* a hyperaccumulator of nickel from New Caledonia. *Science* 193, 579-580.

Jaffré,T. and Schmid,M.(1974) Accumulation du nickel par une Rubiacée de Nouvelle Calédonie: *Psychotria douarrei* (G.Beauvisage) Däniker. *Comptes Rendus de l'Academie des Sciences, Paris, Série D.* 278, 1727-1730.

Kelly,P.C.,Brooks,R.R.,Dilli,S. and Jaffré,T.(1975) Preliminary observations on the ecology and plant chemistry of some nickel-accumulating plants from New Caledonia. *Proceedings of the Royal Society of London Section B* 189, 69-80.

Leblanc,M.,Robinson,B.H. and Brooks,R.R.(1997) Hyperaccumulation de thallium par *Iberis intermedia* (Brassicacée). *Comptes Rendus de l'Academie des Sciences, Paris, Série D.* (in press).

McGrath,S.P.,Sidoli,C.M.D.,Baker,A.J.M. and Reeves,R.D.(1993) The potential for the use of metal-accumulating plants for the *in situ* decontamination of metal-polluted soils. In: H.J.P.Eijsackers and T.Hamers (eds), *Integrated Soil and Sediment Research: a Basis for Proper Protection*. Kluwer Academic Publishers, Dordrecht, pp.673-676.

Minguzzi,C. and Vergnano,O.(1948) Il contenuto di nichel nelle ceneri di *Alyssum bertolonii*. *Atti della Società Toscana di Scienze Naturale* 55, 49-74.

Nicks,L. and Chambers,M.F.(1995) Farming for metals. *Mining Engineering Management* September, 15-18.

Outridge,P.M. and Noller,B.N.(1991) Accumulation of toxic trace elements by freshwater vascular plants. *Reviews of Environmental Contamination and Toxicology* 121, 1-63.

Pichi Sermolli,R.(1948) Flora e vegetazione delle serpentini e delle altre ofioliti dell'alta valle del Tevere (Toscana). *Webbia* 6, 1-380.

Reeves,R.D. and Brooks,R.R.(1983b) Hyperaccumulation of lead and zinc by two
    metallophytes from a mining area of Central Europe. *Environmental Pollution
    Series A*31, 277-287.
Robinson,B.H.,Chiarucci,A.,Brooks,R.R.,Petit,D.,Kirkman,J.H.,Gregg,P.E.H.
    and De Dominicis,V.(1997) The nickel hyperaccumulator plant *Alyssum
    bertolonii* as a potential agent for phytoremediation and phytomining of
    nickel. *Journal of Geochemical Exploration* 59, 75-86.
Severne,B.C. and Brooks,R.R.(1972) A nickel-accumulating plant from Western
    Australia. *Planta* 103, 91-94.
Skipper,H.D. and Turco,R.F.(1995) *Bioremediation: Science and Applications*.
    Soil Science Society of America and others, Madison, 322 pp.
United States Environmental Protection Agency (1993) *Toxics Release Inventory*.
    United States Government Printing Office, Washington D.C. USEPA/745/R-
    93/003.
Utsunomiya,T.(1980) *Japanese Patent Application No.55-729959*.
Yamada,K.,Miyahara,K. and Kotoyori,T.(1975) Studies on soil pollution caused
    by heavy metals. Part IV: soil purification by plants that absorb heavy metals.
    Gamm Ken Nogyo Shienjo Hokuku 15, 39-54.

Chapter two:

Phytochemistry of Hyperaccumulators

R.R. Brooks

Department of Soil Science, Massey University, Palmerston North, New Zealand

Introduction

Plants that hyperaccumulate metals are ideal subjects for phytochemical studies because their elemental concentrations are by definition so high, that milligram quantities of the target organometallic complexes can be isolated and studied outside the plant. Conventional methods of studying metals in plants that do not hyperaccumulate these elements have previously had to rely on use of radioactive tracers or carrying out these studies within the plant or by use of extracts that do not contain enough of the target material for actual separation and storage for future use.

Phytochemical studies of these plants have also been stimulated by the abnormality of such inordinately high metal concentrations that has led us to reassess commonly held assumptions about the mechanisms of metal uptake in the plant kingdom. Once the practical applications of hyperaccumulation had been realised there was further incentive to examine the phytochemistry of hyperaccumulators in order to examine ways of increasing metal uptake to facilitate phytoremediation and phytomining (see Chapters 11-15).

At the time of the establishment of the principles of hyperaccumulation in the late 1970s (Brooks *et al.*, 1977) there had been virtually no phytochemical studies on these plants and for the next decade there was a modest stream of papers emerging mainly from our New Zealand laboratory. Once the commercial potential had been established however, the number of phytochemical studies increased greatly, particularly in the last 3 years after the benchmark paper by McGrath *et al.* (1993) had laid out the parameters for potential phytoremediation of contaminated land by use of hyperaccumulator plants.

Most of the phytochemical studies on hyperaccumulator plants have centred around those that accumulate nickel, partly because "nickel plants" are so much more numerous (see Table 3.3) than those that take up zinc, cobalt and copper. This relative abundance also implies a much wider choice of species selection for commercial applications and a consequent increase in the number of phytochemical studies entailed by this wider choice.

In this chapter, the phytochemical studies will be discussed in relation to the individual elements concerned but before doing so there will be some attention to a wide variety of instrumental methods of chemical analysis that have allowed for the speedy detection of hyperaccumulator plants using very small samples (see also Chapter 4) and which have enabled the identification and quantification of the elements within the organometallic complexes themselves.

Methods of Quantitative Determination of Metals in Plants

Introduction

The correct method of sample preparation is a *sine qua non* for successful measurement of elemental concentrations in plant material. Methods for processing vegetation samples prior to chemical analysis have been reviewed thoroughly by Hall (1995). All too often, the analyst, who may have little contact with the field worker, strives to produce maximum accuracy and reproducibility (perhaps within 1%) for his methodology, whereas the plant sample itself may have been selected so carelessly that replicates could easily give analytical values differing by 50-100%.

The type of sample preparation that should be used is almost entirely dependent on the method of analysis that is to be employed. There are numerous options that include: no sample preparation other than drying, wet ashing, dry ashing, fusion and selective leaching. Each technique has its advantages and disadvantages, but by far the most common procedures are dry or wet ashing. These procedures will be discussed below.

For nearly 100 years, the only instrumental method of trace analysis available for the analyst was emission spectrometry (ES). Until the early 1930s the method was only semiquantitative at best, but the pioneering work of V.M.Goldschmidt in Göttingen led to quantitative spectrometry and during the next 30 years allowed for the determination of at least 60 chemical elements with adequate sensitivity and precision.

The early 1960s saw the birth of a revolution in quantitative analysis, beginning with flame atomic absorption spectrometry (FAAS). For the first time, it became possible to determine single elements at a very fast rate in solutions of plant material with excellent precision and sensitivity. This precision was afforded by the fact that the method made use of solutions rather than inhomogeneous solid samples.

In the late 1970s and early 1980s a further revolution occurred when inductively-coupled plasma emission spectrometry (ICP-ES) was developed. This method allowed for the simultaneous determination of at least 20 elements with about the same precision and limits of detection as FAAS. As if this were not enough, the scientists then produced inductively coupled plasma emission (ICP) mass spectrometry (ICP-MS) whereby the limits of detection were reduced by as much as two orders of magnitude in some cases.

In the period between the late 1950s and up to the present time, the above instrumental developments were parallelled by the evolution of various techniques of neutron activation analysis (NAA) involving the instrumental (INAA) and radiochemical (RNAA) variants of the procedures. These methods will be described below.

Table 2.1 shows the form of vegetation sample required for the main analytical methods in use today.

Table 2.1. Form of vegetation sample required for specific methods of chemical analysis.

Method of analysis	Sample form
Emission spectrometry (ES)	ash
Flame atomic absorption spectrometry (FAAS)	solution
Graphite furnace atomic absorption spectrometry (GFAAS)	solution
Instrumental neutron activation analysis (INAA)	dry or ash
Laser-excited plasma emission spectrometry (ICP-MS)	ash
Plasma emission mass spectrometry (ICP-MS)	solution
Plasma emission spectrometry (ICP-ES)	solution
Radiochemical neutron activation analysis (RNAA)	solution
X-ray fluorescence spectrometry (XRF)	dry or ash

Sample collection

The question of sample collection has been discussed extensively in numerous publications (e.g. Brooks, 1983; Dunn, 1995; Ernst, 1995). The degree to which the sampling procedure should be refined, depends entirely on the use to which the analytical data are to be put. For example, if it is merely required to identify whether or not a specific plant is a hyperaccumulator of a given element, it may only be necessary to select leaves or twigs of any age. If however the data were to be used for biogeochemical prospecting, a much more careful programme should be followed. For example, Ernst (1995) has listed the following variables that should be taken into account in a sampling programme: date of sampling (i.e. season), weather, soil type, vegetation type, sampling design, age of individuals, whether flowering or fruiting, whether senescent, plant organ, sampling height, whether or not infested by fungi, whether injured by herbivores, etc.

There are so many variables that can affect the reliability of the data that it is surprising that any worthwhile work is ever achieved. However, in identifying hyperaccumulators or using them for some specific purpose, the selection of the correct plant organ is paramount. As an illustration of this, a recent project involving a "nickel plant", *Homalium kanaliense* in New Caledonia (Jaffré and Brooks unpublished data) provided a nickel concentration of about 6000 $\mu g/g$

(0.6%) in dried leaves, whereas the bark and trunk wood contained only about 350 μg/g nickel. If only wood or bark had been used in the survey, hyperaccumulation would not have been detected.

Sample preparation

Washing

The question of sample preparation has been reviewed thoroughly by Hall (1995). The first step obviously involves washing the plant material and there is some disagreement as whether to wash or not. Dunn (1992) considers that washing of leaves or needles is not necessary in northern forests unless there is uncovered mineralisation. In desert areas or near mine dumps or smelters where wind-borne dust is a problem, washing is clearly necessary. This problem is far less serious for hyperaccumulator plants than for "normal" species. This is because the metal content of the dried plant is usually higher than that of the soil in which it grows so that any wind-borne dust would have the effect of apparently diluting the true metal concentration in the plant. Nevertheless, there have been several errors in the literature derived from contaminated samples.

There is no evidence that any vascular plant is able to hyperaccumulate or even accumulate chromium. Indeed this element, along with titanium, can be used as an index of sample contamination. Wild (1974) reported 0.24% chromium in a specimen of *Sutera fodina* from the Great Dyke in Zimbabwe. However, Brooks and Yang (1984) found only a mean of 2 μg/g in 3 specimens from the same area. The original sample had been collected near the Noro chromium mine and had probably been contaminated by wind-borne mineralised dust.

Dry ashing

Dry ashing of plant material can easily be carried out by placing the sample in a borosilicate beaker (commonly 10 g of material in a 50 mL container) and heating at 500°C overnight or until the sample is fully combusted. The earlier use of silica, porcelain or platinum vessels is usually quite unnecessary and expensive.

There are several advantages in dry ashing of plant material:

1 - improved sample representivity in that 1 g of ash may correspond to as much as 15-20 g of dry material;

2 - improved sensitivity because the elemental content has been concentrated by a factor of about 15;

3 - avoidance of having to use mineral acids for sample decomposition with consequent contamination of the sample.

The main disadvantage of dry ashing is that some elements such as cadmium, arsenic, selenium, mercury, and lead are volatile and can be lost, at least in part, at 500°C. This latter problem can be addressed to some extent by addition of an *ashing aid* such as magnesium. Loss by volatilisation can be reduced by making sure that chlorides are not present as many elements have volatile chlorides.

The form of the target element can also decide whether there are losses on ignition of the plant material. Hall *et al.* (1990) found that dry ashing caused the form of palladium to change so that it was less soluble in mineral acids. Reheating to 870°C was needed to render it soluble. Gold is also alleged to be lost during high temperature ashing if chlorides are present (Girling *et al.*, 1979), but Hall *et al.* (1991) found no loss of gold at temperatures as high as 870°C.

In the course of studies involving herbarium specimens, my colleagues and I analysed about 20,000 samples during a 10-year period (see also Chapter 4). The procedure that we found to be satisfactory involved ashing about 100 mg of leaf sample in a 10 mL borosilicate test tube, ashing at 500°C, and redissolving the ash in 5 mL of 2M hydrochloric acid followed by analysis with either FAAS or ICP-ES. The procedure was simple, rapid, and relatively free from contamination problems.

Wet ashing

Decomposition by wet ashing is usually carried out to avoid loss of volatiles such as cadmium and lead. There is no wide agreement as to which mixture of acids should be used, but oxidising acids such as nitric or perchloric acids have to be used to destroy the organic matter. Perchloric acid is extremely dangerous and can cause explosions. The usual procedure is therefore to use a mixture of nitric and perchloric acids. The nitric acid begins the oxidation and is removed towards the end of the procedure as it has a lower boiling point than perchloric. The latter completes the final oxidation. Sometimes hydrogen peroxide can be added to nitric or sulphuric acids to improve the oxidation. If the plant material is siliceous (as in the case of *Equisetum* spp.), it may be necessary to use hydrofluoric acid.

The undeniable disadvantage of wet oxidation is that contamination from the mineral acids occurs and blanks therefore have to be performed. If the concentration of the target element is low, the blank can be as high or higher than the signal from the analyte. This does not however usually present a problem with hyperaccumulator plants because they usually have a high concentration of the analyte element.

Selective leaching

Sometimes it is appropriate to determine the form of a given element in plant material by selective leaching. For example, nickel complexes in most "nickel plants" are water-soluble polar compounds. A programme of selective leaching with different solvents can identify the various forms of the target metal. This is illustrated in Table 2.2 that shows a schematic programme for studying nickel in sequential extracts of four species of nickel hyperaccumulators from the island of Palawan in the Philippines (Homer, 1991).

Table 2.2. Distribution of nickel in sequential extracts from four nickel hyperaccumulators from the Philippines.

Species	A	B	C	D	E	F	G	H
Brackenridgea palustris	0.86	67.04	10.02	4.60	6.15	0.23	9.35	1.76
Dichapetalum gelonioides	1.08	77.0	14.64	1.65	3.96	0.03	1.48	0.16
Phyllanthus palawanensis	0.23	40.27	10.06	25.26	10.92	0.25	6.06	6.95
Walsura monophylla	1.04	66.91	10.54	1.34	7.70	0.14	10.89	1.45

A - ethanol extract (small neutral molecules), B - aqueous extract (polar low-molecular weight compounds), C - 0.2M HCl extract (acid-soluble polar compounds), D - precipitate from C in acetone (proteins and pectates), E - residue in 0.5M HClO$_4$ (lignins, cellulose etc.), F - precipitate from E in acetone (nucleic acids), G - residue in 2M NaOH (complex proteins and polysaccharides), H - final residue (cellulose, lignins, and insoluble material in cell walls).
Source: Homer (1991).

Fusion

Fusions are seldom carried out on plant material because other methods are much more effective. However, there are cases where refractory compounds involving chromium, zirconium, hafnium, niobium or boron may be resistant to conventional acid attack. In such cases the plant ash may be fused with an alkaline flux such as sodium peroxide (oxidative), lithium metaborate, or sodium and/or potassium carbonate. The sample is usually fused in a platinum crucible at temperatures from 500-900°C depending on the fusion agent. The melt is then extracted with hot water. The resultant solution may be analysed immediately or perhaps acidified before use.

The main advantage of the fusion process is that virtually all the target elements are brought completely into solution. The main disadvantage is that the large amount of fusion agent (typically 5:1 in relation to the sample) provides a solution with a very high ionic strength so that extensive dilution is required before the solution can be analysed. The second major disadvantage is that the throughput rate is very low and could require a large number of expensive platinum or zirconium crucibles. Graphite crucibles can be used as a cheaper alternative but can only be employed for about five fusions before they are oxidised in the muffle furnace.

Instrumental methods of chemical analysis

Emission spectrometry (ES)

The emission spectrograph was the workhorse of the analyst for nearly 100 years until it became eclipsed by other instrumental methods in the early 1960s. Although the technique is seldom used today, it was the forerunner of the popular plasma emission spectrometry (ICP-ES) which today is one of the commonest instrumental methods. The principles of the two procedures are almost identical.

The emission spectrograph consists of a source, dispersion prism or grating, and readout system that for most of its history was a photographic plate. The source is usually a d.c. arc struck between a graphite counter electrode and a hollow graphite electrode in which the sample is placed. The sample electrode is usually the anode. The emission spectrum is focused by a convex lens through a slit on to a quartz or glass prism which refracts the energy into a series of spectral lines that are recorded on to a photographic plate.

An alternative system replaces the prism by a diffraction grating that produces a series of lines with a better dispersion than the quartz prism. Once again a photographic plate can be used to record the spectrum.

A later development was the emergence of the so-called "direct reader" in which the spectral lines are recorded with an electronic readout permitting reference to standard curves and a direct printout of the elemental abundances. Although this later development greatly speeded up the analytical process, it was not able to compensate for the inherent poor precision (5-10% at best) of ES compared with other methods.

An undeniable advantage of ES was however, the fact that the whole picture was immediately obvious to the viewer of the photographic plate: i.e. some 50 elements would be shown on the plate and could be identified and quantified at leisure. Virtually all instrumental trace element analyses in the period 1930-1960 were carried out by use of this instrument and are a fitting tribute to its contribution to scientific knowledge during this period.

Inductively coupled plasma emission spectrometry (ICP-ES)

The first commercial ICP-ES instrument appeared in the mid 1970s. The procedure is based on the older emission spectrometer except that the d.c. arc source is replaced by a system in which a solution is nebulised and fed into a high-temperature rf (radio frequency) plasma where it is heated to a very high temperature (6000-10,000°C) and produces a complex emission spectrum of a large number of elements. Line dispersion and electronic readout are similar to those of the older ES direct reader except that modern electronic developments permit a much more sophisticated interpretation of the data and readout options.

As summarised by Hall (1995), the main advantages of ICP-ES are as follows:

1 - the ability to measure 20-60 elements simultaneously in a cycle time of 2-3 min;

2 - the long linear dynamic range of 4-6 orders of magnitude that avoids the need for sample dilution (a problem with atomic absorption spectrometry);

3 - superior sensitivity for such elements as boron, phosphorus, sulphur, and refractories such as the rare earth elements;

4 - much reduced chemical interferences in the hot argon plasma.

Two options are offered by manufacturers: the first is the *simultaneous* instrument whereby all the selected channels (typically 20) produce an immediate readout of each analyte. The other (less expensive) option is the *sequential* system whereby the instrument scans the whole wavelength range and records the elemental abundance at each point in the scale. This procedure is much slower than the simultaneous reader. Hall (1995) has estimated that the determination of 30 samples for 30 elements would need 2 hours on a simultaneous system, but 6 hours with a sequential instrument.

Inductively coupled plasma mass spectrometry (ICP-MS)

The first commercial ICP-MS instrument appeared in 1983 (Sciex) in Canada, followed shortly afterwards by the VG instrument in the UK. The principle of the technique involves a plasma torch exactly the same as the source in ICP-ES. The plasma is, however, fed into a quadrupole mass spectrometer. The instrument possesses extraordinary sensitivity that allows for measurement of elemental abundances in the ng/g (ppb) or even pg/g (ppt) range compared with μg/g (ppm) for the older ICP-ES instruments.

Interference problems are not as extensive as with ICP-ES but since ICP-MS measures the m/e quotient (i.e. mass/charge) of individual isotopes, two different isotopes could give the same signal if one were twice the mass of the other and had twice the ionic charge. This problem can be solved by finding another isotope where the interference does not occur.

A new modification of ICP-MS is the use of the so-called *laser ablation* technique whereby a laser beam is used to excite ions from a solid sample. This procedure is not yet well developed and there are problems associated with sample homogeneity because only a small sample size is excited by the laser. Usually the sample is fused into a glass to improve homogeneity and the standards are themselves contained in fused glass.

The greatest disadvantage of ICP-MS is the very great cost of the instruments that are at least twice the price of those of the more conventional ICP-ES. This has tended to restrict use of these more expensive instruments.

It might well be argued that in a discussion of instrumental methods of identifying and determining hyperaccumulator plants, a method of great sensitivity would not be required. This is certainly true for hyperaccumulators of commoner elements such as nickel and zinc, but if hyperaccumulators of the rarer elements such as the noble metals are to be discovered and studied, classical methods of instrumental analysis might not be sensitive enough for this purpose.

Flame atomic absorption spectrometry (FAAS)

Flame atomic absorption spectrometry is based on the pioneering work of A.A.Walsh (Walsh, 1955). The instrument consists basically of a source, sample cell, grating monochromator and electronic readout. The source is a *hollow cathode lamp* that produces a fine-line emission spectrum of the analyte element. This spectrum is focused through the sample cell which comprises a nebulisation system that introduces the sample solution into a long narrow flame of air-acetylene or nitrous oxide-acetylene (the latter providing a higher temperature). When the analyte ions reach the hot flame they are converted into *ground-state atoms* that absorb their own radiation from the hollow cathode source. The degree of this absorption is related to the initial concentration of the analyte in the solution and is recorded by the electronic readout after selection of the appropriate spectral line by the monochromator.

Following their introduction about 1960, FAAS instruments have become the workhorse of the analytical laboratory, few of which are without at least one of them. The instruments are relatively inexpensive and provide a good level of sensitivity. Most of the hyperaccumulator species known today, were identified by means of FAAS. The limit of detection for most elements is in the $\mu g/g$ range on the original solid sample, and in some cases (e.g. cadmium, magnesium and zinc) can be in the ng/g range.

Graphite furnace atomic absorption spectrometry (GFAAS)

The graphite furnace depends on electrothermal excitation to produce ground-state atoms of the analyte in a small narrow graphite tube or furnace that replaces the flame cell in FAAS. The sample is prepared in solution and a few microlitres of solution are fed into the top of the graphite tube. The tube is heated to perhaps 120°C to dry the sample that is then charred at typically 500°C to remove organic matter. The next stage in the heating cycle involves the process of atomisation in which the temperature is raised to anywhere between 1000 and 3000°C depending on the analyte. The ground-state atoms inside the furnace absorb their own radiation from the hollow-cathode lamp exactly as with FAAS.

The limits of detection with GFAAS are very much lower (typically 100 times) than with FAAS because the ground-state atoms remain longer in the graphite tube during the reading. This residence time of seconds compares with the microseconds that a given atom is present in the flame of FAAS.

A further advantage of GFAAS is that very much smaller samples can be used. For example, it would be easily possible to measure the concentration of an analyte in pollen where the sample weight might be in micrograms and the solution volume just a few microlitres.

A major disadvantage of GFAAS is the slower throughput of samples and the serious interference problems that can arise.

Instrumental neutron activation analysis (INAA)

Instrumental neutron activation analysis arose from the post-war development of nuclear technology and involves the use of a reactor. When samples are irradiated with neutrons, some of the stable isotopes of the analyte elements are converted into radioactive nuclides. Many of these are gamma emitters that can be detected and quantified by examination of the gamma spectrum without the need of a sample digestion procedure. This latter point is one of the great advantages of INAA because this avoids dilution of the sample and there is no possibility of contamination from reagents.

After irradiation, samples are allowed to "cool" for predetermined periods. This procedure allows for reduction of interferences. For example, ^{76}As has a half life of 1.1 days whereas ^{198}Au requires 2.7 days to lose half of its activity. The gold and arsenic nuclides are very close to each other in energy and the latter, usually found with gold, will obscure the gold signal. However after a cooling period of 2.7 days, the gold will have lost half of its activity whereas the arsenic signal will have decreased to 18% of its former value.

Modern facilities for INAA are highly automated and commercial companies can carry out determination of a suite of elements at very low cost in vegetation and soil samples.

Radiochemical neutron activation analysis (RNAA)

The principles of radiochemical neutron activation are as follows: the sample is irradiated as before in a nuclear reactor and then mixed with a stable carrier comprising the analyte element. The mixture is then dissolved in an appropriate solvent such as a mineral acid and the analyte (i.e. radioactive isotope plus stable carrier) is then separated by some appropriate procedure such as extraction with an organic solvent or by ion exchange chromatography. At this stage the chemical yield is obtained by measuring the weight of a precipitate for example. The gamma activity of the pure separate is then determined and the concentration in the original sample calculated after making allowance for the chemical yield.

The technique of RNAA can lower detection limits by up to 100-fold, though at the cost of much lower productivity and far greater expense. For example, INAA determination of iridium can be achieved at a cost of about $10 and a limit of detection of perhaps 1 ng/g. With RNAA the cost is increased to about $500 per sample but with a limit of detection of 0.01 ng/g (10 pg/g). A schematic representation of INAA and RNAA determination of iridium (Brooks, 1987a) is given in Fig.2.1.

Comments given above under ICP-MS in relation to the need for ultrasensitive methods of analysis for hyperaccumulator species are equally relevant for RNAA. If the sample is very small and the analyte concentration low, RNAA might still be useful in identifying hyperaccumulation of a rare element. Hyperaccumulation is after all a *relative* rather than *absolute* concept.

X-ray fluorescence spectrometry (XRF)

In a review of instrumental methods of analysis, Hall (1995) pointed out that XRF is only of marginal interest for trace analysis because the limits of detection afforded by this technique are usually in the upper part of the μg/g range. This limitation is of course not particularly serious when dealing with hyper-

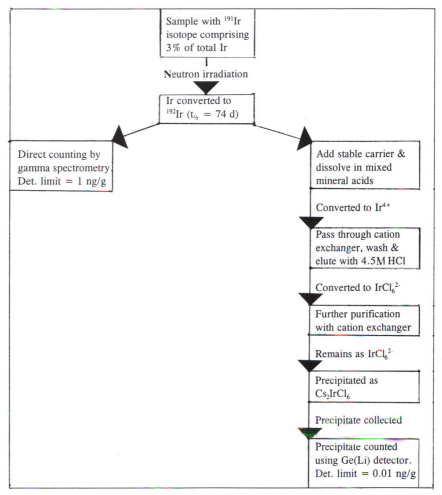

Fig.2.1. Schematic representation of the principles of instrumental and radiochemical neutron activation analysis (INAA and RNAA). Source: Brooks (1987a).

accumulator plants that typically contain analytes in the percent rather than $\mu g/g$ range.

In XRF analysis, the sample (usually in the form of plant ash) is either pressed to a small pellets or fused to a glass with boric acid. The sample is irradiated with primary X-rays from an X-ray source. Secondary X-rays are then produced that are dispersed through a diffracting crystal and their energy and intensity measured by a scintillation counter. These are known as *wavelength dispersive* instruments.

An alternative to the wavelength dispersive instrument is the *energy dispersive* spectrometer. In this case there is no diffracting crystal and the energy of the

Table 2.3. Advantages and limitations of the major analytical techniques.

Technique	Strengths	Weaknesses	Best elements
INAA	*-direct analysis* (little sample preparation, minimal sample contamination) *-multi-element* (large "packages") *-excellent precision and accuracy* *-low cost* (e.g. $C9 for "Au + 34") *-few matrix interferences* *-non-destructive* *-variable sample weight*	*-availability of reactor* *-high capital cost* *-delayed turnaround* (decay periods)	Ag,Au,As,Ba,Br,Ca,Co,Cr, Cs,Fe,Hf,K,Mo,Na,Rb, Sb,Sc,Sr,Ta,Th,U,W, REE,La,Ce,Nd,Sm,Eu,Tb, Yb,Lu. Complementary to ICP-ES
ICP-ES	*-multi-element* (large packages) *-low-cost packages* (e.g. $C6 for 32 elements) *-few matrix interferences*	*-decompn. necessary* (contamin./time loss) *-destructive* *-spectral interferences* (especially from Fe)	Ag,Al,B,Ba,Be,Ca,Cd,Co, Cu,Fe,K,Li,Mg,Mn,Mo,Na, Ni,P,Pb,S,Sr,Ti,V,Y,Zn
ICP-MS	*-superior sensitivity* *-multi-element* *-flexible, versatile* *-few spectral interferences* *-isotopic information*	*-decomposition necessary* *-destructive* *-high cost of analysis* *-limited availability to to date* *-matrix interferences*	All above mass no. 80. Complementary to ICP-MS. Good for low abundance elements, PGE,REE & hydride-forming elements As,Sb,Se,Te,Bi
AAS	*-very few interferences* *-robust and rugged method* *-easy to use* (little training) *-versatile* (e.g. organic solvents to concentrate analytes)	*-single element at a time* *-costly if many elements* *-decomposition necessary* *-small linear dynamic* *-range* (cannot determine major and trace elements in same solution)	As for ICP-ES without P,S,B. Hg by cold vapour is good. GFAAS enhances sensitivity by several orders of magnitude but slower and more expensive
XRF	*-direct solid analysis on powders* *-powders or glass discs can be preserved* *-multi-element (Na-U)* *-instrumentation fully automated* *-excellent precision for major elements*	*-fusion needed for major (light) elements to avoid matrix effects* *-high cost of fused disc pkg.* ($30 for 17 elements) elements) and pressed pellet pkg. ($6 for 1st element & $2 for next etc.) *-matrix interferences* *-min. sample wt. of 2 g for fused disc & 5 g for pellet*	Majors elements & Ba,Cl,Cr Fe,Mn,N,S,P,Rb,Sr,Th,Ti,Y, Zr. Complementary to INAA

Source: Hall (1995).

secondary X-rays is measured simultaneously by a detector. This procedure is much faster than in the case of the wavelength dispersive instruments where the whole spectrum has to be scanned for a period of perhaps 20 minutes.

XRF tends to be used today mainly for whole-rock analyses rather than for vegetation samples, but there is no reason why the technique cannot be used for the latter purpose and there is the further advantage that a whole suite of different elements can be determined at the same time.

Concluding remarks

It will be clear from the above discussion that there is a wide range of different instrumental methods that may be used for the purpose of identifying and analysing hyperaccumulator plants. Each method has specific advantages and disadvantages that are summarised in Table 2.3 above. Perhaps the commonest of these procedures has been FAAS that we have used in our laboratory to identify nearly 400 hyperaccumulators of various elements by analysis of over 20,000 plant samples during a 25-year period.

There are however, some advantages in multi-element analyses. For example, in our survey of zinc and nickel levels in species of *Thlaspi* (Reeves and Brooks, 1983a), we used ICP-ES and inadvertently discovered that *Thlaspi rotundifolium* subsp. *cepaeifolium* is also a hyperaccumulator of lead (Reeves and Brooks, 1983b).

Phytochemical Studies on Hyperaccumulator Plants

Introduction

Most phytochemical studies have been carried out on plants that hyperaccumulate nickel. This is mainly because they greatly outnumber the other hyper-accumulators such as those of copper, cobalt, zinc, and selenium. This section of the chapter will be arranged under the individual elements and there will also be a subsection on a specific group of ligands, the *phytochelatins*, that have been implicated in complexing of heavy metals in plants. The contents of this section represent a considerable update of an earlier review by Brooks (1987b).

Hyperaccumulation of nickel

Inorganic constituents

The first investigation of the chemical composition of a hyperaccumulator of nickel was that of Minguzzi and Vergnano (1948) who determined SiO_2, Fe_2O_3, MgO, CaO and NiO in various organs of *Alyssum bertolonii*, a nickel plant from Tuscany. These data are shown in Table 2.4. A relatively constant Ca/Ni mole quotient was found in all plant parts except the seeds. The authors suggested that the plant is able to compensate for increased nickel levels by increased uptake of calcium.

Table 2.4. Percentage of oxides of nickel and other elements in dry matter of *Alyssum bertolonii*.

Organ	Ash %	SiO_2	Fe_2O_3	MgO	CaO	NiO	Ca/Ni*	Ca/Mg*
Roots	6.89	0.90	0.59	0.69	1.35	0.40	4.85	1.39
Leaves	15.50	0.33	0.18	1.14	5.15	1.55	4.42	3.21
Flowers	8.39	0.08	0.08	0.63	2.39	0.60	4.54	2.67
Fruits	10.00	0.21	0.09	0.84	2.65	0.74	4.71	2.26
Seeds	6.70	0.20	0.17	0.50	1.64	0.78	2.77	2.32

*mole quotients. Source: Minguzzi and Vergnano (1948).

Table 2.5. The distribution of nickel ($\mu g/g$ dry weight) in organs of *Alyssum heldreichii*.

Organ	% of total wt.	Ni content	% of total Ni
Lower roots	11.9	4330	4.5
Upper roots	9.4	9150	7.6
Lower stems	10.0	7190	6.4
Middle stems	5.4	9660	4.6
Upper stems	3.4	16,740	5.0
Lower lateral stems	10.0	17,060	15.1
Lower lateral leaves	27.2	12,500	29.2
Mid-stem leaves	12.9	11,890	13.5
Upper stem leaves	7.6	14,070	9.5
Apical buds	2.2	23,400	4.6
Seeds	negligible	1880	negligible

Source: Morrison (1980).

Morrison (1980) carried out a thorough investigation of the nickel content of various organs of *Alyssum heldreichii*. His findings are summarised in Table 2.5 above. It will be seen that the greatest accumulation of nickel occurred in the leaf material and the least in the roots. The lower stem and middle stem zones had lower nickel levels than the upper stems, and lower lateral stems. This is of some significance because the last two plant parts were green whereas the former were brown and woody. It would seem that nickel can be preferentially accumulated in photosynthetic tissues rather than in non-photosynthetic material. Similar observations were made by Minguzzi and Vergnano (1948) and Vergnano Gambi *et al.* (1977) using *A.bertolonii*. The same workers concluded that the degree of accumulation of nickel was related to the length of the growing period rather than to the nickel content (total or exchangeable) of the soil.

Several studies have been carried out to determine the relationship between hyperaccumulation of nickel and the uptake of other trace elements. In a study on nickel and 15 other elements in endemic plants of the Great Dyke in Zimbabwe, Brooks and Yang (1984) found that the nickel content of leaf material was correlated positively only with other siderophiles (elements of the iron family) such as cobalt, chromium and manganese but was not related to any of the plant nutrients. A similar study by Yang *et al.* (1985) on serpentine-endemic species of the Flacourtiaceae family from New Caledonia showed an almost identical pattern: i.e. nickel in leaves correlated positively only with other siderophiles, though in this case there was also correlation with sodium and zinc.

From the above studies, it does not seem that nickel is able to affect the nutrient balance of serpentinophytes to any marked degree in spite of its hyperaccumulation by some taxa.

Organo-metallic complexes

Carboxylic acids

About 20 years ago, Jaffré *et al.* (1976) determined that the sap of *Sebertia acuminata* (*sève bleue*), a serpentine-endemic tree from New Caledonia, contained an inordinately high nickel content in its blue-green sap (hence the French common name for the tree). The tree is portrayed in Fig.3.6 and was reported by these workers to have a nickel content of 25.7% in the dried sap and 11.2% in the fresh material. The leaves, trunk bark, twig bark, fruits and wood contained respectively: 1.17, 2.45, 1.12, 0.30, and 0.17% nickel. The 11.2% nickel contained in the fresh latex represents by far the highest nickel content recorded for any living material.

Lee *et al.* (1977) isolated and characterised the nickel compound in the sap of *sève bleue* and in leaves of other hyperaccumulators such as *Homalium francii*, *H.guillainii*, *H.kanaliense*, *Hybanthus austocaledonicus* and *H.caledonicus*. Experiments with high-voltage paper electrophoresis (Fig.2.2) showed that at pH 6.5, an aqueous extract of *Sebertia acuminata* gave peaks corresponding to $Ni(H_2O)_6^{2+}$, and a negatively-charged 2:1 citratonickel complex. An extract of *Homalium guillainii* showed only the presence of the citrato complex. The identity of these peaks was further confirmed by a combination of gas-liquid chromatography and mass spectrometry. The counter cation to the nickel citrato complexes was a mixture of $Ni(H_2O)_6^{2+}$ and hydrated Ca^{2+} and Mg^{2+} ions in the case of the *Homalium* and *Hybanthus* plants but was only the aquo complex of nickel in the latex of *S.acuminata*.

The relationship between nickel and citric acid was further investigated by Lee *et al.* (1978) who found both constituents in mature leaves of 15 New Caledonian hyperaccumulators and in two *Alyssum* species as well as in the Zimbabwean nickel plant *Pearsonia metallifera*. This association is shown in Fig.2.3 and it is clear that the two variables are closely related. The same workers also found traces of malic and malonic acids in the extracts, though these were minimal in the latex of *S.acuminata*.

Further work on the composition of nickel complexes was performed by Kersten *et al.* (1980) using both gel and ion exchange chromatography, as well as high performance liquid chromatography (HPLC), and a combination of gas-liquid chromatography and mass spectrometry.

In their work on *Psychotria douarrei*, these workers showed that the nickel was present mainly as a negatively-charged malate complex balanced by a cationic nickel aquo complex. In contrast, the hyperaccumulator *Phyllanthus serpentinus* had its nickel bound as 42% citrate and 40% malate.

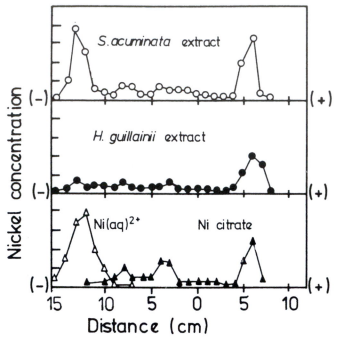

Fig.2.2. Results of high-voltage electrophoresis on extracts of *Sebertia acuminata* and *Homalium guillainii* at pH 6.5. The *Sebertia* extract shows both positively and negatively charged species corresponding to a nickel-aquo complex and a citrato complex respectively. The *Homalium* extract shows only the citrato complex. Source: Lee (1977).

Nickel plants are ideal subjects for phytochemical studies of their organic constituents because the nickel contents are usually so high that it is readily possible to isolate milligram quantities of these complexes rather than the microgram amounts possible with "normal" plants. Despite this obvious advantage, it is surprising to find that until the 1990s, only two other groups outside Massey University, New Zealand, had carried out extensive studies on these interesting plants. I refer to the work of W.Ernst at the Vrije Universiteit in Amsterdam, and that of the late O.Vergnano Gambi and her associates at the University of Florence, Italy, who identified nickel complexes in *Alyssum bertolonii* (Pelosi *et al.*,1974). These workers used gel chromatography to separate soluble nickel complexes and deduced that the nickel was bound mainly to an organic acid. Later, Pelosi *et al.* (1976) purified a nickel complex from the same species, again using gel chromatography. The purified product was examined by a mixture of gas-liquid chromatography and mass spectrometry and was found to contain a mixture of malic and malonic acids. The association of nickel with these acids in *A.bertolonii* and in *A.pintodasilvae* was further investigated by Pancaro *et al.* (1978a - see also Pancaro *et al.*,1978b for an English summary of this work). These workers used as a control, specimens of *A.bertolonii* grown in ordinary non-serpentine garden soil.

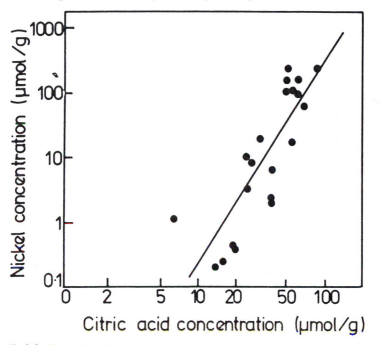

Fig.2.3. The relationship between nickel and citric acid in hyperaccumulators of the genera *Homalium, Hybanthus, Sebertia, Geissois, Psychotria, Alyssum* and *Pearsonia.* Source: Lee (1977).

The results of the above survey are shown in Table 2.6. They found that nickel in leaves of plants grown on serpentine soil was mainly associated with malic and malonic acids present in an approximately 1:1 mole ratio (ca. 200 μmol/g dry mass). Control samples of *A.bertolonii* obtained from plants grown on ordinary garden soil and therefore deficient in nickel (< 40 μg/g), contained malic and malonic acid concentrations an order of magnitude lower. High levels of malic acid (120 μmol/g) related to high nickel concentrations (166 μmol/g) were found in the leaves of *A.pintodasilvae*, though in this case the level of malonic acid was very low. Experiments on purified extracts from *A.bertolonii* confirmed the involvement of the organic acids in the nickel metabolism of the leaf tissues which contained 1400 μmol/g of malonic acid and 800 μmol/g of malic acid. In the seeds of this plant, nickel was bound mainly to malic acid (300 μmol/g).

From the above work, it might be concluded that nickel is bound primarily to

malic and malonic acids in *Alyssum* and to citric acid in many other
hyperaccumulators such as *Sebertia acuminata, Homalium* and *Hybanthus*. There
is however, little direct evidence for a direct association between citric acid and
nickel in *Alyssum*.

Table 2.6. Concentrations (μmol/g) of malic and malonic acids and of inorganic constituents in dried
leaves of *Alyssum bertolonii* (on serpentine soil [A] and on ordinary garden soil [B]]) and
A.pintodasilvae on serpentine soil (C).

Plant	Date	Organic acids			Cations				
		Malic	Malonic	Total	Ni	K	Ca	Mg	Total
A	17/2/75	304	176	480	238	390	1060	175	1863
	7/9/75	257	208	465	213	399	723	269	1604
	11/11/75	153	162	315	272	374	686	238	1570
	11/1/76	161	140	301	341	325	661	245	1572
B	10/7/76	10	38	48	0.56	31	1260	101	1712
	10/1/77	29	44	73	0.65	508	1320	87	1915
C	15/5/77	120	36	156	166	465	1448	124	2203

After: Pancaro *et al.* (1978).

It would be somewhat facile to assume that nickel in hyperaccumulators is
bound only to citric, malic and malonic acids. Morrison (1980) studied the
organic constituents of several *Alyssum* species and found many organic
constituents other than these three organic acids. Some of these were derivatives
of one or more of these acids such as the trimethyl esters of citric and homocitric
acid. The latter parent acid has also been identified in the Zimbabwean *Pearsonia
metallifera* by Stockley (1980).

The phytochemistry of Zimbabwean serpentine-tolerant plants has also been
investigated by Ernst (1972). He studied *Indigofera setiflora* and *Dicoma
niccolifera* from the Great Dyke area. The latter species contains up to 700 μg/g
nickel in dried leaves and does not quite qualify for hyperaccumulator status. It
has, however, been included for the sake of completeness. The same is true of
I.setiflora (415 μg/g nickel). Ernst found 73 μg/mL of nickel in the cell sap of
D.niccolifera and also carried out sequential extraction of root material from this
species and from leaves of *I.setiflora*. These data are shown in Table 2.7. The
proportion of nickel extractable with water from leaves of both species closely
followed the nickel content of the cell sap. It appeared, therefore, that much of
the water-soluble fraction was located within the leaf vacuole system.

As reported by Ernst (1972), about 75% of the nickel in leaves of *I.setiflora*
was relatively tightly bound to the plant material and could only be removed by
solvents with a high exchange capacity such as sodium chloride and citric acid.
Ernst concluded that an appreciable proportion of the nickel in the residue was

bound to the cell walls. He suggested that fixation at these sites was a mechanism whereby the nickel could be detoxified by storage and removed from the plant at leaf fall. When nickel levels in the substrate were high, these storage sites became saturated and the nickel burden in the cells was characterised by an increased water-soluble fraction.

Table 2.7. Percentage of total nickel in various extracts of organs of *Dicoma niccolifera* and *Indigofera setiflora*.

Species	Organ	a	b	c	d	e	f	g
I.setiflora	Leaves	0.2	23.4	3.5	34.3	34.3	4.3	415
D.niccolifera	Root cortex	21.2	6.2	21.1	19.4	32.1	510	-
	Root xylem	29.3	21.2	13.3	20.1	13.4	2.7	238

a - butanol, b - water, c - sodium chloride solution, d - citric acid solution, e - hydrochloric acid, f - insoluble residue, g - total nickel in $\mu g/g$ dry weight. After: Ernst (1972).

Perennial organs such as roots present an entirely different problem compared with leaves. In the latter case, leaf fall can lead to removal of toxic metals, whereas in roots the same mechanism cannot be operative. In roots the capacity to tolerate heavy metals depends on the capacity to render the metals either soluble or insoluble and to retain the insoluble fraction while permitting the soluble fraction to translocate to the leaves for subsequent removal at leaf fall.

The data in Table 2.7 for roots of *Dicoma niccolifera* show that residual nickel is much more abundant in the root cortex than in the woody material. In general the older roots are predominant sites for the inactivation of heavy metals accumulated by these plants. The nickel in the wood was characterised by a high percentage of water-soluble and easily-exchangeable forms.

The phytochemistry of the Zimbabwean hyperaccumulator *Pearsonia metallifera* was investigated by Stockley (1980) using a sequential extraction system (Bowen *et al.*,1962 as modified by Lee, 1977). The percentage of total nickel in each fraction is shown in Table 2.8. It will be noted that over 75% of the nickel was extractable with water. The aqueous fractions were subjected to gel chromatography to separate a green crystalline material which contained most of the nickel. The extract was methylated and passed though a gas-liquid chromatographic column coupled to a mass spectrometer. Two peaks were obtained of which one was the trimethyl ester of citric acid. The other appeared to be the trimethyl ester of 3-hydroxy-3-carboxylhexanedioic acid.

Stockley (1980) also found malonic acid in the crude extract of the plant material and proposed that this causes the tricarboxylic acid cycle (Krebs Cycle) to be inhibited at the step involving conversion of succinate to fumarate. Because of this, the conversion of oxaloacetate to citrate is controlled by the amount of malate produced from back-to-back condensation of acetate units, and by the small residual conversion of succinate to fumarate.

The phytochemistry of *Pearsonia metallifera* is obviously much more complicated than that of other hyperaccumulators such as *Sebertia acuminata*.

Table 2.8. Percentage of total nickel in extracts of leaves of *Pearsonia metallifera*.

Extractant	Percentage nickel
Ethanol (95%)	2.79
Water (first extract)	55.76
Water (second extract)	15.33
Water (third extract)	4.74
Hydrochloric acid (0.2M)	20.35
Acetone-insoluble fraction from above	Trace
Perchloric acid (0.5M)	0.84
Acetone-insoluble fraction from above	0.03
Sodium hydroxide (2M)	0.05
Residue	0.05

Source: Stockley (1980).

A process of sequential extraction was also used by Brooks *et al.* (1981) to study nickel in *Alyssum serpyllifolium* and its close relatives *A.pintodasilvae* and *A.malacitanum*. They found that more than half of the nickel was soluble in water and dilute acid showing that it was present as polar complexes. They also found an association between nickel and citric, malic and malonic acids (cf. Pelosi *et al.*,1974, 1976).

Homer (1991) used a combination of gel chromatography and gas chromatography to study the nature of nickel complexes in the Philippine hyperaccumulators *Dichapetalum gelonioides* subsp. *tuberculatum*, *Phyllanthus palawanensis*, and *Walsura monophylla*. She prepared milligram quantities of purified material and after methylation subjected the samples to gas chromatographic analysis. The samples were found to be composed mainly of nickel, and citric and malic acids. In the case of *D.gelonioides* these three components accounted for 96% of the total mass of the purified extract. The data are shown in Table 2.9.

The data show elevated amounts of malic acid in the *Dichapetalum* and

Table 2.9. The nickel and organic acid contents (% w/w) and mole ratios of purified extracts from three hyperaccumulators from Philippines.

Species	Nickel (A)	Citric (B)	Malic (C)	A:B:C mole ratio
D.gelonioides	18.0	24.0	43.0	1:0.4:1
P.palawanensis	5.0	5.1	4.2	1:0.4:0.4
W.monophylla	3.0	2.7	8.2	1:0.2:1

Source: Homer (1991).

Walsura and comparable concentrations of citric and malic acids in the *Phyllanthus*. They confirm the apparent association of organic acids with nickel complexes in hyperaccumulator plants. The findings of these and other workers should not however be accepted uncritically. Although both nickel and organic acids are found in these complexes it must be remembered that some of these are somewhat unstable. The nickel malate complex, for example, might easily be hydrolysed merely by the process of extraction of the complex with water or dilute acid. To find both malic acid and nickel in the same aqueous extract therefore does not necessarily mean that they were originally complexed together in the plant material. Table 2.10 shows the stability constants for various complexes between nickel and aminoacids and carboxylic acids (Homer *et al.*,1997).

Table 2.10. Stability constants for complexes between nickel and aminoacids and carboxylic acids.

Acid	$\log K_1$	$\log K_2$
AMINOACIDS		
Alanine	5.41	9.89
Aspartic acid	7.16	12.40
Cysteine	9.82	20.07
Glycine	6.18	11.13
Histidine	8.67	15.50
Proline	5.94	10.85
Serine	5.45	9.96
Tyrosine	5.10	9.46
Others	5.20-5.70	9.70-10.90
CARBOXYLIC ACIDS		
Citric acid	5.51	7.84
Malic acid	3.30	?
Malonic acid	3.29	?
Tartaric acid	5.47	7.60

K_1 refers to complexes of the form NiL where L is the organic ligand. K_2 refers to complexes of the structure ML_2. Source: Homer *et al.* (1997).

It is clear from the above table that nickel complexes with aminoacids are considerably more stable than those with carboxylic acids. The next section will therefore discuss the possibility that aminoacids are involved in these complexes.

Aminoacids

It is well known that divalent nickel is able to complex with many aminoacids found in plants. For example Cataldo *et al.* (1988) working on nickel in xylem exudates of soybean found that some of the nickel was associated with aminoacids, some with carboxylic acids, and some with neither. Stability constants (see Table 2.10) for 1:1 nickel-aminoacid complexes show that these are at least as stable as anionic complexes with citric acid and considerably more so than those involving neutral nickel-malate bondings. Where there is an excess of aminoacids, even more stable 1:2 nickel complexes may form. The presence of aminoacids possibly linked with nickel was determined by Kelly *et al.* (1975) and by Farago *et al.* (1980). Bick *et al.* (1982) determined nickel and aminoacids in leaves of *Alyssum bertolonii* and found significant variations in the concentrations of these acids at different seasons.

Homer *et al.* (1997) determined aminoacids in three species of Philippine hyperaccumulators. The results are shown below in Table 2.11.

Table 2.11. Aminoacids and nickel in leaf extracts of the Philippine hyperaccumulators *Walsura monophylla* (A), *Phyllanthus palawanensis* (B) and *Dichapetalum gelonioides* (C).

Material	A	B	C
Nickel in plants (μg/g)	1391	1990	7866
(μmol/g)	23.7	34.3	134
Total aminoacids (μmol/g)			
Alanine	1.78	0.49	0.13
Arginine	0.10	0.18	0.12
Aspartic acid	6.45	0.06	0.05
Cysteine	<0.05	<0.05	<0.05
Glutamine	29.8	0.15	0.12
Glycine	0.28	0.06	0.21
Histidine	0.17	0.23	0.18
Isoleucine	0.13	0.05	0.12
Leucine	0.14	0.05	<0.05
Lysine	0.13	0.09	0.13
Methionine	0.09	<0.05	<0.05
Phenylalanine	0.97	0.41	0.60
Proline	29.0	14.3	3.01
Serine	<0.05	0.16	0.70
Threonine	4.61	0.30	<0.05
Tyrosine	0.23	0.07	0.15
Valine	0.52	0.12	<0.05

Source: Homer *et al.* (1997).

Except for glucosamine, all the aminoacids in the standards were detected in the plant material. Proline appeared to be the most abundant in all three species though glutamine was present at high concentration in *W.monophylla*. The mole ratio of nickel to aminoacid ranged from 0.3 in *Walsura* to 23.4 in the more strongly nickel-accumulating *Dichapetalum*. In the latter case the amount of aminoacid fell far short of that required to complex all of the metal. In view of what is to follow below, it should be noted that the histidine concentration was comparatively low in all three species.

Homer *et al.* (1997) made the following conclusions about the nature of nickel complexes in hyperaccumulator plants:

1 - The high concentrations of nickel in leaves of some plants cannot be accounted for by complexing with molecules of high molar mass. It appears that this element is associated with polar molecules of low molar mass.

2 - Extracts consistently show the presence of carboxylic acids such as citric and malic.

3 - Nickel is usually associated with anionic complexes with at least one cationic form.

4 - There is evidence for the association of amino acids with nickel in xylem exudates of both accumulator and non-accumulator species.

The above work by Homer *et al.* (1997) was submitted and accepted by the journal in May 1995. A later paper by Krämer *et al.* (1996) was submitted in September of the same year but has appeared well before the other. The Krämer paper has made a significant advance in our knowledge of nickel binding in plants and has displayed compelling evidence that nickel is bound to free histidine in *Alyssum* species such as *A. lesbiacum*. The details of the experiments are as follows:

Plants were grown in hydroponic solution and the xylem sap removed from the upper surface of excised material. Over a wide range of nickel treatments there was a linear relationship between the nickel content and level of free histidine as is shown in Fig.2.4.

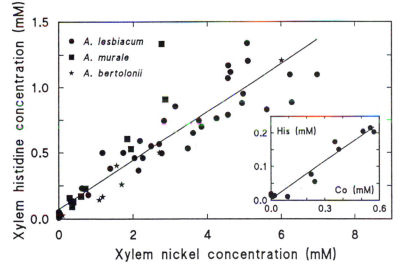

Fig.2.4. Relationship between free histidine and the nickel content of *Alyssum lesbiacum* seedlings grown under hydroponic conditions. The inset shows a similar pattern for cobalt. Source: Krämer (1996).

The proposed association between nickel and histidine was further reinforced by experiments in which histidine was fed to the same hyperaccumulator and also to the non-accumulator *Alyssum montanum*. The results are shown in Fig.2.5 below and show an increase in both biomass and nickel flux in the xylem of both of these two species.

The findings of Krämer (1996) and Krämer *et al.* (1996) beg the question as to the validity of the large volume of previous work that has implicated carboxylic acids in nickel complexing. The two theories are however not in conflict if we consider the separate questions of *transport* and *storage*. The *"sève bleue" Sebertia acuminata* (Fig.3.6) has a blue latex. A blue colour has never been found in complexes involving oxygen atom coordination to nickel and is usually associated with direct nitrogen-metal bonding. All the evidence points to aminoacid-nickel complexes involved in the transport process, whereas carboxylic acids could be implicated in the storage of the nickel.

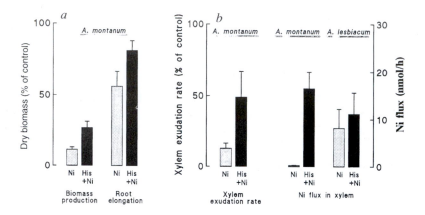

Fig.2.5. Effect of histidine on (a) biomass production and (b) xylem exudation and nickel flux in these exudates. *Alyssum montanum* is a non-accumulator and *A.lesbiacum* is a hyperaccumulator of nickel. Source: Krämer (1996).

The latest data are encouraging and it seems that after 20 years, we are somewhat closer to solving the nature of nickel complexes in hyperaccumulator plants as well as casting fresh light on the transport and storage mechanisms involved. These findings are not only of academic significance. There are very real practical applications in the burgeoning fields of phytoremediation and phytomining (see Chapters 11-15).

Localisation of nickel in plant tissues

The localisation of nickel in plant tissue of hyperaccumulators has been investigated by various South African workers. Przybyłowicz *et al.* (1995) used the Dynamic Analysis method for on-line elemental imaging to study nickel in *Senecio coronatus*, a hyperaccumulator from Natal Province. The technique permitted the determination of nickel, potassium and calcium concentrations in selected parts of the fruits and epidermis of the plant. Resolution was of the order of a few micrometres.

In a further study on the *S.coronatus*, Mesjasz-Przybyłowicz *et al.* (1994) used the proton microprobe to produce detailed high resolution maps of elemental distributions, principally nickel, in lateral and cross-sectional scans in the tissue of this species. The highest nickel concentrations were found in the epidermis of leaves, stems and roots. The scan permitted the determination of other elements in the plant material: namely S, Cl, K, Ca, Ti, Cr, Mn, Fe, Ni, Cu, Zn, Se, Br, Rb and Sr. Nickel data are shown in Table 2.12 below.

Table 2.12. Mean nickel content (μg/g dry weight) in soils and tissues of the hyperaccumulator *Senecio coronatus* using proton microprobe analysis.

Material	Kaapsehoop Mine	Agnes Mine
Epidermis	not determined	15,030
Leaf	12,500	5070
Stem	1600	4420
Root	110	590
Total in soil	3400	520
Available in soil	117	90

After: Mesjasz-Przybyłowicz *et al.* (1994).

Hyperaccumulation of copper and cobalt

Introduction

It is appropriate that phytochemical studies on copper and cobalt should be considered together since plants that are able to hyperaccumulate one element can usually do so in respect of the other. The only true "copper/cobalt plants" are found only in Central Africa (see Chapters 3 and 4) where they grow on mineralised outcrops in the Shaban Copper Arc and Zambian Copperbelt of Zaïre and Zambia respectively (Brooks and Malaisse, 1985; Brooks *et al.*,1992).

Hardly any phytochemical work has been carried out on these interesting plants because of the extreme difficulty in obtaining plant material or seed in view of the current unrest and instability in Zaïre where at present, about one third of the country is controlled by rebels.

However, some phytochemical studies were carried out by Morrison (1980) who worked on the copper/cobalt hyperaccumulators *Haumaniastrum robertii, Aeollanthus biformifolius, Buchnera metallorum, Faroa chalcophila* and *Silene cobalticola*. The initial studies involved fractionation of copper and cobalt by a process of sequential extraction (Bowen *et al.*,1962). The results are shown in Table 2.13.

In contrast to above work on *Alyssum* and the findings of Reilly (1969) for the copper indicator *Becium homblei*, very little of the heavy metals in the five Zaïrean metallophytes were found to be bound with aminoacids. It is obvious that most of the cobalt in the complexes is bound to organic ligands that form a number of polar compounds.

Table 2.13. The total metal content of dry plant tissue and fractionation (%) of copper and cobalt in plant tissue extracts of five Zaïrean metal-tolerant plants.

	I	II	III	IV	V
COBALT					
Total (µg/g)	2380	1510	134	4690	233
A	0.5	0.1	1.2	0.2	0.5
B	33.8	10.3	20.4	42.9	34.9
C	31.7	36.3	48.9	39.7	53.4
D	1.8	1.5	0.9	1.6	1.2
E	22.0	39.6	21.3	9.9	9.0
F	0.9	0.7	0.2	0.7	0.1
G	8.5	7.0	4.4	4.5	0.7
H	0.9	4.7	2.7	0.5	0.4
COPPER					
Total (µg/g)	3920	3520	700	489	33
A	0.8	0.1	1.4	1.7	2.5
B	11.3	9.1	16.6	12.6	22.2
C	47.8	59.3	54.6	39.9	38.8
D	4.2	4.0	0.8	3.0	1.9
E	26.0	20.7	17.4	18.3	23.2
F	1.2	0.6	0.2	2.5	0.5
G	1.8	2.0	3.6	7.1	5.4
H	6.9	4.3	5.3	14.9	5.6

I - *Aeollanthus biformifolius*, II - *Buchnera metallorum*, III - *Faroa chalcophila*, IV - *Haumaniastrum robertii*, V - *Silene cobalticola*.
A - neutral small molecules including aminoacids and pigments, B - water-soluble low molecular weight polar compounds, C - acid-soluble polar compounds and some structural groups, D - proteins and pectates, E - polar compounds and some structural groups, F - nucleic acids, G - remaining proteins and and polysaccharides, H - cellulose, lignin, and immobile fractions of cell walls. Source: Morrison (1980).

Localisation of cobalt in Haumaniastrum robertii

Proton microprobe analysis was performed on a leaf sample of *Haumaniastrum robertii* from Zaïre by Morrison *et al.* (1981). Determinations were made on the spatial distribution of cobalt, manganese, calcium and potassium in the sample. The metals were heterogeneously distributed with high cobalt concentrations associated with low potassium levels and *vice versa*. This is demonstrated in Fig.2.6. The lighter regions represent highest metal concentrations and the darkest the lowest. In Fig.2.7 the results of a scan across the leaf are shown. Manganese, cobalt and calcium follow each other very closely whereas the potassium content is inversely related to the other three metals. It is difficult to account for this pattern of distribution, but Morrison *et al.* (1981) have suggested that cobalt might be coprecipitated in the leaf along with calcium oxalate crystals that are known to occur in plants (Al-Rais *et al.*,1971).

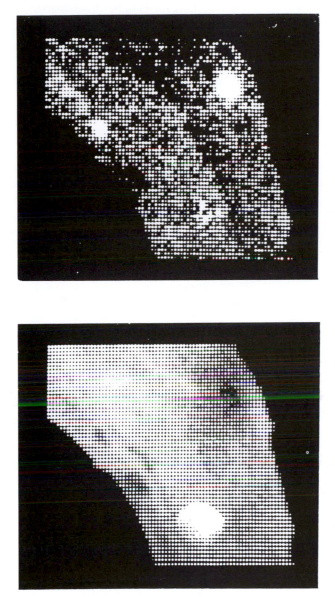

Fig.2.6. Proton microprobe micrographs showing the distribution of cobalt (upper photo) and potassium (lower photo) in leaf material of *Haumaniastrum robertii*. The lighter regions represent greatest concentrations. Source: Morrison (1980).

The possibility that cobalt existed as oxalate within the leaf sample was checked by determining cobalt and oxalate concentrations in fraction C of the extraction programme shown in Table 2.13. It was established that the two components existed in a 1:1.1 mole ratio. Despite the progress so far, a great deal of work remains to be done to establish the true nature of cobalt and copper complexes among the hyperaccumulators of Shaba Province, Zaïre.

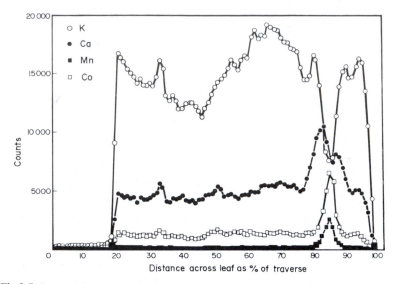

Fig.2.7. Proton microprobe scan of leaf of *Haumaniastrum robertii* showing distribution of potassium, calcium, cobalt and manganese. Region occupied by the leaf is 18-99% of the traverse distance which covers 6.75 mm. Source: Morrison (1980).

Hyperaccumulation of zinc and cadmium

It is appropriate that the phytochemistry of zinc and cadmium should be considered together because both metals can be hyperaccumulated by the same plants. The volume of such phytochemical studies is considerably less than in the case of nickel, perhaps because there are far fewer hyperaccumulators of these elements; 16 for zinc (see Table 3.4) and only one, *Thlaspi caerulescens*, for cadmium.

The phytochemistry of zinc has been reviewed by Jackson *et al.* (1990). This element is specifically required in plants. Metalloenzymes containing zinc include many of the enzymes involved in DNA and RNA synthesis as well as protein synthesis and metabolism. It is a cofactor in over 200 enzymes. At higher concentrations however, this element causes the blockage of xylem elements (Robb *et al.*, 1980).

Zinc-tolerant plants usually accumulate high levels of organic acids (cf. nickel above) and the formation of zinc-citrate complexes has been suggested by Godbold *et al.* (1988). This is analogous to the behaviour of nickel (see above).

Mathys (1977) suggested that zinc binds to malic acid (again similar to the behaviour of nickel).

Most of the phytochemical work involving zinc has been carried on non-accumulator species. Pollard and Baker (1996) studied the genetics of zinc hyperaccumulation by *Thlaspi caerulescens* and found significant differences in biomass and metal content between different populations of this species, but the study was not orientated to phytochemistry.

The phytochemistry of cadmium in plants has been studied to a greater extent than has that of zinc. This is because cadmium toxicity in foodstuffs is of more concern worldwide than is that of zinc.

A recent review by Prasad (1995) has a section on cadmium complexes in plants. He emphasised the role of *phytochelatins* (PC) in complexing with cadmium, but as these compounds have been given a separate section below, they will not be discussed further here. Prasad (1995) did however review the evidence for complexing of cadmium with metallothioneins (MT). The latter are thought to be aggregates of PCs and contain a number of aminoacids with glutamic acid, cysteine and glycine as major constituents. In cases where MT-like proteins were purified, metal ions including cadmium were also bound to low molecular weight peptides, probably PCs.

In addition to MTs and PCs, other intracellular ligands may play a role in complexing with Cd (e.g. glutathione and carboxylic acids). The latter were found to complex in the vacuoles of tobacco plants (Kortz *et al.*, 1989). The above findings are strongly reminiscent of the work on nickel plants where implication of aminoacids or carboxylic acids in complex formation with nickel was indicated.

Localisation of cadmium and zinc in Thlaspi caerulescens

The study of the phytochemistry of hyperaccumulators of cadmium is limited to the only species capable of such hyperaccumulation, *Thlaspi caerulescens*. There have been no detailed investigations into the phytochemistry of cadmium complexes in this species but some work has been carried out on the localisation of cadmium and zinc in the plant.

Vázquez *et al.* (1992) used X-ray microanalysis (EDAX) to examine the

Table 2.14. Mean zinc and cadmium concentrations (μg/g) in roots and shoots of *Thlaspi caerulescens* grown in hydroponic solutions.

Solution Zn	Solution Cd	Root Zn	Root Cd	Shoot Zn	Shoot Cd
1 μM	100 μM	400	19,500	590	13,680
100 μM	1 μM	3950	180	6190	170
100 μM	100 μM	4390	10,780	8950	12,500

Source: Vázquez *et al.* (1992).

localisation of zinc and cadmium in roots of *Thlaspi caerulescens*. The plants were grown hydroponically in solution containing zinc and cadmium in different proportions as shown in Table 2.14.

Large amounts of cadmium and zinc accumulated in both shoots and roots of the plant though zinc levels were always higher in shoots than in roots. The zinc and cadmium concentrations in cell walls, vacuoles and intercellular space are shown in Table 2.15.

Table 2.15. Zinc and cadmium concentrations (μg/g) and relative concentrations (counts per sec/background counts) of calcium and iron in cortex and epidermal and subepidermal cells as determined by EDAC in *Thlaspi caerulescens* roots exposed to varying zinc and cadmium concentrations (μM).

Element	[Zn]	[Cd]	Cell wall	Vacuole	Intercellular space
EPIDERMAL AND SUBEPIDERMAL CELLS					
Cadmium	100	1	n.d.	n.d.	n.d.
Zinc	100	1	258	1026	n.d.
Calcium	100	1	19.00	8.43	30.10
Iron	100	1	1.76	16.80	0.60
Cadmium	1	100	800	100	860
Zinc	1	100	350	90	180
Calcium	1	100	16.1	9.06	15.30
Iron	1	100	22.2	3.25	7.50
CORTEX					
Cadmium	1	100	500	350	2270
Zinc	1	100	190	230	700
Calcium	1	100	8.3	11.6	10.9
Iron	1	100	10.9	11.2	37.7

n.d. - not detected.
After: Vázquez *et al.* (1992).

The above authors concluded that cadmium is stored principally in the apoplast and to a lesser extent in the vacuoles where this element occurs together with calcium and iron. Zinc was mainly accumulated in vacuoles and to a lesser extent in cell walls. It shows a localisation pattern similar to other species not capable of storing this metal in an innocuous form.

Leblanc (pers. comm.) has recently studied the localisation of zinc and other elements in plant tissues of *Thlaspi caerulescens*. He used scanning electron microscopy (SEM) together with an energy-dispersive X-ray spectrometer (EDS). There was no evidence of surface contamination in any of the specimens but there were small (10 μm) hemispherical bodies at the surface of some of the leaves (Fig.2.8). These convex smooth bodies coat the small hairs (2-3 μm) on the leaf surface and adhere by capillary attraction. They are essentially compounds of carbon and oxygen with remarkable concentrations of zinc and sulphur. There are locally numerous (10-100/mm^2) on the surface of mature leaves. The bodies have the appearance of viscous droplets and have a chemical composition similar to metallothionines. Leblanc proposed that these bodies represent xylem exudates produced by excision of the leaf samples. Their presence at the leaf surface does however afford a means of studying this material by SEM.

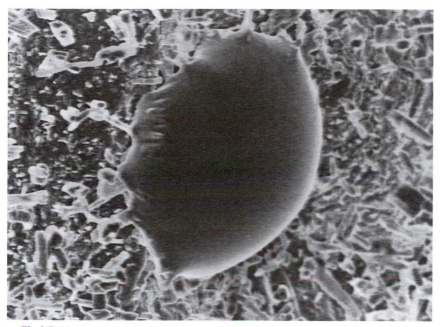

Fig.2.8. Electron micrograph of metal-rich viscous droplets (10 μM) adhering to fine hairs on the surface of a leaf of *Thlaspi caerulescens*. Photo by M.Leblanc.6

Hyperaccumulation of selenium

Selenium hyperaccumulators (Table 3.5) were among the first to be discovered and were certainly the first to undergo phytochemical research. This research was stimulated by the very practical considerations of finding an answer to stock losses caused by ingestion of seleniferous plants. Brown and Shrift (1982) and Jackson *et al.* (1990) have reviewed the plant chemistry of these species and some of their findings are given below. Selenium is complexed in plants largely as selenoaminoacids such as selenocysteine and selenomethionine. Selenoproteins also exist and are typified by glutathione peroxidise extracted from the marine diatom *Thalassiosira pseudonana*.

A notable feature of the selenoaminoacids and selenoproteins is the way in which selenium is able to mimic and replace sulphur in the selenium hyperaccumulators. It is not however a perfect mimic and the compounds produced are not able to replace the sulphur compounds in their physiological functions. This accounts for the high toxicity to animals of seleniferous plants such as the "*loco weed*" *Astragalus* (see Chapter 3).

One of the major selenium compounds present in hyperaccumulators of this element is Se-methylselenocysteine (Trelease *et al.*, 1960). Sulphur-methylcysteine is also found in *Astragalus* species but both of the above compounds are absent in non-accumulators of selenium.

Selenium hyperaccumulators are capable of excluding selenoaminoacids from their proteins (Peterson and Butler, 1967; Brown and Shrift, 1981).

Phytochelatins

Phytochelatins deserve separate treatment in this chapter because they have received a great deal of attention from researchers. During the past decade, discovery of the true nature of metal complexes in hyperaccumulator plants has almost seemed to workers in this field as being a present-day equivalent of the *philosopher's stone* of ancient alchemy where every new discovery does not seem to quite achieve finality of the solution of the problem.

Phytochelatins (PCs) are glutathione-derived peptides with the general structure of $(\gamma\text{-Glu-Cys})_n\text{Gly}$, where n varies from 2 to 11. The term was first proposed by Grill *et al.* (1985) and used to cover small cysteine-rich peptides that are biosynthesised by plants exposed to heavy metal stress. Another term, *phytochelates* was first proposed by Crowley *et al.* (1987) to replace the earlier term *phytosiderophores*. The latter term implied specificity for iron but these compounds chelate many elements other than iron. Phytochelates are not in fact phytochelatins. The former represents a more general expression to define all organic compounds capable of chelating metals, including phytochelatins themselves.

A variety of metal ions such as Cu(II), Cd(II), Pb(II) and Zn(II) induce PC synthesis in plants and may or may not complex with the metals themselves. Recently, Mehra *et al.* (1996) have shown that silver also binds with phytochelatins with values of n = 2,3 and 4.

Copper phytochelatins have been isolated from the metal-tolerant plant *Mimulus guttatus* by Salt *et al.* (1989). The copper-containing peptide contained 39.1% glutamic acid, 39.1% cysteine and 13.4% glycine. This suggested a phytochelatin with a ratio of 3:3:1 Glx:Cys:Gly.

The role of phytochelatins in plant physiology has been reviewed by Kinnersley (1993). Phytochelatins function by complexing with heavy metals in the plant and thus detoxifying them. If the plant fails to synthesise PCs the result is either growth inhibition or death (Steffens, 1990). High levels of PCs have been found in plants that display tolerance to heavy metals (Robinson, 1990). Grill *et al.* (1987) investigated the effect of copper upon growth and PC synthesis in *Rauvolfia serpentina*. As shown in Fig.2.9, hydroponic cultures of this plant grew rapidly without addition of Cu^{2+} and had negligible amounts of PCs. Upon addition of 50 μM Cu^{2+}, growth ceased for about 10 h because of the toxicity effect. At the same time, PC synthesis commenced. Within 2 days the intracellular PC content reached a limiting value and 80 % of the copper had been removed from the solution. Rapid cell growth then resumed. These experiments represent perhaps the greatest evidence so far obtained for the role of phytochelatins in metal tolerance strategies adopted by plants.

Huang *et al.* (1987) selected tomato varieties that were able to grow in the presence of 5 mM Cd^{2+}. This was more than 10 times the lethal concentration (400 μM) of unselected cells. The cadmium tolerance was associated with synthesis and accumulation of cadmium in these selected cells. The same cells also showed increased tolerance to copper.

Gekeler *et al.* (1989) showed that more than 200 plant species had the ability to form PCs. Extrapolating from his smaller number of samples, he concluded that over 300,000 plant species should have the ability to form these chelates as a detoxification strategy.

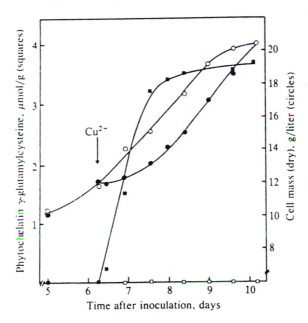

Fig.2.9. Growth retardation and phytochelatin synthesis in cells of *Rauvolfia serpentina* after metal stress following addition of 50 μM Cu^{2+} at which point phytochelatins (solid squares) began to be formed. Growth of copper-treated cells (solid circles) was inhibited for about 10 hours compared with the control (open circles). Thereafter, growth resumed. Source: Grill *et al.* (1987).

Accumulation and Transport Mechanisms of Heavy Metal Hyperaccumulation

So far, the discussion has been centred around the identity of heavy metal complexes in hyperaccumulators and little attention has been paid to the mechanisms of accumulation and transport within the plant. It may be appropriate therefore to end this chapter with a few general observations in this subject with particular reference to the large number of so-called "nickel plants."

It would seem that most, if not all, of these nickel plants are *facultative* rather than *obligate* since they will usually grow quite well in non-serpentinic soils, though they are liable to fungal attack and will not usually tolerate much competition from other plants. Nickel uptake may also well be a defence mechanism against insect predation (see Chapter 8). Even the unusual *Sebertia acuminata* from New Caledonia (see also Chapter 3) with its latex of almost pure nickel citrate is probably not obligate on serpentine. The latex seems most likely as an efficient agent in transporting phytotoxic nickel from the root systems to the leaves for storage until leaf fall.

All the above studies have shown that uptake of nickel by plants is a selective process not accompanied to any marked extent by accumulation of other elements to an inordinate degree. Still and Williams (1980) have discussed this point and proposed that the accumulation of nickel is due to a selective transport ligand in the root membrane. This "selector" is restricted to the membrane, so that other organic compounds such as citric and malic acids would be needed to act as "transport" ligands. A mechanism of hyperaccumulation based on the work of Still and Williams (1980) has been proposed by Morrison (1980). When the former examined the transport of nickel in plants they were unable to decide whether the transporting ligand was able to cross the root membrane as part of a selector-transport-nickel complex or became complexed after the nickel had been released after crossing the membrane.

Morrison (1980) proposed that the transport ligand forms a ternary (mixed ligand) complex with the selector-nickel complex on the internal surface of the root membrane. Thus the method proposed that the selector ligand (S) complex with the aquonickel(II) ions of the soil solution on the external surface of the root membrane. The selector-nickel complex (SNi) then moves through the membrane to the inner surface where a ternary complex (SNiT) is formed with the transport ligand (T) which is most probably an oxygen donor.

Sigel (1973) has shown that not only are mixed ligand systems generally more stable than binary systems, but mixed nitrogen-oxygen systems are more stable than nitrogen- or oxygen-dominated systems. Still and Williams (1980) have proposed that since the ligand has a nitrogen donor, the transport ligand is most favourably an oxygen donor. It is the breakdown of the ternary complex which releases the transport-nickel complex into the xylem. The selector ligand is then free to move back across the root membrane to repeat the process. This proposal of nickel complexation via a selector-transport-nickel complex has one particularly important advantage: free aquonickel(II) ions, the cause of nickel toxicity to plants, are not formed internally. Within the plant cells the nickel is always in a complexed form. The transport-nickel complex moves through the xylem to the leaf cells. Here it crosses the plasmalemma, cytoplasm and tonoplast to enter the vacuole. In the vacuole, the transport-nickel complex reacts with a terminal "acceptor" ligand (A) to form the acceptor-nickel complex (NiA), and releases the transport ligand. The acceptor nickel complex accumulates within the vacuole where it cannot interfere with the cell's physiological processes. The tonoplast must, therefore, be impermeable to this complex. The transport ligand moves out of the vacuole, the tonoplast being permeable to this ligand, through the cytoplasm and plasmalemma into the phloem and hence to the roots where it diffuses back into the xylem. It is not impossible that in some cases (e.g. citric acid), the transport and acceptor ligands could be the same species. The combination of roles would mean that a non-cyclic system is developed with the transport-nickel complex moving through to the vacuole where it accumulates while a fresh supply of ligand is always made available to the roots.

A great deal of research still remains to be done on hyperaccumulators of nickel. Fruitful avenues of research on these interesting plants should be centred on mechanisms of nickel uptake. Such work may well lead to progress in colonising serpentine areas with suitable crop plants and help to raise the standard of living of many developing countries such as the Philippines where so much potentially-arable land is unexploited due to the presence of serpentine and lateritic soils.

References

Al-Rais,A.H.,Myers,A. and Watson,L.(1971) The isolation and properties of oxalate crystals from plants. *Annals of Botany* 35, 1213-1218.

Bick,W.,De Kock,P.L. and Vergnano Gambi,O.(1982) A relationship between free amino acid and nickel contents in leaves and seeds of *Alyssum bertolonii*. *Plant and Soil* 66, 117-119.

Bowen,H.J.M.,Cawse,P. and Thick,J.(1962) The distribution of some inorganic elements in plant tissue extracts. *Journal of Experimental Botany* 13, 257-267.

Brooks,R.R.(1983) *Biological Methods of Prospecting for Minerals*. Wiley, New York, 322 pp.

Brooks,R.R.(1987a) Analytical chemists and dinosaurs. *Canadian Journal of Chemistry* 65, 1033-1041.

Brooks,R.R.(1987b) *Serpentine and its Vegetation*. Dioscorides Press, Portland, 454 pp.

Brooks,R.R.,Baker,.J.M. and Malaisse,F.(1992) Copper flowers. *National Geographic Research and Exploration* 8, 338-351.

Brooks,R.R.,Lee,J.,Reeves,R.D. and Jaffré,T.(1977) Detection of nickeliferous rocks by analysis of herbarium specimens of indicator plants. *Journal of Geochemical Exploration* 7, 49-77.

Brooks,R.R. and Malaisse,F.(1985). *The Heavy Metal Tolerant Flora of South-central Africa*. Balkema, Rotterdam, 199 pp.

Brooks,R.R.,Shaw,S. and Asensi Marfil,A.(1981) The chemical form and physiological function of nickel in some Iberian *Alyssum* species. *Physiologia Plantarum* 51, 167-170.

Brooks,R.R. and Yang,X.H.(1984) Elemental levels and relationships in the endemic serpentine flora of the Great Dyke, Zimbabwe, and their significance as controlling factors for this flora. *Taxon* 33, 392-399.

Brown,T.A. and Shrift,A.(1981) Exclusion of selenium from proteins of selenium-tolerant *Astragalus* species. *Plant Physiology* 67, 1051-1053.

Brown,T.A. and Shrift,A.(1982) Selenium toxicity and tolerance in higher plants. *Biological Reviews* 57, 59-84.

Cataldo,D.A.,McFadden,K.M.,Garland,T.R. and Wildung,R.E.(1988) Organic constituents and complexation of nickel (II), iron (III), cadmium (II) and plutonium (IV) in soybean xylem exudates. *Plant Physiology* 86, 734-738.

Crowley,D.E.Reed,C.P.P. and Szanizlo,P.J.(1987) Microbial siderophores as iron sources for plants. In: Winkelmann,G.,Van der Helm,D. and Neilands,J.B.(eds), *Iron Transport in Microbes, Plants and Animals*. VCH Publishers, New York, pp. 375-386.

Dunn,C.E.(1992) Biogeochemical exploration for deposits of the noble metals. In: Brooks,R.R.(ed.), *Noble Metals and Biological Systems: their Role in Medicine, Mineral Exploration and the Environment*. CRC Press, Boca Raton, pp. 47-89.

Dunn,C.E.(1995) A field guide to biogeochemical prospecting. In: Brooks,R.R., Dunn,C.E. and Hall,G.E.M.(eds), *Biological Systems in Mineral Exploration and Processing*, Ellis Horwood, Hemel Hempstead, pp.345-370.

Ernst,W.(1972) Ecophysiological studies on heavy metal plants in south Central Africa. *Kirkia* 8, 125-145.

Ernst,W.(1995) Sampling of plant material for chemical analysis. *The Science of the Total Environment* 176, 15-24.

Farago,M.E.,Mahmoud,I.E.D.A.W. and Clark,A.J.(1980) The amino acid content of *Hybanthus floribundus*, a nickel-accumulating plant and the difficulty of detecting nickel amino acid complexes by chromatographic methods. *Inorganic and Nuclear Chemistry Letters* 16, 481-484.

Gekeler,W.,Grill,E.,Winnacker,E.L. and Zenk,M.M.(1989) Survey of the plant kingdom for the ability to bind heavy metals through phytochelatins. *Zeitschrift für Naturforschung* 44c, 361-369.

Girling,C.A.,Peterson,P.J. and Warren,H.V.(1979) Plants as indicators of gold mineralization at Watson Bar, British Columbia, Canada. *Economc Geology* 74, 902-907.

Godbold,D.L.,Fritz,E.L. and Hütterman,A.(1988) Aluminum toxicity and forest decline. *Proceedings of the National Academy of Science USA* 85, 3888-3892.

Grill,E.,Winnacker,E.L. and Zenk,M.H.(1985) Phytochelatins: the principal heavy-metal complexing peptides of higher plants. *Science* 230, 674-676.

Grill,E.,Winnacker,E.L. and Zenk,M.H.(1987) Phytochelatins, a class of heavy-metal-binding peptides from plants, are functionally analogous to metallothioneins. *Proceedings of the National Academy of Sciences* 84, 439-443.

Hall,G.E.M.(1995) Sample preparation and decomposition. In: Brooks,R.R., Dunn,C.E. and Hall,G.E.M.(eds), *Biological Systems in Mineral Exploration and Processing*, Ellis Horwood, Hemel Hempstead, pp.427-442.

Hall,G.E.M.,Pelchat,J.C. and Dunn,C.E.(1990) The determination of Au,Pd, and Pt in ashed vegetation by ICP-mass spectrometry and graphite furnace atomic absorption spectrometry. *Journal of Geochemical Exploration* 37, 1-23.

Hall,G.E.M.,Rencz,A.N. and MacLaurin,A.I.(1991) Comparison of analytical results for gold in vegetation with and without high-temperature ashing. *Journal of Geochemical Exploration* 41, 291-307.

Homer,F.A.(1991) Chemical Studies on Some Plants That Hyperaccumulate Nickel. PhD Thesis, Massey University, New Zealand, 267 pp.

Homer,F.A.,Reeves,R.D. and Brooks,R.R.(1997) The possible involvement of aminoacids in nickel chelation in some nickel-accumulating plants. *Current Topics in Phytochemistry* 14, 31-33.

Huang,B.,Hatch,H. and Goldsbrough,P.B.(1987) Selection and characterization of cadmium tolerant cells in tomato. *Plant Science* 52, 211-221.

Jackson,P.J.,Unkefer,P.J.,Delhaize,E. and Robinson,N.J.(1990) Mechanisms of metal tolerance in plants. In: Katterman,F.(ed.) *Environmental Injury to Plants*. Academic Press, New York, pp. 231-255.

Jaffré,T.,Brooks,R.R.,Lee,J. and Reeves,R.D.(1976) *Sebertia acuminata* a nickel-accumulating plant from New Caledonia. *Science* 193, 579-580.

Kelly,P.C.,Brooks,R.R.,Dilli,S. and Jaffré,T.(1975) Preliminary observations on the ecology and plant chemistry of some nickel-accumulating plants from New Caledonia. *Proceedings of the Royal Society (London) Section B* 189, 69-80.

Kersten,W.J.,Brooks,R.R.,Reeves,R.D. and Jaffré,T.(1980) Nature of nickel complexes in *Psychotria douarrei* and other nickel-accumulating plants. *Phytochemistry* 19, 1963-1965.

Kinnersley,A.M.(1993) The role of phytochelatins in plant growth and productivity. *Plant Growth Regulation* 12, 207-218.

Kortz,R.M.,Evangelou,B.P. and Wagner,G.J.(1989) Relationships between cadmium, zinc, Cd-peptide and organic acid in tobacco suspension. *Plant Physiology* 91, 780-787.

Krämer,U.(1996) Nickel Hyperaccumulation in the Genus *Alyssum* L. PhD Thesis, University of Oxford.

Krämer,U.,Cotter-Howells,J.D.,Charnock,J.M.,Baker,A.J.M. and Smith,J.A.C. (1996) Free histidine as a metal chelator in plants that accumulate nickel. *Nature* 379, 635-638.

Lee,J.(1977) Phytochemical and Biogeochemical Studies in Nickel Accumulation by some New Caledonian Plants. PhD Thesis, Massey University, New Zealand, 181 pp..

Lee,J.,Reeves,R.D.,Brooks,R.R. and Jaffré,T.(1977) Isolation and identification of a citrato complex of nickel from nickel-accumulating plants. *Phytochemistry* 16, 1503-1505.

Lee,J.,Reeves,R.D.,Brooks,R.R. and Jaffré,T.(1978) The relationship between nickel and citric acid in some nickel-accumulating plants. *Phytochemistry* 17, 1033-1035.

Mathys,W.(1977) The role of malate, oxalate and mustard oil glucosides in the evolution of zinc-resistance in herbage plants *Physiologia Plantarum* 40, 130-136.

McGrath,S.P.,Sidoli,C.M.D.,Baker,A.J.M. and Reeves,R.D.(1993) The potential for the use of metal-accumulating plants for the *in situ* decontamination of metal-polluted soils. In: Eijsackers,H.J.P. and Hamers,T.(eds), *Integrated Soil and Sediment Research: a Basis for Proper Protection.* Kluwer Academic Publishers, Dordrecht, pp. 673-676.

Mehra,,R.K.,Tran,K.,Scott,G.W.,Mulchandani,P. and Saini,S.S.(1996) Ag(I)-binding to phytochelatins. *Journal of Inorganic Biochemistry* 61, 125-142.

Mesjasz-Przybyłowicz,J.,Balkwill,K.,Przybyłowicz,W.J. and Annegarn,H.J. (1994) Proton microprobe and X-ray fluorescence investigations of nickel distribution in serpentine flora from South Africa. *Nuclear Instruments and Methods in Physics Research* B89, 208-212.

Minguzzi,C. and Vergnano,O.(1948) Il contenuto di nichel nelle ceneri di *Alyssum bertolonii* Desv. *Memorie dagli Società Toscana di Scienze Naturali Serie A* 55, 49-74.

Morrison,R.S.(1980) Aspects of the Accumulation of Cobalt, Copper and Nickel by Plants. PhD Thesis, Massey University, New Zealand, 287 pp.

Morrison,R.S.,Brooks,R.R.,Reeves,R.D.,Malaisse,F.,Horowitz,P.,Aronson,M. and Merriam,G.R.(1981) The diverse form of heavy metals in tissue extracts of some metallophytes from Shaba Province, Zaïre. *Phytochemistry* 20, 455-458.

Pancaro,L.,Pelosi,P.,Vergnano Gambi,O. and Galoppini,C.(1978a) Ulteriori indagini sul rapporto tra nichel e acidi malico e malonico in *Alyssum. Giornale Botanico Italiano* 112, 141-146.

Pancaro,L.,Pelosi,P.,Vergnano Gambi,O. and Galoppini,C.(1978b) Further contribution on the relationship between nickel and malic and malonic acids in *Alyssum. Giornale Botanico Italiano* 112, 282-283.

Pelosi,P.,Galoppini,C. and Vergnano Gambi,O.(1974) Sulla natura dei composti del nichel presenti in *Alyssum bertolonii* Desv. Nota 1. *Agricoltura Italiana* 29, 1-5.

Pelosi,P.,Fiorentini,R. and Galoppini,C.(1976) On the nature of nickel compounds in *Alyssum bertolonii* Desv. *Agricultural and Biological Chemistry* 40, 1641-1642.

Peterson,P.J. and Butler,G.W.(1967) Significance of selenocystathionine in an Australian selenium-accumulating plant, *Neptunia amplexicaulis. Nature* 213, 599-600.

Pollard,A.J. and Baker,A.J.M.(1996) Quantitative genetics of zinc hyperaccumulation in *Thlaspi caerulescens. New Phytologist* 132, 113-118.

Prasad,M.N.V.(1995) Cadmium toxicity and tolerance in vascular plants. *Environmental and Experimental Botany* 35, 525-545.

Przybyłowicz,W.J.,Pineda,C.A.,Prozesky,V.M. and Mesjasz-Przybyłowicz,J. (1995) Investigation of Ni hyperaccumulation by true elemental imaging.

Nuclear Instruments and Methods in Physics Research B 104, 176-181.

Reeves,R.D. and Brooks,R.R.(1983a) Hyperaccumulation of lead and zinc by two metallophytes from a mining area in Central Europe. *Environmental Pollution Series A* 31, 277-287.

Reeves,R.D. and Brooks,R.R.(1983b) European species of *Thlaspi* L. (Cruciferae) as indicators of nickel and zinc. *Journal of Geochemical Exploration* 18, 275-283.

Reilly,C.(1969) The uptake and accumulation of copper by *Becium homblei* (De Wild.)Duvign. and Plancke. *New Phytologist* 68, 1081-1087.

Robb,J.,Busch,L. and Rauser,W.E.(1980) Zinc toxicity and xylem vessel alteration in white bean. *Annals of Botany (London)* 46, 43-50.

Robinson,N.J.(1990) Metal binding polypeptides in plants. In: Shaw,A.J.(ed.) *Heavy Metal Tolerance in Plants*. CRC Press, Boca Raton, 195-214.

Salt,D.E.,Thurman,D.A.,Tomsett,A.B. and Sewell,A.K.(1989) Copper phytochelatins of *Mimulus guttatus*. *Proceedings of the Royal Society (London) B236, 79-89*.

Sigel,H.(1973) Structural aspects of mixed-ligand complex formation in solution. In: Sigel,H.(ed.) *Metal Ions in Biological Systems. v.2*. Dekker, New York.

Steffens,J.C.(1990) The heavy-metal binding peptides of plants. *Annual Reviews of Plant Physiology and Plant Molecular Biology* 41, 553-575.

Still,E.R. and Williams,R.J.P.(1980) Potential methods for selective accumulation of nickel II ion in plants. *Journal of Inorganic Biochemistry* 13, 35-40.

Stockley,E.(1980) Biogeochemical Studies on the Nickel Complex Contained in the Nickel-accumulating Legume *Pearsonia metallifera* from the Great Dyke Area, Zimbabwe. BSc Hons Thesis, Massey University, New Zealand.

Trelease,S.F.,Disomma,A.A. and Jacobs,A.L.(1960) Selenoamino acid found in *Astragalus bisulcatus*. *Science* 132, 3427.

Vázquez,M.D.,Barceló,J.,Poschnrieder,C.,Mádico,J.,Hatton,P.,Baker,A.J.M. and Cope,G.H.(1992) Localization of zinc and cadmium in *Thlaspi caerulescens* (Brassicaceae), a metallophyte that can hyperaccumulate both metals. *Journal of Plant Physiology* 140, 350-355.

Vergnano Gambi,O.,Pancaro,L. and Formica,C.(1977) Investigations on a nickel-accumulating plant *Alyssum bertolonii* Desv. I Nickel, calcium and magnesium content and distribution during growth. *Webbia* 32, 175-188.

Walsh,A.A.(1955) The application of atomic absorption spectra to chemical analysis. *Spectrochimica Acta* 7, 108-117.

Wild,H.(1974) Indigenous plants and chromium in Rhodesia. *Kirkia* 9, 233-241.

Yang,X.H.,Brooks,R.R.,Jaffré,T. and Lee,J.(1985) Elemental levels and relationships in the Flacourtiaceae of New Caledonia and their significance for the evaluation of the "serpentine problem". *Plant and Soil* 75, 281-292.

Chapter three:

Geobotany and Hyperaccumulators

R.R. Brooks

Department of Soil Science, Massey University, Palmerston North, New Zealand

Introduction

Biological methods of prospecting (Brooks *et al.*,1995) are extremely varied and include several subdisciplines such as *biogeochemistry, geobotany, geozoology* and *geomicrobiology*. Only the first two, as well as geomicrobiology in part (see Chapter six), really impinge on the use of hyperaccumulator plants; so there will be no further mention of geozoology beyond the statement that it involves the use of animals in mineral prospecting.

Biogeochemical procedures (Brooks *et al.*,1995) date back only to 1938 when the Soviet scientist S.M.Tkalich (Tkalich, 1938) observed that the iron content of tundra plants accurately reflected the concentration of this element in the substrate. This topic will be discussed in greater detail in Chapter 4.

In contrast to biogeochemistry that depended on advances in analytical chemistry before it could be developed, geobotany depends only on visual observation of the vegetation cover and therefore has a much longer history, dating back at least to Roman times.

The famous mediaeval metallurgist and miner, Georgius Agricola (Georg Bauer) published his celebrated *De Re Metallica* (Of Metallic Matters) in 1556. He can really be acknowledged as the world's first practical geobotanist. In the English translation (Hoover and Hoover, 1950) we read that:

> Over concealed ore bodies the soil will produce only small and pale-coloured plants... in a place where there is a multitude of trees, if a long row of them at an unusual time lose their verdure and become black or discoloured and frequently fall by the violence of the wind, beneath this spot there is a vein...likewise along a course where a vein extends, there grows a certain herb or fungus that is absent from the adjacent space....

In the same century, Thalius (1588) determined that *Minuartia verna* was an indicator of ores in the Harz Mountains, Germany. Even as early as Roman times Vitruvius in the reign of Augustus Caesar observed in 10 BC that:

...the slender bulrush, the wild willow, the alder, the agnus castus, reeds, ivy and the like...usually grow in marshy places...but water is to be sought in those regions and soils other than marshes, in which such trees are found naturally and not artificially planted.

The target was water rather than minerals, but the principle was the same. Geobotanical methods of mineral exploration are concerned with the detection of subsurface mineralisation by an interpretation of its vegetative cover. This can involve several distinct fields of methodology and includes:
    1 - study of indicator plants;
    2 - study of ore-indicating plant communities;
    3 - study of morphological changes to plants;
    4 - aerial photography and/or satellite imagery.
Though geobotany was the earliest of the biological methods of exploration and predated biogeochemistry by as much as 2000 years, the former techniques involving solely ground observations have become less popular with the passing years, perhaps because so many of the world's obvious or near-surface mineral deposits have already been discovered. It is in satellite imagery where the greatest geobotanical interest lies today and will do so for the foreseeable future. In spite of this trend, it is clear that ground observations will still be important because it is "ground truth" that must be employed to evaluate anomalies indicated by aerial photography or satellite imagery.

Fig.3.1. *Aeollanthus biformifolius*, a hyperaccumulator of cobalt and copper from Zaïre. Photo by F.Malaisse.

As this book is concerned with hyperaccumulation of metals by plants, it is clear that the broad geobotanical approach involving satellite imagery for example, will hardly be appropriate for the small number of rare plants that will need to be investigated by field operations. However, a good example of successful field operations can be provided by the story of mineral exploration in Central Africa which reached a peak in the 1950s, although it began at the turn of the century. The prospectors of this period soon realized that small herbs such as *Haumaniastrum katangense* and *Aeollanthus biformifolius* (hyperaccumulators of copper and cobalt - see Fig.3.1) and *Becium homblei* (a non-accumulator), both members of the Lamiaceae (mint family), were confined to soils containing in excess of 100 μg/g (ppm) copper and were able to use these plants to map these metalliferous soils. These and other plants became known as *copper flowers* (Anon. 1959; Brooks *et al.*,1992a; Duvigneaud,1958; 1959).

Despite the emphasis on individual indicator plants, implicit in a work on hyperaccumulator taxa, these unusual plants must not be considered in isolation. They usually form part of a specific community containing many non-accumulating companions and associated plants and therefore the whole community must be considered and studied in a manner recommended by geobotanists. This involves plant mapping and the recognition of specific phytosociological groupings. These topics will now be considered below.

Mapping Techniques in Indicator Geobotany

Fundamental principles

The subject of plant mapping has been covered by Brooks (1983) and Brooks *et al.* (1995). In many cases the distinction between two different plant communities is so pronounced that a mere visual observation is all that is needed to observe the geological boundaries. This is particularly the case for serpentine floras. Even when the rock types are relatively similar, pronounced differences in vegetation can occur.

In cases where purely visual observation is not sufficient, recourse must be made to geobotanical mapping. Such an operation should ideally be carried out by a skilled botanist or ecologist since the amateur can readily confuse plants which are superficially similar and might have difficulty in distinguishing different ecotypes (even if they were indeed distinguishable).

Plant mapping involves the selection of a number of sample plots known as *quadrats*. There is no general agreement as to the best method of selection of these quadrats. Some workers believe that they should be selected in a random manner, whereas others believe that they should be chosen in a definite pattern. This latter procedure, though frequently used, is open to criticism insofar as it makes the assumption that the vegetation associations are already known.

If the non-random approach is to be used, the following procedure should be adopted. Every plot should have the utmost uniformity it is possible to find in the area concerned; not only with regard to plant species, but also with consideration of such factors as aspect, slope, drainage, relief and altitude. Particular care must be taken that the quadrats do not include two or more different associations as may occur with plots of non-uniform slope.

The size of the quadrat now has to be established. As a general rule, the size should be the minimum needed to include most of the plants of the association and will obviously be related to the homogeneity of the community. In assessing the size of the *minimal area*, the law of diminishing returns will obviously apply: i.e. successive increases in size of the sample area will give progressively smaller amounts of additional information. The concept of minimal area has been reviewed by Goodall (1952). A species-area plot at first rises sharply and then becomes flatter, although never completely horizontal because the whole area would have to be included in the quadrat to be sure of including every single species present. Greig-Smith (1964) has discussed the problem of minimal area at some length, and there appears to be no general agreement as to a universal criterion to determine this area. As a general rule however, the following procedure may be adopted in the field.

Begin with a small quadrat of perhaps 5 m², note the species within it and then increase the size of the plot progressively (10, 50, 100 m² etc.) noting at each stage additional species encountered. When there is an appreciable drop in the rate of increase of new species, the optimum quadrat size will have been found. When the size and position of the test plots have been established, the next procedure will involve an evaluation of the *density* of individual species and their *spacing* (reciprocal of density). In determining density, direct counting or a scale of numbers may be used. The scale is somewhat arbitrary but can give good results in the hands of a skilled field worker. The system is as follows: 1-very rare, 2-rare, 3-infrequent, 4-abundant, 5-very abundant. One disadvantage of this system is that data are heavily dependent on the personal assessment of one individual and are not always comparable with data collected by other workers. The use of this scale is nevertheless justified on the grounds of speed and practicality. It is also a useful system when the vegetation cover is dense so that counting of absolute numbers of individuals would have been impossible. A geobotanical map of an area need contain nothing more than the density or spacing data enumerated above. If other parameters are added, a more meaningful map can be compiled.

The space demand of a species introduces another concept, that of *cover*. It is assumed that the entire shoot system of a plant is projected on the ground and that this area (equal to the area of shade if the sun were directly overhead), represents the cover. Cover is usually expressed as a percentage. The total of all species will often exceed 100% due to overlap.

A quick method of assessing cover involves measuring the total length of interception by plants on line transects. The proportion of the total length of the transect intercepted by a given species is a measure of its cover. Another procedure is the *points-quadrat method*, a strategy particularly suitable for small plants in a small quadrat of perhaps 1 m². The quadrat is divided into a grid of perhaps 400 squares each 5 cm × 5 cm. The interception of each plant species with one of the squares is recorded and the cover hence calculated. A typical grid of this type is shown in Fig.3.2 which represents a 1 m^2 experimental quadrat over serpentine soils at Murlo in Italy. The main species present are *Armeria maritima* and the nickel hyperaccumulator *Alyssum bertolonii*. The reader is referred to Greig-Smith (1964) for further details of these and other means of measurement.

Fig.3.2. Experimental plot over serpentine soils in the Murlo district of Italy. This shows *Armeria maritima* and the hyperaccumulator *Alyssum bertolonii*. Photo by A.Chiarucci.

Some mention should also be made of the concept of *layering*. The vegetation may be considered not only laterally but also vertically. In the vertical concept, several distinct layers are recognized. These are the tree layer, shrub layer, herb layer and moss layer. Clearly, the denser the upper tree layer, the greater will have to be the tolerance towards reduced light intensity by the lower members of the community. Mosses, as might be expected, will tolerate the least light intensity.

There are other criteria of plant communities which can also be employed in geobotanical mapping. These are however somewhat outside the scope of this book and will only be mentioned briefly.

Sociability expresses the space relationship of individual plants and can be expressed in terms of a simple scale (Braun Blanquet, 1932) as follows: soc.1, growing in one place singly; soc.2, grouped or tufted; soc.3, in troops, small patches or cushions; soc.4, in small colonies in extensive patches or forming carpets; soc.5, in great crowds.

Vitality is a measure of how a plant prospers in the community and can be expressed by a number of conventional symbols (Braun Blanquet, 1932) as follows: ● well developed and regularly completing life cycle; ◉ strong and increasing but usually not completing life cycle; ☉ feeble but spreading and never completing life cycle; ○ occasionally germinating but not increasing.

Periodicity is a measure of the regularity or absence of rhythmic phenomena in plants such as flowering, fruiting etc. A study of this criterion involves continuous and systematic research and is outside the scope of this book.

If mean values for density, spacing, cover and other parameters are determined for each species and averaged over several quadrats (preferably chosen randomly), it will be possible to characterize the plant associations in the test area and to attempt to correlate this with the geological environment.

Although the use of quadrats is the most usual approach in indicator geobotany, a different approach will be needed for plants growing over narrow ore bodies. In such cases, the use of line transects or belt transects is recommended. Line transects consist of parallel straight lines run through an area with the aid of tape measures and compasses, whereas belt transects consist of lines of continuous quadrats running across the profile of the area.

Representation of plant densities in each quadrat can be made in various ways. A simple method is to record the number of individuals in each 30m × 30m (a common dimension) quadrat or else to record each species as a percentage of the total number of individuals of all species in each quadrat. As mentioned above, it is much easier to carry out plant mapping in arid areas than in thick forest. In the latter environment, the line transect is usually more appropriate. If a multi-storey vegetation assemblage is involved, it is hard to evaluate the extent to which each plant or plant grouping intersects the line. In such cases an alternative method consists of laying a tape measure through the community and recording every individual of a specified size (e.g. trunk diameter) within a predetermined distance of the tape. There are of course many variants of this procedure which can be adapted to specific conditions.

Another type of plant mapping is more qualitative. Figure 3.3 represents work carried out on the copper-cobalt deposits of Shaba Province, Zaïre (Malaisse *et al.*, 1979). Representation is exaggerated both in regard to altitude and the height of vegetation. This form of mapping nevertheless gives a better feeling of reality than the more quantitative scheme. Hyperaccumulators seldom exist in isolation and form part of a specific *phytosociological community*. This discipline is not well understood in the English-speaking world since it is hardly taught in its universities and colleges. The basic elements of the subject will therefore now be outlined in order to assist in its understanding.

Phytosociological classification

In phytosociology, constituents of a plant community are grouped into a series of hierarchial classificatory groups that when published as a table seem extremely complicated. The highest hierarchial group is the *order* (Ordnung).

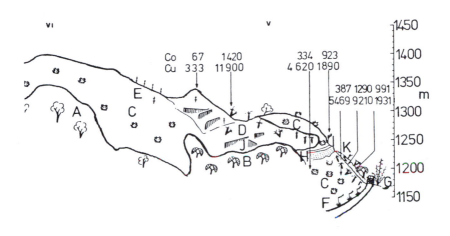

Fig.3.3. Map of vegetation over copper mineralization at Fungurume,
Shaba Province, Zaïre. Source: Malaisse *et al.* (1979).

As an example of an order we can refer to the order *Violetalia calaminariae* to which, according to Ernst (1974), all Western European metal tolerant plant communities belong. The order is named after *Viola calaminaria*, the well known hyperaccumulator of zinc with over 1% of this element in its dry tissue (Baumann, 1885). The endings *-etalia,-ariae* are typical of the naming of orders.

The next grouping in the hierarchial succession is the so-called *alliance* (Verband). Referring once again to Ernst (1974) we find that the *Violetalia calaminariae* order can be subdivided into three alliances: *Galio anisophylli-Minuartion vernae, Thlaspion calaminaris* (based on another zinc hyperaccumulator, *Thlaspi calaminare*) and *Armerion halleri*. The ending *-ion* is usually found in the nomenclature and names such as *Armerion halleri* refer to key species such as *Armeria halleri*.

The alliance is further subdivided into *associations* and *subassociations* each with their *differentials* (Trennarten) and *companions* (Begleiter). The nomenclature for associations is along the lines of *Violetum dubyanae* with the typical ending *-etum*. This association is part of the *Galio anisophylli-Minuartion vernae* alliance of the *Violetalia calaminariae* order. Because of the extreme complexity of the hierarchial system, it is seldom possible to present a table of data to show the complete progression from order down to subassociation. It is therefore the custom to present tables of only the association. This is demonstrated in Table 3.1 below which is a condensed version of the *Asperulo-Staehelinetum baeticae* association of serpentine rocks in Andalusia, Spain (López González, 1975). The community is host for the nickel hyperaccumulator *Alyssum* *malacitanum*. The plus sign indicates that the species is present. Each column refers to an individual series of observations in a transect across the community. A view of this community is shown in Fig.3.4.

Fig.3.4. View of the *Asperulo-Staehelinetum baeticae* association between Alora and Carratraca over serpentine in Andalusia, Spain. The large mass in the centre is the nickel hyperaccumulator *Alyssum malacitanum*. Photo by R.R.Brooks.

Phytosociology is not a particularly exact science insofar as two workers studying the same area might well produce quite different evaluations of the groupings that are present. Nevertheless, the discipline is extremely useful in being able to summarize in tables, a semiquantitative evaluation of the communities that are present. Such data can be of great value for geobotanical mineral prospecting, particularly if they can be combined with aerial photographs or satellite imagery.

The above discussion leads quite naturally to a consideration of some of the major floras that contain important hyperaccumulator plants.

Plant Communities with Hyperaccumulator Plants

Introduction

It is clear from the above discussion that many mineralised areas apparently have a characteristic flora that may be unique to that particular locality or may be specific for all discrete areas of mineralisation in that region. It is among such

Table 3.1. The *Asperulo-Staehelinetum baeticae* association over serpentine soils in Andalusia, Spain.

Relevé Number	1	2	3	4
Height above sea level (m)	700	600	550	600
Percentage cover	80	90	95	95
Exposure	SW	SW	S	NW
Slope	30°	60°	40°	40°
Area of quadrat (m²)	100	600	100	80
Number of species in the inventory	45	28	33	33
Characteristic species of the association				
Asperula asperrima var. *asperrima* - -	-	+		
Staehelina baetica	2.3	1.2	0.2	1.2
Characteristic species of the *Staehelino-Ulicion baetici* alliance				
Centaurea carratracensis	1.2	+	1.2	2.1
Ulex parviflorus subsp. *funkii*	3.3	3.2	2.2	3.3
Teucrium haenseleri	1.2	1.1	-	-
Scorzonera baetica	2.1	2.2	1.1	1.1
Serratula baetica var. *baetica*	1.2	-	2.2	1.2
Characteristic species of the *Phlomiditalia purpurea* order				
Phlomis purpurea subsp. *purpurea*	3.3	3.2	1.2	1.2
Elaeoselinum tenuifolium	-	1.2	1.2	-
Genista umbellata subsp. *equisetiformis*	-	+	-	-
Characteristic species of the *Ononido Rosmarinetea* class				
Cistus salviefolius	2.1	-	1.2	2.2
*Alyssum malacitanum**	2.2	2.2	1.2	2.3
Companions				
Asphodelus cerasiferus	1.2	1.2	-	-
Dactylis glomerata var. *hispanica*	2.2	2.2	1.2	-
Brachypodium distachyum	1.2	+	1.2	-
Fumana thymifolia subsp. *glutinosa*	1.2	-	+	2.2

* - Hyperaccumulator of nickel, + - present but rare, numerals - relative abundances. After López Lopez González (1975).

characteristic floras that most, if not all, metal-hyperaccumulating plants are to be found. The phytosociological communities that comprise these floras have been described by Ernst (1974). They include:

    1 - ultramafic (serpentine) floras;

    2 - the floras of copper/lead/zinc sulphide mineralisation;

    3 - selenium floras;

    4 - the copper/cobalt floras of Central Africa.

Each of these groupings will be considered individually below. Before doing so however, it will be appropriate to detail the abundances of specific elements in "normal" and hyperaccumulating plant species as is shown in Table 3.2. The table shows background concentrations, the metal content of non-hyperaccumulators growing in mineralisation, and the mean metal content of typical hyperaccumulators. The values in the last column are not the highest concentrations that can be encountered in hyperaccumulators.

Table 3.2 Typical mean elemental concentrations ($\mu g/g$ [ppm] dry weight) in "normal" and hyperaccumulator plants.

Element	Normal in background	Normal in mineralisation	Hyperaccumulator plant
Cadmium	0.1	2	100
Cobalt	1	3	5000
Copper	10	20	5000
Manganese	400	1000	10000
Nickel	3	20	5000
Selenium	0.1	1	1000
Zinc	70	100	10000

After: Brooks (1983).

Ultramafic ("serpentine") floras

Of all morphological changes produced in vegetation by the substrate, those found in serpentine floras are probably the most extreme. So great is the differentiation between these floras and those of adjacent substrates that geological boundaries are readily observable.

Figure 3.5 shows the boundary between ultramafic (serpentine - right side of photo) and schistose (left side) rocks at Red Mountain, South Island, New Zealand. The thick vegetation over the schist is replaced by an almost totally barren plant cover over ultramafics.

Until fairly recently there were no textbooks dealing with serpentine ecology in general and the world literature was confined to a few classical papers on specific regions such as Finland (Lounamaa, 1956), Greece (Krause, 1958), Italy (Pichi Sermolli, 1948), New Caledonia (Jaffré, 1980), Portugal (Pinto da Silva, 1970), Sweden (Rune, 1953), and Zimbabwe (Wild, 1965). This deficiency has however been remedied by the recent appearance of four books (Brooks 1987; Roberts and Proctor 1992; Baker et al.,1992; Brooks et al.,1995) that have covered the field of serpentine ecology. It is to be expected that further books will appear in the near future due to the interest generated by these earlier volumes.

The ecology and physiognomy of serpentine floras have often been studied from the standpoint of the so-called *serpentine problem* in which it has been sought to establish the reason for the establishment of this unusual characteristic flora over soils normally considered to be highly phytotoxic.

A serpentine community will usually show a general sparseness of vegetation with a shortage of species as well as individuals. There is usually a high degree of accompanying endemism. For example, the serpentine flora of New Caledonia has an endemism as high as 90% (Jaffré, 1980), whereas in Britain there may be only one or two species that are truly endemic to this type of substrate.

Fig.3.5. Vegetation boundary between schistose rocks (left) and ultramafics (right) at Red Mountain, South Island, New Zealand. Photo by L.F.Molloy.

Serpentine plant communities are hosts to one of the most important groups of hyperaccumulator, the so-called *nickel plants* (Brooks *et al.*,1977) that have over 1000 μg/g (ppm) nickel in their dry mass. Perhaps one of the most astonishing examples of this hyperaccumulation is afforded by the New Caledonian *Sebertia acuminata* (*sève bleue* i.e. "blue sap") that exudes a blue sap containing over 11% nickel (wet weight). This is illustrated in Fig.3.6.

Serpentine soils are considerably different from "normal" soils in that they are very rich in chromium, cobalt, iron, magnesium and nickel as well as being deficient in the nutrients calcium, molybdenum, nitrogen, phosphorus and potassium (Brooks, 1987).

Lounamaa (1956) and Robinson *et al.* (1935) concluded that the unusual serpentine flora results from excessive amounts of chromium, cobalt and nickel. Other workers (Minguzzi and Vergnano, 1948; Rune, 1953) claimed that it was nickel that was mainly responsible for the specialized vegetation.

The low calcium content of serpentine soils led Kruckeberg (1985), Walker (1954), and Walker *et al.* (1955) to conclude that the flora of these soils is unusually tolerant of low calcium/magnesium ratios in the substrate. Yet another theory (Krause, 1958; Paribok and Alekseeva-Popova, 1966; Sarosiek, 1964) is that survival of plants on serpentine soils depends on their ability to adapt, at least partially, to all the adverse edaphic factors of these soils, unlike non-serpentine plants, and in this way counteract to some extent, the deficiency of calcium in ultramafic soils.

The possible role of nickel in the control of serpentine floras has been discussed by Lee (1974) who observed that some New Zealand serpentine-endemic plants often had a higher nickel content than did non-endemic species. For example, the endemic *Pimelea suteri* contained over 400 μg/g nickel in its dried leaves compared with no more than 50 μg/g in other plants growing over the same substrate. In the case of serpentine-endemic *Alyssum* species (Brooks *et al.*,1979) the difference is even more pronounced as nickel contents of over 1 % (10,000 μg/g) have been found in some species.

Fig.3.6. The New Caledonian *Sebertia acuminata* (sève bleue) that exudes a sap containing 11 % nickel (wet weight). Photo by T.Jaffré

Perhaps one of the most comprehensive surveys of the vegetation of a single ultramafic region is a study by Jaffré (1980) of the serpentine flora of New Caledonia. It is clear however, that the *serpentine problem* is far from being solved and much work remains to be done before it can be established with certainty why this peculiar plant community is to be found over ultramafic rocks. It is even possible that the main factors influencing the development of a serpentine flora vary in different areas and that no universal factors exist.

Ultramafic (serpentine) soils have provided by far the greatest number of hyperaccumulators, mainly of nickel and some of manganese. Brooks *et al.* (1995) have reported 188 nickel plants but the total is now well over 300 following the discovery of about 128 more in Cuba by Reeves *et al.* (1996) and Reeves (pers. comm. 1997). The list of nickel hyperaccumulators is given in Table 3.3 below and does not include the last 48 discovered in Cuba. In order to save space, individual taxa within a particular genus are not named except in the case of the recently discovered Cuban plants (Reeves *et al.*,1996).

Table 3.3. Hyperaccumulators of nickel (maximum concentrations in μg/g [ppm] dry mass).

FAMILY/Genus	Location	Ni Conc.
ACANTHACEAE		
Blepharis acuminata	Zimbabwe	2000
Justicia lanstyakii	Brazil	2690
Lophostachys villosa	Brazil	1890
Ruellia geminiflora	Brazil	3330
ADIANTACEAE		
Adiantum sp.	Brazil	3540
ANACARDIACEAE		
Rhus wildii	Zimbabwe	1600
ASTERACEAE		
Berkheya coddii	South Africa	11,600
B. zeyheri	South Africa	17,000
Chromolaena meyeri	Brazil	1100
Dicoma niccolifera	Zimbabwe	1500
Leucanthemopsis alpina	Italy	3200
Senecio coronatus	South Africa	24,000
S. pauperculus	Newfoundland	1900
Solidago hispida	Newfoundland	1020
BORAGINACEAE		
Heliotropium sp.	Brazil	2020
BRASSICACEAE		
Alyssum 48 taxa	S.Europe/Turkey	1280-29,400
Bornmuellera 6 taxa	Balkans/Turkey	11,400-31,200
Cardamine resedifolia	Italy	3270
Cochlearia aucheri	Turkey	17,600
C. sempervivum	Turkey	3140
Peltaria emarginata	Greece	34,400
Streptanthus polygaloides	California	14,800
Thlaspi 23 taxa	Worldwide	2000-31,000
BUXACEAE		
Buxus aneura	Cuba	1450
B. baracoensis	Cuba	1590
B. crassifolia	Cuba	8350-12,250
B. excisa	Cuba	2150
B. flaviramea	Cuba	4500-8360
B. foliosa	Cuba	1320
B. gonoclada	Cuba	2610
B. heterophylla	Cuba	3480-8740
B. historica	Cuba	4810
B. imbricata	Cuba	1940
B. moana	Cuba	1100-1760
B. pilosula	Cuba	4870-9200
B. pseudoaneura	Cuba	1240

Table 3.3 continued

B. retusa	Cuba	310-10,310
B. revoluta	Cuba	7870-15,630
B. serpentinicola	Cuba	10,410
B. vaccinioides	Cuba	25,420
CAMPANULACEAE		
Campanula scheucheri	Italy	1090
CARYOPHYLLACEAE		
Arenaria 3 species	USA/Canada	2300-2370
Minuartia laricifolia	Italy	2710
M. verna	Italy	1390
CONVOLVULACEAE		
Merremia xanthophylla	Zimbabwe	1400
CUNONIACEAE		
Geissois 7 species	New Caledonia	1000-34,000
Pancheria engleriana	New Caledonia	6300
DICHAPETALACEAE		
Dichapetalum gelonioides		
subsp. *tuberculatum*	Philippines	26,600
subsp. *andamanicum*	Andaman Is.	3160
DIPTEROCARPACEAE		
Shorea tenuiramulosa	Sabah	1000
ESCALLONIACEAE		
Argophyllum grunowii	New Caledonia	1380
A. laxum	New Caledonia	1900
EUPHORBIACEAE		
Baloghia sp.	New Caledonia	5380
Cleidion viellardii	New Caledonia	9900
Cnidoscolus bahianus	Brazil	1020
Leucocroton acunae	Cuba	10,140
L. angustifolius	Cuba	6790-19,160
L. anomalus	Cuba	13,330
L. baracoensis	Cuba	2260
L. bracteosus	Cuba	11,660
L. brittonii	Cuba	5800
L. comosus	Cuba	6470-11,740
L. cordifolius	Cuba	2040-19,620
L. cristalensis	Cuba	4970-8070
L. discolor	Cuba	7670
L. ekmanii	Cuba	4610-8550
L. flavicans	Cuba	6710-15,500
L. incrustatus	Cuba	4260
L. linearifolius	Cuba	13,310-27,240
L. longibracteatus	Cuba	3850

Table 3.3 continued

L. moaensis	Cuba	9770-15,510
L. moncadae	Cuba	15,330
L. obovatus	Cuba	5070-9980
L. pachyphylloides	Cuba	5800-18,050
L. pachyphyllus	Cuba	693-9220
L. pallidus	Cuba	10,760
L. revolutus	Cuba	8910-17,240
L. sameki	Cuba	13,080
L. saxicola	Cuba	10,820-18,480
L. stenophylla	Cuba	12,090-24,500
L. subpeltatus	Cuba	13,890
L. virens	Cuba	5630-24,360
L. wrightii	Cuba	7410-12,600
Phyllanthus 16 taxa	New Caledonia	1090-38,100
P. chamaecristoides		
subsp. *chamaecristoides*	Cuba	18,530
subsp. *baracoensis*	Cuba	3400-31,740
P. chryseus	Cuba	10,790-13,790
P. cinctus	Cuba	11,510-21,870
P. comosus	Cuba	9340-19,380
P. comptus	Cuba	7260
P. cristalensis	Cuba	4200-8750
P. discolor	Cuba	13,670-31,499
P. ekmanii	Cuba	12,060-19,060
P. formosus	Cuba	7400
P. incrustatus	Cuba	10-1582
P. microdictyus	Cuba	4950-19,750
P. mirificus	Cuba	4480-7690
P. myrtilloides		
subsp. *alainii*	Cuba	14,330
subsp. *erythrinus*	Cuba	16,940-33,240
subsp. *myrtilloides*	Cuba	8490-9970
subsp. *shaferi*	Cuba	7910-21,710
subsp. *spathulifolius*	Cuba	5780-8900
P. nummularioides	Cuba	12,240-22,930
P. orbicularis	Cuba	4140-10,950
P. × *pallidus (=discolor* × *orbicularis)*	Cuba	15,390-60,170
P. phlebocarpus	Cuba	4890-19,400
P. pseudocicca	Cuba	9460-22,670
P. scopulorum	Cuba	13,650-21,930
P. williamioides	Cuba	232-18,100
FABACEAE		
Anthyllis sp.	Italy	4600
Pearsonia metallifera	Zimbabwe	10,000
Trifolium pallescens	Italy	1990
FLACOURTIACEAE		
Casearia silvana	New Caledonia	1490
Homalium 7 species	New Caledonia	1160-14,500
Xylosma 11 species	New Caledonia	1000-3750

Table 3.3 continued

JUNCACEAE		
Juncus lutea	Italy	2050
MELIACEAE		
Walsura monophylla	Philippines	7090
MYRISTICACEAE		
Myristica laurifolia	Indonesia	1100
OCHNACEAE		
Brackenridgea palustris		
subsp. *foxworthyi*	Philippines	7600
subsp. *kjellbergii*	Sulawesi	1050
ONCOTHECACEAE		
Oncotheca balansae	New Caledonia	2500
POACEAE		
Trisetum distichophyllum	Italy	1710
RANUNCULACEAE		
Ranunculus glacialis	Italy	1260
RUBIACEAE		
Mitracarpus sp.	Brazil	1000
Psychotria douarrei	New Caledonia	19,900
SAPOTACEAE		
Sebertia acuminata	New Caledonia	11,700
SAXIFRAGACEAE		
Saxifraga 3 species	Italy	2970-3840
SCROPHULARIACEAE		
Esterhazya sp.	Brazil	1060
Linaria alpina	Italy	1990
STACKHOUSIACEAE		
Stackhousia tryonii	Queensland	21,500
TILIACEAE		
Trichospermum kjellbergii	Sulawesi	1600
TURNERACEAE		
Turnera subnuda	Brazil	6130
VELLOZIACEAE		
Vellozia sp.	Brazil	3080
VIOLACEAE		
Agatea deplanchei	New Caledonia	2500

Table 3.3 continued

Hybanthus 5 taxa	New Caledonia	3000-17,600
H. floribundus	W.Australia	10,000
Rinorea bengalensis	SE Asia	17,500
R. javanica	Kalimantan	2170

After: Brooks *et al.* (1995); Reeves *et al.* (1996).

NB - The list does not include the last 48 Cuban species discovered by Reeves (pers. comm. 1997).

The geographical distribution of hyperaccumulators of nickel is quite extraordinary. In descending order, the numbers of separate taxa are: Cuba (128), Southern Europe and Anatolia 92, New Caledonia 56, Southeast Asia 12, Brazil 11, Southern Africa 8, North America 6, Australia 2 and Dominican Republic 1. Both Cuba and New Caledonia have by far the greatest number of nickel plants within a single territory. This is probably because of their isolation from major land masses during a time frame of 10-30 million years.

In contrast to Cuba, North America has furnished very few hyperaccumulators of any metal. This is almost certainly a temporal factor because there are no unequivocal records of hyperaccumulators within areas that have been previously glaciated.

Fig.3.7 shows the extent of previous glaciation in the world. The last Ice Age ended about 10,000 years ago, and this time frame is clearly not sufficient to have allowed for the development of hyperaccumulating plants.

There are many hyperaccumulators of nickel that have been, or could be, used for mineral prospecting. It should however be pointed out that there are far more indicator plants (i.e. thousands rather than a few hundred) that could also be used for this purpose.

Considering the Cuban flora alone, there are 920 species that are endemic to nickel-rich serpentine soils (Reeves *et al.*,1996). Each of these can therefore indicate the presence of nickel-rich soils. If the discussion is restricted to hyperaccumulators, all of the 48 *Alyssum* species with this character, indicate ultramafic substrates and by inference the presence of nickel, chromium, and cobalt that are associated with this rock type.

Brooks and Wither (1977) predicted the nature of the rock type upon which *Rinorea bengalensis* was growing, merely by analysing herbarium specimens and not having visited any of the countries from which the plants has been collected.

A nickel hyperaccumulator in Western Australia, *Hybanthus floribundus* has been used for geobotanical prospecting in that territory (Severne, 1972; Severne and Brooks, 1972; Cole, 1973).

The remarkable hyperaccumulation of nickel by *Hybanthus floribundus* was first reported in the open literature (it had previously been discovered by M.M.Cole, but not reported until 1973) by Severne and Brooks (1972). At that time it was only the third hyperaccumulator of nickel to have been discovered (the first two were *Alyssum bertolonii* and *A.murale*). They reported 1600 μg/g nickel in dried leaves compared with only 700 μg/g in the associated soils. The species has several different varieties and forms that are discussed later in Chapter 4.

Severne (1972) studied the distribution of this species over a nickel deposit in the Kurrajong region of Western Australia and compared its frequency of occurrence (number of individuals per 900 m^2 quadrat) and related its density to the nickel content of the soil (Fig.3.8). He concluded that this species is restricted to laterised ultramafic outcrops and creek beds draining these areas.

Zinc (base metal) floras

Plant communities growing over copper/lead/zinc (base metal) deposits have a certain similarity with serpentine floras. Plant growth is retarded and stunted, and broadleaf plants are absent. Endemic forms are often found if the area of mineralisation is large enough (Baumeister, 1954; Schwanitz and Hahn, 1954; Schwickerath, 1931).

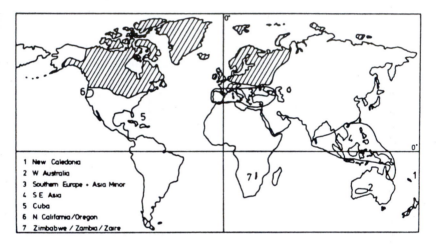

Fig.3.7. Extent of world-wide glaciation at the end of the last Ice Age, 10,000 years ago. The location of hyperaccumulators of nickel is also indicated. After: Brooks (1987).

In many cases it cannot be established whether or not the characteristic flora is due to the presence of one or all of the three base metals mentioned above since all three are usually found together in areas of sulphide mineralisation. There has been a tendency to classify such communities as *zinc* or *Galmei* floras since zinc is usually the main constituent. The true zinc floras are found in western Germany and eastern Belgium where the soils are rich in zinc and do not contain inordinately high levels of copper or lead. Galmei floras have been known for well over a century when Baumann (1887) reported the very high concentration of zinc (>1% dry weight) in *Viola calaminaria* (Fig.3.9). Early miners were guided to these zinc deposits by the appearance thereon of this specific flora. These vegetation communities belong to the *Violetum calaminariae westfalicum* association (Ernst, 1964).

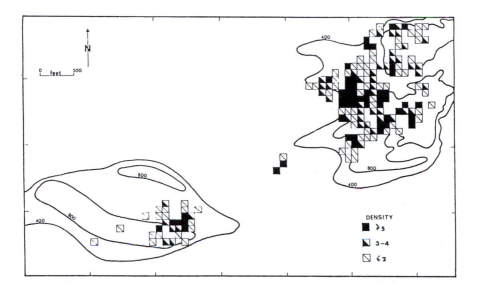

Fig.3.8. The relationship between the abundance of *Hybanthus floribundus* and the nickel content of the soil (μg/g shown in contours) in the Kurrajong area of Western Australia. Source: Severne (1972).

Viola calaminaria is not the only hyperaccumulator plant species in this community. Linstow (1929) reported that *Thlaspi calaminare* was also a hyperaccumulator of zinc and contains ten times as much of this element in the leaves as in the roots.

Studies by Reeves and Brooks (1983a) have shown that at least ten other species of *Thlaspi* contain comparable amount of zinc. They also found that one of these, *Thlaspi rotundifolium* subsp. *cepaeifolium* (Reeves and Brooks, 1983b) also appeared to have the ability to hyperaccumulate lead.

A bibliography of zinc and other heavy metal floras has been compiled by Ernst (1967) who listed 96 references. Further references are given in his book on heavy-metal floras (Ernst, 1974).

Table 3.4 lists hyperaccumulators of zinc, and Fig.3.10 shows a typical view of a zinc/lead flora near St Laurent le Minier, some 40 km north of Montpellier in southern France. The main species are *Thlaspi caerulescens*, *Armeria maritima* and *Minuartia verna*. The latter is not a hyperaccumulator but is confined to metal-rich soils throughout the temperate zones of Europe. It is noteworthy that in northern Europe it is found in the tundra where it is not confined to mineralisation.

Another species that is frequently found in the copper/lead/zinc community is *Armeria maritima* whose distribution is somewhat unusual. It is common either over mineralisation or near the sea shore where it is subject to salt spray.

Fig.3.9. *Viola calaminaria*, the first known hyperaccumulator of zinc; recognised as such in 1885.

Table 3.4. Hyperaccumulators of zinc (maximum concentrations in % dry weight).

Species	Location	%Zn
Arenaria patula - CARYOPHYLLACEAE	USA	1.31
Cardaminopsis halleri - BRASSICACEAE	Germany	1.36
Haumaniastrum katangense - LAMIACEAE	Zaïre	1.98
Noccaea eburneosa - BRASSICACEAE	Switzerland	1.05
Silene cucubalus - CARYOPHYLLACEAE	USA	0.47
Thlaspi alpestre - BRASSICACEAE	UK	2.50
T. brachypetalum	France	1.00
T. bulbosum	Greece	1.05
T. caerulescens	Germany	2.73
T. calaminare	Germany	3.96
T. limosellifolium	France	1.10
T. praecox	Bulgaria	2.10
T. rotundifolium subsp. *cepaeifolium*	Italy	2.10
T. stenopterum	Spain	1.60
T. tatraense	Slovakia	2.70
Viola calaminaria - VIOLACEAE	Germany	1.00

Source: Brooks *et al.* (1995)

Cardaminopsis halleri is another hyperaccumulator common in the Galmei floras of Central and Western Europe. It is not found in the southern mineral exposures. Perhaps the most extensively studied species within the Galmei flora of Western Europe has been *Thlaspi calaminare* a species very closely related to

Thlaspi caerulescens. The latter has being trialed for experiments on *phytoremediation* in which the plant is grown over polluted soils and harvested to remove zinc and cadmium (McGrath *et al.*,1993 - see Chapter 12).

Selenium floras

The identification of the elevated selenium content of plants from the Colorado Plateau in the western United States represents a piece of brilliant detective work by O.A.Beath and his colleagues at Laramie, Wyoming (Beath *et al.*,1939a, 1939b, 1940; Trelease and Beath, 1949). It had been known for the previous 100 years that cattle and sheep grazing in this part of the United States suffered serious disease caused by eating the "loco weed" *Astragalus* spp. Lack of suitable methods of analysis prevented identification of the causal agent until O.A. Beath suggested that it might be selenium and devised a method for determination of this element in various forage crops including *Astragalus*. The analytical method that was used at that time was a very complicated and time-consuming fluorimetric method, so it is indeed surprising that Beath and his associates were able to analyse many hundreds of plant samples and reported hyperaccumulation of selenium (> 1000 μg/g dry weight) in at least 13 species of *Astragalus* as well as in a few other taxa such as *Aster venusta, Atriplex confertifolia, Oonopsis wardii,* and *Stanleya pinnata*. One of these plants, *Astragalus bisulcatus*, is illustrated in Fig.3.11.

Fig.3.10. View of the zinc/lead metal-tolerant flora of the Mine des Malines, near St. Laurent le Minier, north of Montpellier, showing *Minuartia verna* (small flowers) surrounded by *Armeria maritima* subsp. *halleri* (large flowers). Photo by R.R.Brooks.

Selenium hyperaccumulation in Queensland was later reported by McCray and Hurwood (1963) who reported a mean of 2661 μg/g in *Neptunia amplexicaulis* and 1121 μg/g in *Acacia cana*.

Fig.3.11. The selenium hyperaccumulator *Astragalus biulcatus* with about 4000 μg/g Se (dry weight). Photo by H.L.Cannon.

Although selenium plants contain high concentrations of this element, they are probably capable of growth without it (Shrift, 1969). They are able to substitute selenium for sulphur in their metabolism without adverse effects (Shrift, 1969). Uranium deposits on the Colorado Plateau are primarily *carnotite* containing appreciable quantities of selenium. The presence of carnotite results in greater availability of selenium to plants. Cannon (1959, 1960, 1964) has been able to use the *Astragalus* species for indirect prospecting for uranium because the plants tend to grow in areas of maximum total or available selenium in the soils. Her classic work represents one of the most successful known applications of the geobotanical method and resulted in the discovery of several uranium deposits in the Colorado Plateau.

Some of the taxa in a selenium flora are able to absorb huge quantities of this element (over 0.5% of the dry weight of some *Astragalus* species) so that the garlic-like odour of volatile seleniferous compounds can often be detected in the plants themselves, sometimes even from a fast-moving vehicle.

It is clear from the above discussion that the advantage of selenium plants lies not so much in the discovery of selenium deposits, but rather in the indirect indication of uranium mineralisation. There is however another field in which these plants are of supreme interest. This is *hypervolatilisation*, the direct antithesis of hyperaccumulation. As will be discussed in Chapter 13, the *Astragalus* and other species are able to hyperaccumulate selenium and convert it to a volatile form that can be liberated into the atmosphere. There is here a potential for decontaminating soils polluted with this element.

Table 3.5. Hyperaccumulators of selenium and their elemental content (μg/g dry weight).

Species (No.)	Location	Mean	Range
Acacia cana (1)	Queensland	1121	
Aster venusta (1)	United States	2070	
Astragalus beathii (3)	United States	1906	1034-3135
A. bipinnata (1)	United States	1456	
A. bisulcatus (16)	United States	2276	1144-5330
A. haydenianus (3)	United States	2147	1916-2377
A. limatus (1)	United States	2175	
A. osterhoutii (2)	United States	2017	1356-2678
A. pattersoni (23)	United States	2696	1006-5993
A. pectinatus (16)	United States	2397	1330-6801
A. praelongus (11)	United States	2531	1030-4500
A. preussi 3)	United States	1189	1000-1438
A. racemosa (8)	United States	2145	1330-3920
A. sabulosus (3)	United States	1989	1734-2210
A. scobinatulus (1)	United States	1282	
Atriplex confertifolia (1)	United States	1260	
Neptunia amplexicaulis (14)	Queensland	2661	1143-4164
Oonopsis wardi (1)	United States	1422	
Stanleya pinnata (4)	United States	1510	1110-2490

NB - Values reported are only for individuals with > 1000 μg/g Se (dry weight). Individuals can usually be found with only a few μg/g Se.

Copper/cobalt floras

Although many base metal deposits are covered by ubiquitous metal-tolerant plants such as *Minuartia verna* or *Thlaspi alpestre* (Ernst, 1974), there is only one region in the world where a true copper flora exists. This is in Shaba (formerly Katanga) province in southeastern Zaïre. The *Shaban Copper Arc* extends from Kolwezi in the west to Lubumbashi in the east and is notable for its production of copper, which until the recent troubles in the country accounted for some 5% of the world total. Cobalt production is also important. The deposits take the form of bare hills (see Fig.3.12) that are islands of refuge for one of the world's richest and most unusual metal-tolerant floras. The degree of endemism is extremely high and there are deposits of only a few hectares in area that contain one or more endemic species found there and nowhere else. An example of this is the prostrate herb *Vigna dolomitica* that is confined to a few dozen hectares of the Mine de l'Etoile near Lubumbashi.

The flora of the Shaban Copper Arc has been described by Brooks and Malaisse (1985) and by Brooks *et al.* (1980, 1992a). They have reported the presence of over 50 species endemic to copper/cobalt deposits including several individual plants that have been used for geobotanical prospecting.

Fig.3.12. Aerial view of bare copper deposits in Zaïre. The foreground is Swambo with Mindigi in the background. Photo by R.R.Brooks.

A view of a typical copper-tolerant community (at Shinkolobwe) is shown in Fig.3.13 where distinct communities are developed over substrates containing different concentrations of copper. It is surprising that these plants have developed such a high tolerance to an element that is normally considered to be highly phytotoxic. Not only are the plants tolerant to copper, but in some cases they can hyperaccumulate this element to such a high degree that there are cases of taxa having over 1% (10,000 μg/g) copper (or cobalt) in dried tissue. For example, *Aeollanthus subacaulis* var. *linearis* has been recorded as having a copper content of 1.37% in the dried whole plant (Malaisse *et al.*,1978).

A total of 26 hyperaccumulators of cobalt and 24 of copper have now been identified in Zaïre. Of these, 9 hyperaccumulate both metals. They are listed in Table 3.6 below.

The copper/cobalt flora of Zaïre is perhaps the world's finest example of a geobotanical expression of mineralisation. It is under great threat from opencast mining operations that have already caused the extinction of several taxa and it might therefore be argued that accelerated research and plant collecting for herbaria should be undertaken before this great flora remains but a distant memory. Unfortunately, even this option is hardly viable at present because of the current state of anarchy in that unfortunate country and it will be many years before the political situation will have improved significantly. A recent botanical expedition to the Shaban Copper Arc sponsored by the National Geographic Society (Brooks *et al.*,1992a) might well be the last this century.

A review of geobotanical studies in Africa has been given by Ernst (1974). Most of the work has been centred on southeastern Zaïre (Brooks and Malaisse, 1985) and southern Africa.

As mentioned previously, prospecting work in the 1950s was centred around the distribution of the famous "copper flowers" *Becium homblei* and the copper/cobalt hyperaccumulator *Haumaniastrum katangense*.

Fig.3.13. View from the summit of a copper/cobalt-mineralised hill at Shinkolobwe, Zaïre. Photo by R.R.Brooks.

Most of the geobotanical work in Zimbabwe is due to the pioneering work of the late Hiram Wild who carried out an excellent inventory of the serpentine flora of the Great Dyke (Wild, 1965). He identified two hyperaccumulators of nickel in this flora, *Dicoma niccolifera* and *Pearsonia metallifera*, both indicate nickel-rich soil. The vegetation of copper-rich soils has been described by Ernst (1974).

The Geographical Distribution of Hyperaccumulators

Introduction

The worldwide distribution of hyperaccumulators of various elements is of great interest. In the case of nickel, the nickel plants are very widely distributed over

Table 3.6. Zaïrean hyperaccumulators of copper and cobalt (maximum concentrations in μg/g dry weight).

Species	Copper	Cobalt
Aeollanthus biformifolius - LAMIACEAE	3920	2820
A. saxatilis	-	1000
Alectra sessiliflora - SCROPHULARIACEAE		
var. *sessiliflora*	-	2782
var. *senegalensis*	1590	-
A. welwitschii	-	1561
Anisopappus davyi - ASTERACEAE	2889	2650
A. hoffmanianus	1065	-
Ascolepis metallorum - CYPERACEAE	1200	-
Becium aureoviride - LAMIACEAE		
subsp. *lupotoense*	1135	-
Buchnera henriquesii - SCROPHULARIACEAE	3520	2435
Bulbostylis mucronata - CYPERACEAE	7783	2130
Celosia trigyna - AMARANTHACEAE	2051	-
Commelina zigzag -COMMELINACEAE	1210	-
Crassula alba - CRASSULACEAE	-	1712
C. vaginata	-	1405
Crotalaria cobalticola - FABACEAE	-	3010
Cyanotis longifolia - COMMELINACEAE	-	4200
Eragrostis boehmii - POACEAE	2800	-
Gutenbergia cupricola - ASTERACEAE	5095	2309
Haumaniastrum homblei - LAMIACEAE	-	2633
H. katangense	8356	2240
H. robertii	2070	10,200
H. rosulatum	1089	-
Hibiscus rhodanthus - MALVACEAE	-	1527
Icomum tuberculatum - LAMIACEAE	-	1429
Ipomoea alpina - CONVOLVULACEAE	12,300	-
Lindernia damblonii - SCROPHULARIACEAE	-	1113
L. perennis	9322	2300
Monadenium cupricola EUPHORBIACEAE	-	1234
Pandiaka metallorum - AMARANTHACEAE	6260	2139
Rendlia cupricola - POACEAE	1560	-
Silene cobalticola - CARYOPHYLLACEAE	1660	-
Sopubia dregeana - SCROPHULARIACEAE	-	1767
S. metallorum	-	1742
S. neptunii	-	2476
Striga hermontheca - SCROPHULARIACEAE	1105	-
Triumfetta digitata - TILIACEAE	1060	-
T. welwitschii var. *descampii*	-	2201
Vernonia petersii - ASTERACEAE	1555	-
Vigna dolomitica - FABACEAE	3000	-
Xerophyta retinervis var. *retinervis* - VELLOZIACEAE	-	1520

Source: Brooks *et al.* (1995).

ultramafic rocks in all of the continents except Antarctica. As has been mentioned above (see Fig.3.7), they never occur in terrain that has previously been glaciated, perhaps because 10,000 years is not sufficiently long for the hyperaccumulation character to have evolved. Brooks (1987) has described the distribution of the nickel hyperaccumulators, and these are by far the most numerous. The worldwide distribution of these plants is shown in Fig.3.7.

Fig.3.14. Worldwide distribution of hyperaccumulators of selenium and copper/cobalt.

The geographical distribution of hyperaccumulators of selenium, and copper/cobalt is shown in Fig.3.14. These plants have a much more limited distribution than the nickel plants since the latter are found over the relatively common ultramafic (serpentine) rocks, whereas the selenium plants are virtually confined to the western United States and Queensland, and the copper/cobalt plants are found only in Shaba Province, Zaïre (Brooks and Malaisse, 1985). Zinc hyperaccumulators that grow over lead/zinc sulphide mineralisation are found in most of the temperate zones of Eurasia and have a very scattered distribution because of the sporadic nature of their host substrates. They are considered separately below under the heading of "*Thlaspi*" since this is the most common of the hyperaccumulating taxa and has the character of being able to accumulate cadmium, nickel and zinc.

Alyssum

It is apparent that all of the nickel plants so far discovered have been found in one or more of seven distinct regions: 1 - New Caledonia, 2 - Western Australia, 3 - southern Europe and Asia Minor, 4 - Southeast Asia, 5 - Cuba, 6 - western United States, 7 - Zimbabwe (Great Dyke), and 8 - South Africa. It will also be noted that hyperaccumulators are never found over previously glaciated terrain (see Fig.3.7).

Fig.3.15. *Alyssum robertianum*, a nickel hyperaccumulator from Corsica. Drawing by M.Conrad.

It might perhaps be argued that the distributions shown above are fortuitous rather than real and may be derived from extensive studies of selected parts of the world.   However in the original herbarium survey of Brooks *et al.* (1977) involving the genera *Hybanthus* and *Homalium*, hyperaccumulators were found in only two of the above eight regions despite a relatively uniform distribution of samples throughout the world. The genus with among the greatest number of individual hyperaccumulators of nickel (see Table 3.3 above) is *Alyssum* which contains 48 species of nickel plant. A drawing of a typical plant of this type, *A.robertianum* from Corsica is portrayed in Fig.3.15.

The distribution of hyperaccumulating *Alyssum* species is unusual. They are confined to ultramafic substrates in southern Europe and Asia Minor stretching from Portugal in the west to the Iraq/Turkey/Iran border areas in the east. Anatolia is the site of their maximum multiplicity and diversity. The distribution of these plants in southern Europe is shown in Fig.3.16. A total of 13 of these taxa were identified by Brooks and Radford (1978). Later, another nickel plant, *Alyssum malacitanum* (Brooks *et al.*,1981) has been identified from Andalusia in southeastern Spain.

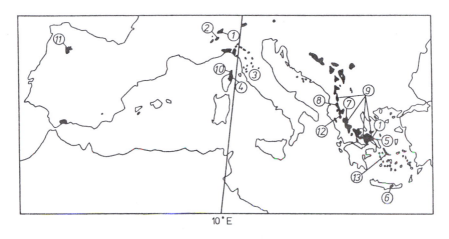

10°E

Fig.3.16. Ultramafic rocks (black) and the distribution of *Alyssum* nickel hyperaccumulators in Southern Europe. 1 - *A.alpestre*, 2 - *A.argenteum*, 3 - *A.bertolonii*, 4 - *A.corsicum*, 5 - *A.euboeum*, 6 - *A.fallacinum*, 7 - *A.heldreichii*, 8 - *A.markgrafii*, 9 - *A.murale*, 10 - *A.robertianum*, 11 - *A.pintodasilvae*, 12 - *A.smolikanum*, 13 - *A.tenium*, 14 - *A.malacitanum*. After: Brooks (1987).

A similar plant was discovered on serpentine near Santiago de Compostela (Galicia) in the northwest of the same country and found to be a hyperaccumulator of nickel. It has now been described as being *Alyssum pintodasilvae*, formerly described from ultramafics near Bragança in northeast Portugal.

The above Iberian species of Alyssum give an excellent illustration of the concept of evolutionary *neoendemism*. The hyperaccumulators *A.pintodasilvae* (Portugal and northwest Spain) and *A.malacitanum* (Andalusia) are all close relatives of the ubiquitous *A.serpyllifolium* found throughout the Iberian Peninsula and southern France. The latter surrounds "serpentine islands" of related nickel plants that have probably evolved from it.

Fig.3.17 shows the distribution of Anatolian species of *Alyssum* from section Odontarrhena (which contains all the hyperaccumulating taxa of this genus). The figure classifies the plants into four categories depending on their nickel content: > 10000 μg/g, 1000-10000 μg/g, 100-999 μg/g, and < 100 μg/g. The distributions are shown in relation to Turkish *vilayets* (administrative districts). The distribution of the plants corresponds exactly to the location of ultramafic rocks in Anatolia. There is hardly a single serpentine area in the country, however small, which does not contain at least one distinctive hyperaccumulator of nickel of the genus *Alyssum*.

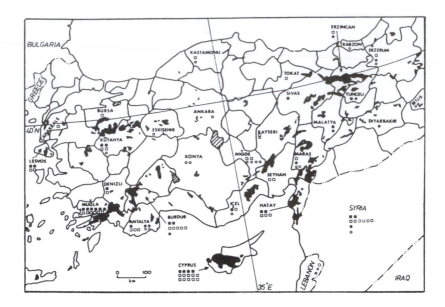

Fig.3.17. Ultramafic rocks (black) and the distribution of *Alyssum* hyperaccumulators in Anatolia
and the eastern Mediterranean. Nickel concentrations in plants (μg/g dry weight) are as
follows: Solid squares = >10,000, open squares = 1000-10,000, solid circles =
100-999, open circles = <100. Source: Brooks *et al*. (1979).

Although species of *Alyssum* extend through central Asia and even across the
Bering Strait to Alaska and the Yukon, hyperaccumulators are not found beyond
a point just east of Turkey, despite the fact that there are plenty of potential
serpentine soils in Asia and northwestern North America. Only *A.penjwinensis*
(northwestern Iran) and *A.singarense* (northeastern Iraq) have the ability to
hyperaccumulate nickel anywhere to the east of Turkey.

An interesting observation can be made about *A.corsicum*. This taxon was
once thought to be endemic to a small ultramafic occurrence of a few hectares in
extent, situated at the outskirts of Bastia in Corsica. Later work however showed
that this species originated in western Anatolia and was later spread to Corsica,
Cyprus and Crete as weed seeds in grain carried to those islands by the Venetian
traders in the 15th century. The seeds fell fortuitously on literally "stony ground"
as they required an ultramafic substrate upon which to germinate and flourish (see
also Chapter 7).

It has been proposed by Brooks *et al*. (1979) that the extent of distribution of
a nickel plant is inversely related to its nickel content. For example, of the 18
species of *Alyssum* in section Odontarrhena containing over 1% nickel in dry
leaves, only one (*A.constellatum*) has a wide distribution, and even that is
confined to eastern Turkey and northern Iraq. *Alyssum* species containing
1000-5000 μg/g nickel have a much wider distribution. Examples of this are
A.alpestre and *A.obovatum*. In *Alyssum* species there is therefore a relationship

between diversity, proliferation and endemism on the one hand, and extraordinarily high concentrations of nickel on the other.

Multiplicity and diversity of species recognized by morphological discontinuities, together with a high level of endemism, have often been associated with ancient disturbed floras. A third characteristic may now be added: hyperaccumulation of nickel. This cannot be regarded as a universal characteristic of ancient floras however, because of the comparative rarity of ultramafic rocks which are obviously a prerequisite for extreme nickel uptake.

It would seem that hyperaccumulation of nickel, like endemism, is an evolutionary adaptation typical of many ancient floras. By their extraordinary ability to accumulate massive concentrations of normally phytotoxic nickel, some genera such as *Alyssum* have adjusted phylogenetically to very hostile edaphic conditions. The development of this physiological tolerance may perhaps be a survival or defence strategy against competition from other taxa. Certainly within section Odontarrhena of *Alyssum* there is no question that the hyperaccumulators of nickel have been enormously successful. This triumph over the environment is illustrated by several taxa that occur on serpentine outcrops as extensive or nearly pure populations with an almost total absence of any other competing species. Examples of such "weedy" colonisers and superadapters are *A.murale* which occurs throughout the Balkans and particularly in the Pindus mountains of Greece, and *A.corsicum* and *A.cypricum* throughout western Anatolia.

The nickel plants of New Caledonia

An example of the clear relation between nickel hyperaccumulation and endemism and multiplicity of taxa is shown by the flora of New Caledonia. At least 46 hyperaccumulators of nickel from 9 genera and 6 different plant families have been reported from this Pacific island. These New Caledonian hyperaccumulators (Table 3.3) belong to the following genera and families: *Casearia* (Flacourtiaceae), *Geissois* (Cunoniaceae), *Homalium* (Flacourtiaceae), *Hybanthus* (Violaceae), *Lasiochlamys* (Flacourtiaceae), *Phyllanthus* (Euphorbiaceae), *Psychotria* (Rubiaceae), *Sebertia* (Sapotaceae), and *Xylosma* (Flacourtiaceae). Among the above genera, the largest number of nickel hyperaccumulators is found in *Homalium*.

Out of the 16 New Caledonian species of this genus recognized by Sleumer (1974), seven possess this unusual accumulatory character. Not only are these seven taxa endemic to New Caledonia, but they are entirely restricted to ultramafic substrates.

The nickel plants of Southeast Asia and Oceania

Several hyperaccumulators of nickel have reported from Southeast Asia. Brooks and Wither (1977) found hyperaccumulation of nickel by *Rinorea bengalensis* which grows throughout the region and extends from Sri Lanka to Queensland.

Although not all specimens of the herbarium material analysed contained > 1000 μg/g nickel, it was clear that such levels were attained or exceeded whenever the plant was growing on ultramafic rocks.

Wither and Brooks (1977) identified hyperaccumulation of nickel by *Planchonella oxyedra* and *Trichospermum kjellbergii* and Baker *et al.* (1992) discovered seven new hyperaccumulators of nickel from the island of Palawan in the Philippines (see Table 3.3). Two of these plants, *Phyllanthus palawanensis* and *Walsura monophylla* carried a green sap with a very high nickel content analogous to *Sebertia acuminata* from New Caledonia (see above).

Only two hyperaccumulators of nickel have been reported from Australia. Bationoff *et al.* (1990) and Bationoff and Specht (1992) reported up to 4.13% nickel in dried leaves of *Stackhousia tryonii* from Queensland, among the highest ever reported for any plant species worldwide. At the other end of Australia in Western Australia, Severne and Brooks (1972) and Cole (1973) reported hyperaccumulation of nickel by *Hybanthus floribundus* (Chapter 4).

The nickel plants of The Americas

Until fairly recently, there had been no determinations of hyperaccumulation of nickel in Cuban plants. This was of course due to the political isolation of that country. However, Berezaín Iturralde (1981, 1992) has reported 4500 and 7700 μg/g (dry weight) in *Buxus flaviramea* and *Leucocroton flavicans* respectively. This earlier work led Reeves *et al.* (1996) to analyse a large number of species in both genera. They found over 80 hyperaccumulators in these two genera (see Table 3.3). The logic of their work had been to investigate further the two genera in which Berezaín Iturralde (1981, 1992) had already found nickel plants. Further studies (Reeves, pers. comm. 1997) has led to the discovery of a further 20 hyperaccumulators in Cuba.

In sharp contrast to Cuba, North America is host to only four hyper-accumulators of nickel. These are confined to the states of California and Oregon. In this region, Reeves *et al.* (1981, 1983) discovered hyperaccumulation of nickel by *Strepanthus polygaloides* and by three varieties of *Thlaspi montanum* (var. *californicum*, var. *montanum* and var. *siskiyouense*). The evolution of serpentine-tolerant forms of *Thlaspi montanum* is another example of *neoendemism*. The existence of forms intermediate between *T.montanum* and var. *siskiyouense* has been observed at Waldo, Oregon by Holmgren (1971). A taxon in this intermediate category contained 12,850 μg/g (1.28%) nickel. *Streptanthus polygaloides*, the only true native species, has been used for experiments in phytomining (see Chapter 14).

Until a few years ago, it had not been considered that South America would be host for hyperaccumulators of nickel until Brooks *et al.* (1990, 1992b) discovered a total of 11 new nickel plants in the serpentine flora of Goiás State, Brazil. One of these, *Cnidoscolus* sp. contained a nickel-rich green latex reminiscent of that of the New Caledonian *Sebertia acuminata* (see above).

Investigations over ultramafic rocks in Argentina and Paraguay have not turned up further nickel plants.

The nickel plants of Southern Africa

The Great Dyke of Zimbabwe and the Barberton Sequence of South Africa are hosts to a number of hyperaccumulators of nickel. Wild (1970) reported the presence of the hyperaccumulator *Pearsonia metallifera* over the Great Dyke and Brooks and Yang (1984) working in the same area, found a further three nickel plants (*Blepharis acuminata*, *Merremia xanthophylla* and *Rhus wildii*).

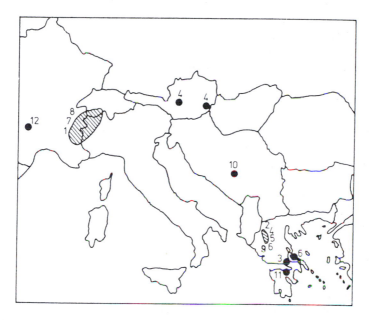

Fig.3.18. Distribution of *Thlaspi* and related *Noccaea* throughout southern and central Europe.
1 - *T.alpestre*, 2 - *T.graecum*, 3 - *T.bulbosum*, 4 - *T.goesingense*, 5 - *T.epirotum*,
6 - *T.ochroleucum*, 7 -*T.alpestre* subsp. *sylvium*, 8 - *T.rotundifolium*,
9 - *Noccaea tymphaea*, 10 - *N.aptera*, 11 - *N.boetica*, 12 -
N.firmiensis. Source: Reeves and Brooks (1983a).

Until fairly recently, no hyperaccumulators of nickel had been recorded from South Africa. However, Morrey *et al.* (1992) report hyperaccumulation of nickel in the following species, all in the Asteraceae, and all from the Barberton Sequence in northeastern South Africa: *Berkheya coddii*, *B.zeyheri* subsp. *rehmannii* var. *rogersiana*, *Senecio coronatus* and *S.lydenburgensis*.

Thlaspi

Like *Alyssum*, the genus *Thlaspi* contains a number of hyperaccumulators of nickel and tends to occupy the same ecological niches in central and southern Europe. Unlike *Alyssum* however, *Thlaspi* nickel plants are not represented in Anatolia. For about a century, one species of *Thlaspi* (*T. calaminare*) has had the reputation of hyperaccumulating zinc. Baumann (1885) reported 17.1% zinc in the ash of specimens growing over calamine deposits in western Germany.

More recent work by Ernst (1968) continues to demonstrate the extraordinary ability of this taxon to hyperaccumulate zinc. He showed that normal zinc levels were around 1% in dried leaves of the plant.

Reeves and Brooks (1983b) have reported hyperaccumulation of zinc and lead in *T. rotundifolium* subsp. *cepaeifolium* from the Cave del Predil region of northern Italy and have also shown (Reeves and Brooks, 1983a) that several of the southern European taxa of this genus can hyperaccumulate nickel. They found at least 12 species with this character. The centre of maximum diversity and multiplicity appears to be in Greece (Fig.3.18) where there are seven taxa which may be classified as nickel plants. Other hyperaccumulators of nickel in the genus *Thlaspi* are to be found in the Switzerland/Italy/France alpine border area, in southcentral France, and in Austria and the former Yugoslavia. American varieties of *T. montanum* have also been found to hyperaccumulate nickel and have already been discussed above. It will be noted from Table 3.3 that several hyperaccumulators are classified as *Noccaea* but were originally part of *Thlaspi* until the genus was revised by Meyer (1973). Reeves and Brooks (1983a) showed that 38 out of 54 taxa studied were also able to hyperaccumulate zinc.

General Discussion

This chapter has reviewed the occurrence and geobotany of a selected number of hyperaccumulators of only six elements. It has not touched on a whole range of other elements such as vanadium, manganese, cadmium, and even the noble metals. This is because very little is known about the uptake of these elements by plants and because they are not common pollutants, it is more difficult to envisage that plants could be used to remediate soils (phytoremediation - see Chapter 12) contaminated with them. It must however be remembered that the degree of "pollution" from these rarer elements is likely to be orders of magnitude less than in the case of common metals such as zinc or nickel. Thresholds of hyperaccumulation for these rarer metals have not yet been formulated, but it could easily be envisaged that a concentration of say 1 μg/g of gold in plant material might well represent hyperaccumulation since this level is probably 200 times the normal background concentration of this element in vegetation.

The emphasis in this chapter has been on geobotany in mineral exploration, and it is quite clear that plants will never be able to indicate the rarer elements directly because their concentrations in soils will be so low that they will have virtually no direct effect on the distribution or morphology of the vegetation. If however, pathfinders such as arsenic can be used, this restriction is no longer valid. As will be revealed in later chapters, the potential uses of hyperaccumulators extend far beyond the boundaries of geobotany alone and will have an ever-increasing importance in new and burgeoning fields such as phytoremediation and even phytomining.

References

Anon.(1959) A flower that led to a copper discovery. *Horizon (RST Group), Ndola* 1, 35-39.

Baker,A.J.M.,Proctor,J. and Reeves,R.D. (eds)(1992) *The Vegetation of Ultramafic (Serpentine) Soils*. Intercept, Andover, 509 pp.

Baker,A.J.M.,Proctor,J.,van Balgooy,M.M.J. and Reeves,R.D.(1992) Hyperaccumulation of nickel by the flora of the ultramafics of Palawan, Republic of the Philippines. In: Baker,A.J.M.,Proctor,J. and Reeves,R.D. (eds), *The Vegetation of Ultramafic (Serpentine) Soils*. Intercept, Andover, pp.291-304.

Bationoff,G.N.,Reeves,R.D. and Specht,R.L.(1990) *Stackhousii tryonii* Bailey: a nickel-accumulating serpentine-endemic species of Central Queensland. *Australian Journal of Botany* 38, 121-130.

Bationoff,G.N. and Specht,R.L.(1992) Queensland (Australia) serpentine vegetation. In: Baker,A.J.M.,Proctor,J. and Reeves,R.D.(eds), *The Vegetation of Ultramafic (Serpentine) Soils*. Intercept, Andover, pp.109-128.

Baumann,A. (1885) Das Verhalten von Zinksalzen gegen Pflanzen und im Boden. *Landwirtschaftliche Versuchsstation* 31, 1-53.

Baumeister,W.(1954) Der Einfluss von Zink in *Silene inflata* Smith. *Berichte der Deutschen Botanischen Gesellschaft* 67, 205-213.

Beath,O.A.,Gilbert,C.S. and Eppson,H.F.(1939a) The use of indicator plants in locating seleniferous areas in western United States. I. General. *American Journal of Botany* 26, 257-269.

Beath,O.A.,Gilbert,C.S. and Eppson,H.F.(1939b) The use of indicator plants in locating seleniferous areas in western United States. II. Correlation studies by states. *American Journal of Botany* 26, 296-315.

Beath,O.A.,Gilbert,C.S. and Eppson,H.F.(1940) The use of indicator plants in locating seleniferous areas in western United States. III. Further studies. *American Journal of Botany* 27, 564-573.

Berezaín Iturralde,R.(1981) Reporte preliminar de plantas serpentinicolas acumuladores e hiperacumuladores de algunos elementos. *Revista del Jardín Botánico Nacional* 2, 48-59.

Berezaín Iturralde,R.(1992) A note on plant/soil relationships in the Cuban serpentine flora. In: Baker,A.J.M.,Proctor,J. and Reeves,R.D.(eds), *The Vegetation of Ultramafic (Serpentine) Soils*. Intercept, Andover, pp.97-99.

Braun Blanquet,J.(1932) *Plant Sociology*. McGraw Hill, New York, 439 pp.

Brooks,R.R.(1983) *Biological Methods of Prospecting for Minerals*. Wiley, New York, 322 pp.

Brooks,R.R.(1987) *Serpentine and its Vegetation*. Dioscorides Press, Portland, 454 pp.

Brooks,R.R.,Baker,A.J.M. and Malaisse,F.(1992a) Copper flowers. *National Geographic Research* 8, 338-351.

Brooks,R.R.,Dunn,C.E. and Hall,G.E.M.(1995) *Biological Systems in Mineral Exploration and Processing*. Ellis Horwood, Hemel Hempstead, 539 pp.

Brooks,R.R.,Lee,J.,Reeves,R.D. and Jaffré,T.(1977) Detection of nickeliferous rocks by analysis of herbarium specimens of indicator plants. *Journal of Geochemical Exploration* 7, 49-57.

Brooks,R.R. and Malaisse,F.(1985) *The Heavy Metal Tolerant Flora of Southcentral Africa*. Balkema, Rotterdam, 199 pp.

Brooks,R.R.,Morrison,R.S.,Reeves,R.D.,Dudley,T.R. and Akman,Y.(1979) Hyperaccumulation of nickel by *Alyssum* Linnaeus (Cruciferae). *Proceedings of the Royal Society (London) Section B* 203, 387-403.

Brooks,R.R. and Radford,C.C.(1978) Nickel accumulation by European species of the genus *Alyssum*. *Proceedings of the Royal Society (London) Section B* 200, 217-224.

Brooks,R.R.,Reeves,R.D. and Baker,A.J.M.(1992b) The serpentine vegetation of Goiás State, Brazil. In: Baker,A.J.M.,Proctor,J. and Reeves,R.D. (eds) *The Vegetation of Ultramafic (Serpentine) Soils*. Intercept, Andover, pp.67-81.

Brooks,R.R.,Reeves,R.D.,Baker,A.J.M.,Rizzo,J.A. and Diaz Ferreira,H. (1990) The Brazilian Serpentine Plant Expedition (BRASPEX), 1988. *National Geographic Research* 6, 205-219.

Brooks,R.R.,Reeves,R.D.,Morrison,R.S. and Malaisse,F.(1980) Hyper-accumulation of copper and cobalt - a review. *Bulletin de la Société Royale Botanique de Belgique* 113, 166-172.

Brooks,R.R.,Shaw,S. and Asensi Marfil,A.(1981) The chemical form and physiological function of nickel in some Iberian *Alyssum* species. *Physiologia Plantarum* 51, 167-170.

Brooks,R.R. and Wither,E.D.(1977) Nickel accumulation by *Rinorea bengalensis* (Wall.) O.K. *Journal of Geochemical Exploration* 7, 295-300.

Brooks,R.R. and Yang,X.H.(1984) Elemental levels and relationships in the endemic serpentine flora of the Great Dyke, Zimbabwe and their significance as controlling factors for this flora. *Taxon* 33, 392-399.

Cannon,H.L.(1959) Advances in botanical methods of prospecting for uranium in the western United States. In: *Proceedings of the First International Geochemical Exploration Symposium at the International Geological Congress, Mexico City*, pp. 235-241.

Cannon,H.L.(1960) The development of botanical methods of prospecting for uranium on the Colorado Plateau. *United States Geological Survey Bulletin* 1085-A, 1-50.

Cannon,H.L.(1964) Geochemistry of rocks and related soils and vegetation in the Yellow Cat area, Grand County, Utah. *United States Geological Survey*

Bulletin 1176, 1-127.

Cole,M.M.(1973) Geobotanical and biogeochemical investigations in the sclerophyllous woodland and shrub associations of the Eastern Goldfields area of Western Australia with particular reference to the role of *Hybanthus floribundus* (Lindl.) F.Muell. as a nickel indicator and accumulator plant. *Applied Ecology* 10, 269-320.

Dudley,T.R.(1964) Synopsis of the genus *Alyssum*. *Journal of the Arnold Arboretum* 45, 358-373.

Duvigneaud,P.(1958) La végétation de Katanga et de ses sols métallifères. *Bulletin de la Société Royale Botanique de Belgique* 90, 127-186.

Duvigneaud,P.(1959) Plantes cobaltophytes dans le Haut Katanga. *Bulletin de la Société Royale Botanique de Belgique* 91, 113-134.

Ernst,W.(1964) Ökologisch-Soziologische Untersuchungen der Schwermetall-pflanzengesellschaften Mitteleuropas unter Einschluß der Alpen. PhD Thesis, University of Münster. 70 pp.

Ernst,W.(1967) Bibliographie der Arbeiten über Pflanzengesellschaften auf schwermetallhaltigen Böden mit Ausnahme des Serpentins. *Excerpta Botanica* 8, 50-61.

Ernst,W.(1968) Zur Kenntnis der Soziologie und Ökologie der Schwermetall-vegetation Grossbritaniens. *Berichte der Deutschen Botanischen Gesellschaft* 81, 116-124.

Ernst,W.(1974) *Schwermetallvegetation der Erde*. Fischer, Stuttgart, 194 pp.

Goodall,D.W.(1952) Quantitative aspects of plant distribution. *Biological Reviews* 27 194-245.

Greig-Smith,P.(1964) *Quantitative Plant Ecology 2nd edn*, Butterworths, London, 256 pp.

Holmgren,P.K.(1971) A biosystematic study of North American *Thlaspi montanum* and its allies. *Memoirs of the New York Botanical Garden* 21, 1-106.

Hoover,H.C. and Hoover,L.H.(1950) *De Re Metallica by Georgius Agricola* (English translation of original Latin text). Dover, New York.

Jaffré,T.(1980) *Etude Ecologique du Peuplement Végétal des Sols Dérivés de Roches Ultrabasiques en Nouvelle Calédonie*. ORSTOM, Paris, 273 pp.

Krause,W.(1958) Andere Bodenspezialisten. In: W.Rühland *et al.* (eds), *Handbuch der Pflanzenphysiologie*. Vol.4. Springer, Berlin, pp. 755-806.

Kruckeberg,A.R.(1985) *California Serpentines: Flora, Vegetation, Geology, Soils and Management Problems*. University of California Press, Berkeley, 180 pp.

Lee,J.(1974) Biogeochemical Studies on some Nickel-accumulating Plants from New Zealand and New Caledonia. MSc Thesis, Massey University, Palmerston North, New Zealand, 117 pp.

Linstow,O.von(1929) Bodenanzeigende Pflanzen. *Abhandlungen des Preußischen Geologischen Landesamts* 114.

López González,G.(1975) Contribución al estudio floristico y fitosociológico de Sierra de Aguas. *Acta Botanica Malacitana* 1, 81-205.

Lounamaa,K.J.(1956) Studies on the content of iron, manganese and zinc in macro-lichens. *Annales Botanicae Fennicae* 2, 127-137.

McCray,C.W.R. and Hurwood,I.S.(1963) Selenium in northwestern Queensland associated with a Cretaceous Formation. *Queensland Journal of Agricultural Science* 20, 475-498.

McGrath,S.P.,Sidoli,C.M.D.,Baker,A.J.M. and Reeves,R.D.(1993) The potential for the use of metal-accumulating plants for the *in situ* decontamination of metal-polluted soils. In: H.J.P.Eijsackers and T.Hamers (eds), *Integrated Soil and Sediment Research: a Basis for Proper Protection.* Kluwer Academic Publishers, Dordrecht, pp.673-676.

Malaisse,F.,Grégoire,J.,Morrison,R.S.,Brooks,R.R. and Reeves,R.D.(1978) *Aeolanthus biformifolius*: a hyperaccumulator of copper from Zaïre. *Science* 199, 887-888.

Malaisse,F.,Grégoire,J.,Morrison,R.S.,Brooks,R.R. and Reeves,R.D.(1979) Copper and cobalt in vegetation of Fungurume, Shaba Province, Zaïre. *Oikos* 33, 472-478.

Meyer,F.K.(1973) Conspectus der *Thlaspi* Arten Europas, Afrikas und Vorderasiens. *Feddes Repertorum* 84, 449-470.

Minguzzi,C. and Vergnano,O.(1948) Il contenuto di nichel nelle ceneri di *Alyssum bertolonii*. *Atti della Società Toscana di Scienze Naturale* 55, 49-74.

Morrey,D.R.,Balkwill,K.,Balkwill,M.-J. and Williamson,S.(1992) The review of the serpentine flora of Southern Africa. In: Baker,A.J.M.,Proctor,J. and Reeves,R.D. (eds) *The Vegetation of Ultramafic (Serpentine) Soils.* Intercept, Andover, pp.67-81.

Paribok,T.A. and Alekseeva-Popova,N.V.(1966) Content of some chemical elements in the wild plants of the Polar Urals as related to the problem of serpentine vegetation (in Russ.). *Botanichesky Zhurnal SSSR* 31, 339-353.

Pichi Sermolli,R.(1948) Flora e vegetazione delle serpentine e delle altre ofioliti dell'alta valle del Tevere (Toscana). *Webbia* 6, 1-380.

Pinto da Silva,A.R.(1970) A flora e a vegetaçao das areas ultrabasicas do Nordeste Transmontana. *Agronomia Lusitana* 30, 175-364.

Reeves,R.D.,Baker.A.J.M.,Borhidi,A. and Berezain,R.(1996) Nickel-accumulating plants from the ancient soils of Cuba. *New Phytologist* 133, 217-224.

Reeves,R.D. and Brooks,R.R.(1983a) European species of *Thlaspi* L. (Cruciferae) as indicators of nickel and zinc. *Journal of Geochemical Exploration* 18, 275-283.

Reeves,R.D. and Brooks,R.R.(1983b) Hyperaccumulation of lead and zinc by two metallophytes from a mining area of Central Europe. *Environmental Pollution Series A* 31, 277-287.

Reeves,R.D.,Brooks,R.R. and McFarlane,R.M.(1981) Nickel uptake by Californian *Streptanthus* and *Caulanthus* with particular reference to the hyperaccumulator *S.polygaloides* Gray (Brassicaceae). *American Journal of Botany* 68, 708-712.

Reeves,R.D.,McFarlane,R.M. and Brooks,R.R.(1983) Accumulation of nickel and zinc in western North American genera containing serpentine-tolerant species. *American Journal of Botany* 70, 1297-1303.

Roberts,B.A. and Proctor,J. (eds)(1992) *The Ecology of Areas with Serpentinized Rocks*. Kluwer, Dordrecht, 427 pp.

Robinson,W.O.,Edgington,G. and Byers,H.G.(1935) Chemical studies of infertile soils derived from rocks generally high in magnesium and generally high in chromium and nickel. *United States Department of Agriculture Bulletin* 471, 1-29.

Rune,O.(1953) Plant life on serpentine and related rocks in the north of Sweden. *Acta Phytogeographica Suecica* 31, 1-135.

Sarosiek,J.(1964) Ecological analysis of some plants growing on serpentine soil in Lower Silesia (in Pol.). *Monographiae Botanicae* 18, 1-105.

Schwanitz,F. and Hahn,H.(1954) *Genetischentwicklungsphysiologische Untersuchungen an Galmeipflanzen* I and II. *Zeitschrift der Botanik* 42, 179-190, and 459-471.

Schwickerath,M.(1931) Das Violetum calaminariae der Zinkböden in der Umgebung Aachens. Eine Pflanzensoziologische Studie. *Beiträge zur Naturdenkmalspflege* 14, 463-503.

Severne,B.C.(1972) Botanical Methods for Mineral Exploration in Western Australia. PhD Thesis, Massey University, Palmerston North, New Zealand, 254 pp.

Severne,B.C. and Brooks,R.R.(1972) A nickel-accumulating plant from Western Australia. *Planta* 103, 91-94.

Shrift,A.(1969) Aspects of selenium metabolism in higher plants. *Annual Reviews of Plant Physiology* 20, 475-494.

Sleumer,H.(1974) A concise revision of the Flacourtiaceae of New Caledonia and the Loyalty Islands. *Blumea* 22, 123-147.

Thalius,J.(1588) *Sylvia Hercynia, Sive Catalogus Plantarum Sponte Nascentium in Montibus*. Frankfurt am Main.

Tkalich,S.M.(1938) Experience in the investigation of plants as indicators in geological exploration and prospecting (in Russ.). *Vestnik dal'nevostochnovo Filiala Akademii Nauka SSSR* 32, 3-25.

Trelease,S.F. and Beath,O.A.(1949) *Selenium: its Geological Occurrence and its Biological Effects in Relation to Botany, Chemistry, Agriculture, Nutrition and Medicine*. Trelease and Beath, New York, 292 pp,

Walker,R.B.(1954) The ecology of serpentine soils. II Factors affecting plant growth on serpentine soils. *Ecology* 35, 259-266.

Walker,R.B.,Walker,H.M. and Ashworth,P.R.(1955) Calcium-magnesium nutrition with reference to serpentine soils. *Plant Physiology* 30, 214-221.

Wild,H.(1965) The flora of the Great Dyke of Southern Rhodesia with special reference to the serpentine soils. *Kirkia* 5, 49-86.

Wild,H.(1970) Geobotanical anomalies in Rhodesia 3. - The vegetation of nickel-bearing soils. *Kirkia* 7, 1-62.

Wither,E.D. and Brooks,R.R.(1977) Hyperaccumulation of nickel by some plants of Southeast Asia. *Journal of Geochemical Exploration* 8, 579-583.

Chapter four:

Biogeochemistry and Hyperaccumulators

R.R. Brooks

Department of Soil Science, Massey University, Palmerston North, New Zealand

Historical and General Introduction

C.E.Dunn (Dunn, 1995) has given an excellent historical outline of the development of biogeochemical exploration that has a much shorter history than that of geobotany because it involves the chemical analysis of plant tissue before relating the abundance of elements in plant material to the presence of mineralisation in the substrate. This introductory section is based on the above paper.

Published accounts on the relationship of plant chemistry to mineralised rock date back to the turn of the century, but were scarce until the late 1930s. In an 1898 study at the Omai gold mine in Guyana, ash samples of baromalli (genus *Catostemma,* from the family Bombacaceae) and "ironwood" were analysed and found to contain 0.3 and 3.0 $\mu g/g$ (ppm) gold, respectively (Lungwitz, 1900).

Kovalevsky (1987) reported that work on biogeochemical exploration in the former USSR began in the 1920s when Aleksandrov discovered an increased concentration of vanadium, radium and uranium in plant ash at a vanadium-uranium deposit, compared to concentrations in plants outside the ore zone. Subsequently, in the 1930s, V.I.Vernadsky supported the establishment of a *Biogeochemical Laboratory of the USSR Academy of Sciences*, to which A.P.Vinogradov made significant contributions.

Development of fundamental theory and application of biogeochemical methods commenced in the 1930s. The first published report of biogeochemical methodology was by the Russian worker Tkalich (1938), who cited the discovery that a deposit of arsenopyrite in Siberia could be traced by the iron content of overlying vegetation. At about the same time, the Swedish scientist Nils Brundin related biogeochemical data to tungsten in Cornwall, England, and to a vanadium deposit in Sweden (Brundin, 1939).

In the early 1940s the North American *father of biogeochemical prospecting*, Harry Warren, commenced more than half a century of studies from his base at the University of British Columbia in Canada. During the first 30 years of his work, Warren and his associates did much to document the relationships between plant chemistry and concealed mineralisation, and to raise the credibility of biogeochemistry from "...general disbelief, through benevolent scepticism to general acceptance that..." when used properly, it can be a viable exploration tool (e.g. Warren and Delavault, 1950; Warren et al.,1968).

Concurrently with the Canadian studies, work was advancing in the United States where Harbaugh (1950) reported the results of a base metal study of soils and vegetation in the mid-western states. Then significant interest stemmed from studies at the U.S. Geological Survey involving the geobotanical and biogeochemical expressions of uranium roll-front deposits in Colorado and the surrounding states (Cannon, 1957, 1960, 1964). At this time Hans Shacklette and his coworkers were conducting laboratory and field experiments on the metal uptake of plants, and they produced a number of USGS publications on the biogeochemical behaviour of different elements (Shacklette et al.,1970), followed by several compilations of baseline data on background concentrations in vegetation (Shacklette and Connor, 1973; Connor and Shacklette, 1975).

In Scandinavia, Nils Brundin and his coworkers were continuing to investigate biogeochemical methods. A monograph on trace elements in plants from Finland provided valuable semiquantitative information on the composition of a wide range of plants growing on different geological substrates (Lounamaa, 1956). Baseline data were presented for lichens, mosses, ferns, conifers, deciduous trees and shrubs, dwarf shrubs, grasses and herbs.

During the above period, Russian workers published numerous papers (mostly in Russian) that described methods in general terms, and results of biogeochemical surveys over various types of mineral deposit. Of importance to the international community was the milestone publication of the first book in English translation that was devoted exclusively to biogeochemical prospecting for minerals (Malyuga, 1964).

For the past 40 years Alexander Kovalevsky, from his base in Ulan-Ude, Siberia, has published the results of a wide range of biogeochemical studies in many papers, some of which have been translated into English. Other leading Russian workers on biogeochemical methods during this period included Grabovskaya (1965); Polikarpochkin and Polikarpochkina (1964); Talipov (1966); Tkalich (1959, 1961, 1970); Vernadsky (1965) and Vinogradov (1954). The landscape geochemistry studies of Perel'man (1961, 1966) also contributed to the literature on biogeochemical methods.

In the mid 1960s the Geological Survey of Canada commenced some fundamental investigations into the use of biogeochemistry in mineral exploration. A trailer was fitted out to be a mobile biogeochemical laboratory containing basic analytical equipment, a muffle furnace, balances and an emission spectrograph. This was taken into the field and the analytical work carried out close to the survey areas (Fortescue and Hornbrook, 1967, 1969; Hornbrook, 1969, 1970a,b). These studies developed protocols for when and how to sample, and suggested a systematic approach to biogeochemical research. They recommended a "visit" lasting one or two days to assess plant distribution and the environment in general (i.e. a brief orientation survey). Then either a "pilot project" of up to 30 days to conduct a comprehensive survey of an area; or, if mining was imminent, a "quick project" to obtain sample material prior to disruption, and potential contamination, of an area by mining.

The 1960s saw New Zealand emerge as a contributor to biogeochemical methods in exploration through the work of my colleagues, students and I at Massey University in Palmerston North. This culminated in 1972 with the first book on geobotany and biogeochemistry in mineral exploration to be published outside the former Soviet Union (Brooks, 1972). A second updated book (Brooks, 1983) followed a decade later and was again updated by a successor (Brooks *et al.*,1995) that included sections on geomicrobiology and on mineral processing by the use of bacteria.

It was in the mid-1970s that biogeochemical studies were in a position to take a big step forward through significant advances in analytical instrumentation. Inductively coupled plasma emission spectrometry (ICP-ES) and instrumental neutron activation analysis (INAA) were developed to provide accurate and precise multi-element data at very low element concentrations, and at remarkably low cost. INAA, in particular, was a great breakthrough since many elements could be determined at the ng/g and even pg/g (ppb and ppt) levels from a single irradiation of dry vegetation. During the boom in uranium exploration in the late 1970s, delayed neutron counting was used for determining uranium concentrations in vegetation (Dunn, 1983a,b; Dunn *et al.*,1984). Later, as exploration interest moved from uranium to gold, INAA was used for the routine determination of gold down to 0.1 ng/g in dry vegetation (Hoffman and Booker, 1986).

The 1979 translation into English of the book *Biogeochemical Exploration for Mineral Deposits* by A.L.Kovalevsky provided new insight into the Russian experience and approach to the science. Subsequently, a revised and extended second edition has been published (Kovalevsky, 1987). A third book is as yet available only in Russian (Kovalevsky, 1991).

Over the last decade the main focus of biogeochemical exploration has been its application to precious metals. A conference held in Los Angeles in 1984 was another milestone in the evolution of the science, bringing together expertise from many disciplines and environments (Carlisle *et al.*,1986). Also, papers in recent years by Erdman and Olsen (1985); Erdman *et al.* (1985); Dunn (1986, 1989, 1992); Dunn *et al.* (1989); Cohen *et al.* (1987); Hall *et al.* (1990) and Kovalevsky and Kovalevskaya (1989) have provided case histories and overviews of procedures and appropriate sample media. In 1992 the book *Noble Metals and Biological Systems* (Brooks, 1992) covered a broad spectrum of current knowledge of these metals.

In China, in pursuance of the 1990s philosophy of "a more open China", information on developments in biogeochemical exploration methods in that country came to light at a meeting of the Association of Exploration Geochemists in Beijing (September, 1993). Systematic biogeochemical studies began in the 1970s, and over the past decade the focus has been mainly on the search for deeply buried mineralisation in arid regions. Researchers claim that from the analysis of willow twigs they can delineate Pb-Zn stratabound deposits beneath more than 100 m of loess and over 100 m of redbeds (Dunn, 1993). At this time the Chinese consider that their biogeochemical studies are still in the experimental stage.

In Australia also, biogeochemical studies are mostly in their early stages of development because of the complexities that arise in dealing with the multitude of species comprising the two main genera of plant - the eucalypti and the acacias. Some of the more significant biogeochemical studies were carried out in Western Australia (Cole, 1973; Severne, 1972).

During the 1980s, a resurgence of interest in biogeochemical methods of exploration took place in North America, especially in the southwestern United States. Big sagebrush (*Artemisia tridentata*) became a common sample medium to assist in the exploration for gold in this arid terrain, whilst in Canada the preferred sample media are commonly black spruce (*Picea mariana*), balsam fir (*Abies balsamea*), douglas fir (*Pseudotsuga menziesii*), western hemlock (*Tsuga heterophylla*) and lodgepole pine (*Pinus contorta*). Work by S.Clark Smith and J.A.Erdman contributed to the biogeochemical data base. The Association of Exploration Geochemists has been instrumental in disseminating biogeochemical exploration methods and data by sponsoring short courses and publishing the course notes (Parduhn and Smith, 1991; Dunn *et al.*,1992; Dunn *et al.*,1993).

C.E.Dunn is arguably the leading proponent of biogeochemistry in the Western World and has brought the technique to a high level of proficiency and respect. Credit must be given to chemists for major developments over the past 10-15 years in analytical instrumentation and methods. This has permitted rapid, cost-effective, accurate and precise quantification of trace levels of elements in biological materials, and provided new insight into element concentrations and distributions in trees and shrubs. As a result of these developments, there has been an exponential increase in biogeochemical knowledge over the past decade.

Biogeochemical Prospecting with Hyperaccumulators

The suitability of hyperaccumulators for biogeochemical prospecting

When dealing with the biogeochemistry of hyperaccumulators of metals, an immediate problem is encountered. With most good biogeochemical indicator plants there is an approximately linear relationship between the elemental content of the plant and the concentration of the same element in the soil. This is an obvious *sine qua non* for the biogeochemical method of exploration. In the case of hyperaccumulators however, the target element tends to exist in an "all or nothing" environment: i.e. the plant will not grow in non-mineralised soils and will have a relatively constant high metal content when it does grow over minerals. This is illustrated in Fig.4.1 which shows the abundance of nickel in a large number of *Alyssum* species.

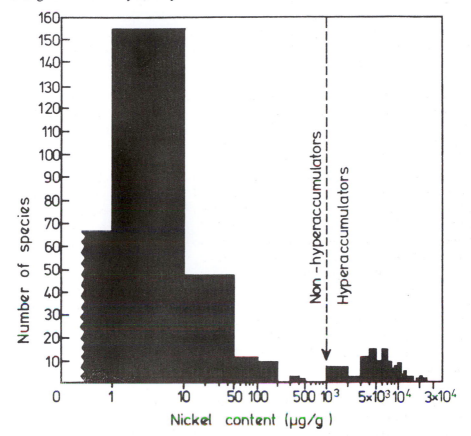

Fig.4.1. Histograms showing the abundance of nickel in *Alyssum* species. All values >1000 μg/g (0.1%) are from Section Odontarrhena of the genus. Source: Brooks (1987).

The concentration of 1000 μg/g (0.1%) dry weight, used to classify hyperaccumulation is clearly not arbitrary. There are two distinct populations. All of the hyperaccumulators are from Section Odontarrhena of the genus. The very presence of these *Alyssum* nickel plants does of course indicate the existence of nickel-rich soils in which they grow, but is probably not representative of the degree of this mineralisation. It must be stated that it is only my personal opinion that hyperaccumulators will not always be suitable for biogeochemical prospecting. This supposition has seldom been tested in the field because most of this type of prospecting has been carried out in previously glaciated boreal parts of the world, particularly Canada and Russia, where hyperaccumulator plants are not found. There is clearly a need for biogeochemical experiments with this type of plant, experiments that could well begin in the Balkan Peninsula, where

Alyssum species such as *A.murale* are found both on and off nickeliferous soils.

Although nickel concentrations in *Alyssum* species are "quantised" into two distinct populations, this is not always the case elsewhere. Reeves (1992) has shown that for some tropical hyperaccumulating plants (such as in Brazil - Brooks *et al.*,1990) there is a continuum of values ranging from 1 to 5000 $\mu g/g$. This is illustrated in Fig.4.2. It is probable therefore that hyperaccumulating plants from this Brazilian serpentine flora could be used in biogeochemical prospecting.

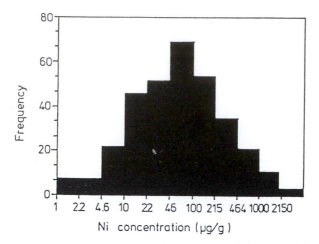

Fig.4.2. Histograms showing the abundance of nickel in serpentine plants (hyperaccumulators and non-accumulators) from Goiás State, Brazil. Source: Reeves (1992).

Strictly speaking, biogeochemistry involves the chemical analysis of vegetation to predict the metal content of the substrate, whereas geobotany relies on identification of plant species or communities to detect ore bodies without predicting the degree of this mineralisation. In some cases the presence of plant species can indicate not only the presence of mineralisation in the substrate, but also the approximate concentration of the target metals. Two examples of this will now be given under the present heading of biogeochemistry rather than earlier in Chapter 3.

Biogeochemical prospecting for copper and cobalt in Zaïre

Table 4.1 records the plant communities growing over a copper/cobalt deposit at Fungurume, Shaba Province, Zaïre. None of the plants growing over the more weakly mineralised part of the area (<350 $\mu g/g$ Co and <12,000 $\mu g/g$ [1.2%] Cu) is a hyperaccumulator. However, for the other, more heavily mineralised communities there are nine such plants, some of which hyperaccumulate both metals. An experienced geobotanist/biogeochemist is thereby able to predict the chemical composition of the substrate without even carrying out the usually obligatory chemical analyses.

Table 4.1. Plant communities over a copper/cobalt deposit in Zaïre.

Community and metal content of soil	Component species
A - Open forest (<280 μg/g Cu in soil)	*Brachystegia bussei, Albizzia adanthifolia, Cussonia arborea*
B - Tall thickets (up to 250 μg/g Co and 350 μg/g Cu in soil)	*Uapaca robynsii, Loudetia superba, Phragmanthera rufescens*
C - Steppe-savanna (up to 350 μg/g Co and 5000 μg/g Cu in soil)	*Haumaniastrum rosulatum, Loudetia simplex, Polygala petitiana, P.usafuensis*
D - Velloziacae steppe (up to 1500 μg/g Co and 12,000 [1.2%] Cu in soil)	*Xerophyta retinervis*, X.demeesmaekeriana, Crassula alba*, Pandiaka metallorum**
E - Commelinaceae steppe (up to 700 μg/g Co and 25,000 μg/g [2.5%] Cu)	*Commelina zigzag*, Ipomoea alpina*, Cyanotis longifolia*, Sopubia dregeana*, Anisopappus davyi*, Bulbostylis mucronata**

*Hyperaccumulator of cobalt and/or copper.

Biogeochemical prospecting for nickel in Western Australia

The second example of a geobotanical/biogeochemical approach to the problem of identification of the degree of mineralisation is provided by the work of Cole (1973); Severne (1972) and Severne and Brooks (1972) in Western Australia (see also Chapter 3) in which they studied the geobotany and biogeochemistry of the nickel hyperaccumulator *Hybanthus floribundus* from Western Australia (Fig.4.3).

In her study of the ecology of *Hybanthus floribundus*, Cole (1973) concluded that it is restricted to soils with a high nickel content in the Widgiemooltha, Mount Thirsty, Mission Ridge, and Kurnalpi regions of Western Australia near Kalgoorlie. Four occurrences of this plant were observed over outcropping serpentinite and at the contact of quartz felspar porphyry and hypersthenite dykes with serpentinite at Widgiemooltha and Doordie Rocks.

Cole (1973) concluded that this species is an indicator of a nickeliferous environment, though it does not necessarily delineate a nickel sulphide ore body within the ultramafics. She established that there was some relationship between the nickel content of the plant and the level of this element in the supporting substrate. Her work involved the analysis of numerous specimens of *H.floribundus*. The nickel content ranged from 131-13,794 μg/g (0.013-1.379%) in dry material. The overwhelming majority of the specimens contained several thousand μg/g of nickel.

It has already been shown in Chapter 3 that *Hybanthus floribundus* indicates nickeliferous substrates, though not necessarily economic mineralisation below the soils. Before considering the possibility of biogeochemical prospecting with this plant, it must be appreciated that it exists in several different forms, each with its own typical nickel signature, and each of which must be identified before a viable prospecting project can be undertaken. Bennett (1969) has recognised three subspecies of this genus (Fig.4.3) as well two forms within one of them. These different taxa, as well as their nickel content and that of the soil, are summarised below in Table 4.2.

Table 4.2. Mean nickel content (μg/g dry weight) of species, subspecies and forms of *Hybanthus* and that of their supporting soils. Number of specimens shown in parentheses.

Species	Plant	Soil
Hybanthus floribundus subsp. *curvifolius* form A (34)	7025	800
H. floribundus subsp. *curvifolius* form B (3)	3000	900
H. floribundus subsp. *adpressus* (12)	1270	134
H. floribundus subsp. *floribundus* (4)	263	50
H. epacroides subsp. *bilobus* (3)	200	10

Source: Severne (1972).

Severne (1972) carried out correlation analysis of nickel concentrations in various plant parts of *Hybanthus floribundus* as a function of the nickel content of the soil. He found no statistically significant relationship between the two variables, though there were some correlations involving nickel and other metals. Although the initial studies with this species showed that it gave greater promise for geobotanical exploration than for the biogeochemical method, the work of Severne and Cole represents perhaps the most serious attempt to use a hyperaccumulator for biogeochemical prospecting. Much more work remains to be carried out in this important mining region to determine the true value of *H.floribundus* for biological exploration methodology.

Biogeochemical prospecting in New Caledonia

Very little biogeochemical prospecting has been carried out in New Caledonia because most of the nickel deposits have an obvious surface expression. The volume of conventional geochemical prospecting by means of soil and rock analyses is truly vast and has left little need for other methodologies. However, Lee *et al.* (1977) did explore the potential of two nickel plants for biogeochemical prospecting. Their work involved the use of *Hybanthus austrocaledonicus* and *Homalium kanaliense* (see Fig.3.3) that can contain up to 1.76% and 0.94% nickel respectively in their dry leaves. The former is a forest species, whereas the *Homalium* is found in the *maquis* shrub community of the Plaine des Lacs region.

Lee *et al.* (1977) analysed a total of 147 specimens of both species together with the soils in which each was found, and determined 11 elements including total nickel in the plants and soils, and extractable (ammonium oxalate) nickel in the soils only. The elemental abundance data were subjected to correlation analysis and several correlations were observed, as is shown in Table 4.3.

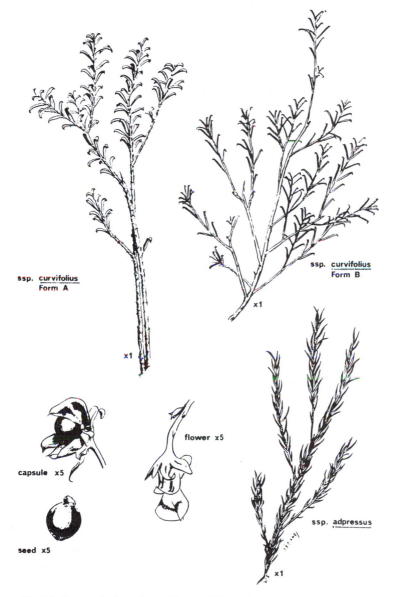

Fig.4.3. Sketch of subspecies and forms of *Hybanthus floribundus*. Source: G.Galloway.

In contrast to *H.kanaliense,* the nickel content of *H.austrocaledonicus* was related not only to the extractable nickel content of the soil, but also to the total nickel concentration. This species therefore shows a better potential for biogeochemical prospecting for nickel. Both species gave a good correlation with the magnesium content of the soils and could presumably be used in determining the more magnesium-rich soils, and by inference, their probable nickel content.

Table 4.3. Plant/soil correlations for various elements in soils and dried leaves of *Homalium kanaliense* (A) and *Hybanthus austrocaledonicus* (B).

		Ca	Co	Cr	Cu	Fe	K	Mg	Mn	Ni	P	Zn	XNi
						SOILS							
Plant A	Ca									+			
	Co	+		-					-				-
	Cr	-						-				-	+
	Cu			+									
	Fe	-								-		-	
	K	-	-	+	+	+	-	-			-		
	Mg	+	+	-		-	+	+	-	+			
	Mn	-	-	+	+	-	-						
	Ni	-						+					+
	P		-		-								
Plant B	Ca							+					
	Co		+										
	Cr							-				+	
	Mg							+				-	
	Mn					+			+				
	Ni							-		+	+	+	
	P											+	
	Zn											+	

XNi = extractable nickel (ammonium oxalate buffer), + = highly significant positive relationship (P < 0.01), - = highly significant negative relationship (P < 0.01). Source: Lee *et al.* (1977).

Biogeochemical Prospecting in the Herbarium

Introduction

The world's herbaria (plant museums) are a repository for about 200 million dried plant samples. There are nearly 2000 herbaria recorded in the Index Herbariorum (Stafleu, 1974). Perhaps the oldest of these is the Paris Herbarium which was founded in 1635 by Louis XIII and has about 6 million specimens. Similar numbers are held by such famous herbaria as those of Kew and the British Museum.

Herbarium specimens of flowering plants, ferns and the larger algae are customarily dried between sheets of heavy blotting paper under pressure so that the dried specimen is flat. Most specimens will dry in about a week if the blotters are changed several times. The drying can be hastened by interspersing ventilators of corrugated cardboard among the blotters and by the gentle application of heat. Although the colours of flowers and leaves eventually fade, herbarium specimens last indefinitely, some being over 200 years old.

A herbarium is a most essential tool for research in plant taxonomy and serves as a permanent record of the distribution of each species, its range of variability and the correlation of that variability with the geography and habitat. A botanist writing a flora of a region, relies largely on herbarium specimens which he studies against a background of his own observations on living plants. In more critical studies of smaller groups of species, cultural and breeding experiments, determination of chromosome number and other detailed analyses are also customary. The herbarium however remains the essential foundation for the work, because no one individual can expect to see even a small group of species in all localities and all habitats in which they occur and under the varying conditions of different years and seasons.

Although herbaria serve as samples of natural populations of plants, they must be interpreted with caution because they tend to be biased towards the unusual. When a region is first explored, collectors have their time cut out, merely to make records of the common species. Thereafter, the less common species and unusual forms of common species, are likely to be collected out of proportion to their abundance. Thus the usefulness of a specimen may be greatly enhanced by some indication on the label as to the frequency of occurrence of the species in the wild.

In the late 1940s, Minguzzi and Vergnano (1948) discovered hyper-accumulation of nickel by *Alyssum bertolonii* and had thought of analysing herbarium material to see if this character was also possessed by other species of this genus. Their approach to the curator of the Florence Herbarium evoked such consternation (O.Vergnano Gambi - pers. comm.) that they desisted in making further requests. The reason for this reaction was not hard to find. At that time, using emission spectrography as an analytical method, virtually the entire specimen would have had to have been destroyed for a single analysis.

The first person to analyse herbarium material was probably E.M.Chenery (Chenery, 1948, 1951) who determined aluminium in representatives of all of the 259 recognized families of dicotyledons. His work was not however orientated towards mineral exploration and consisted of semi-quantitative colorimetric tests of leaf samples of about 6 cm^2 in area.

Herbarium material has been used in geobotanical exploration. For example, Persson (1956) noted the collection localities of Swedish herbarium specimens of a "copper moss" and discovered three localities with anomalous copper levels in the substrate, of which one was an existing copper mine. Later Cole (1971) in the course of geobotanical investigations in southern Africa, identified plant indicators of copper and referred back to a herbarium for other collection localities of these species.

In recent years, new techniques of chemical analysis have resulted in the minimum sample size of vegetation being reduced to a few milligrams or even

less, for the determination of specific elements. Even the plant dust between
herbarium sheets can be successfully analysed without problems. Indeed the
limiting factor is not the size of the sample needed for analysis, but rather the
extent to which this small sample is representative of the entire plant specimen.

During the past few years, my colleagues and I have analysed some 20,000
herbarium specimens (mainly leaf samples of area 1 cm^2 or less) in order to
identify hyperaccumulators and to investigate the possibility of prospecting for
minerals in countries which we have never visited. This proposition might at first
sight appear to be untenable, but it will be shown below that analysis of
herbarium material for this purpose is indeed a viable method for obtaining
biogeochemical data of use for future field exploration. The use of herbarium
materials for biogeochemical prospecting has also been reviewed by Brooks
(1983) and Brooks *et al.* (1995). Much of the material below is based on these
two references.

Analysis of herbarium material

The first step in the analysis of herbarium material is to obtain the specimens.
Provided that requests are modest and that only small samples (1 cm^2 or less) are
required, most herbarium curators are happy to donate material. The following
information is needed along with the sample:

1. The binomial and authorities: e.g. *Rinorea bengalensis* (Wall.) O.K. The authorities are part
of the name of the plant. In this example, the genus is *Rinorea* and the species is *bengalensis*. O.K.
stands for Otto Kuntze. Only for well known botanists are initials used. Perhaps the best known of
these is "L." (Linnaeus). The authorities are part of the name of the plant. In the above example,
"(Wall.)" tells us that Wallich originally described the species but that it was later reclassified by Otto
Kuntze.

2. The collector's number or herbarium accession number. The first is more useful because you
can then check to see whether you have had a duplicate of these taxa from some other herbarium.

3. The collection locality. Full details should be sought. Early (19th century) material often has
a poor description of collection localities: e.g. "near Vienna", whereas modern collectors usually give
very detailed information from which it should be possible to pinpoint the locality to within a few
hundred metres.

Herbaria are repositories for the sacrosanct *type material* which is a standard
upon which names of species are based. A scientific name to be accepted must
be properly published together with a description, but it is the type specimen that
permanently determines the application of the name. In many herbaria, types are
segregated from the rest of the collection in order to minimize the risk of
damage. Curators will not normally supply type material for analysis and it is
ill-mannered to request it.

The second step in chemical analysis is to weigh about two-thirds of the
sample (i.e. up to 50 mg) into a 5 mL borosilicate test tube using a five-figure
balance. Place the tubes in squat borosilicate beakers and ignite them at 500°C
in a muffle furnace for a period of 2-3 hours. After cooling, add a known volume

(1-2 mL) of 2M hydrochloric acid prepared from redistilled constant-boiling (6M) reagent. Warm gently to dissolve the ash if it does not dissolve immediately. With care, four elements can be determined in a 1 mL sample using atomic absorption spectrometry. If plasma emission (ICP) spectrometry is used for the analysis, about 5 mL of solution will be required. The data should be expressed on a dry weight rather than ash weight basis. This is logical because the dry weight is a measure of the true weight of the sample in its natural state.

Replicate analyses of the same original sample should provide reasonably reproducible data (8-10%) but the question is often asked as to the inter-sample variability of different samples from the same specimen. This problem was examined by Wither (1977) who obtained circular disc samples (6 mg) from four separate leaves of the same specimen of the nickel hyperaccumulator *Rinorea bengalensis*. Twenty separate samples (each containing one disc) gave 552 ± 12 $\mu g/g$ nickel. Fifteen separate samples (each containing two discs) gave 536 ± 20 $\mu g/g$ nickel. Ten separate samples of 3 discs each, gave 518 ± 27 $\mu g/g$ of this element. From these experiments it appeared that the sampling error for samples as small as 6 mg, was only 2-5%.

Analysis of herbarium material was the means of identifying well over 300 hyperaccumulators of nickel (Table 3.3), and numerous others of copper, cobalt, and zinc. Herbarium material has also been used, as mentioned above, for "prospecting" for minerals and the host rocks. Some case histories of these studies will now be presented below.

Some case histories of biogeochemical prospecting in the herbarium

Nickel

One of the first herbarium studies was the determination of the nickel content of some 2000 specimens of *Hybanthus* and *Homalium* (Brooks *et al.*,1977a). At that time only four hyperaccumulators of nickel had been recorded. This survey was undertaken because a hyperaccumulator of nickel (i.e. > 1000 $\mu g/g$ in dried leaves), *Hybanthus floribundus*, had been discovered in Western Australia (Severne and Brooks, 1972) and it seemed to be a reasonable assumption that others might be found in the same genus. Since Jaffré and Schmid (1974) had determined that the New Caledonian plant *Homalium guillainii* was also a hyperaccumulator of nickel, it seemed to be logical to also investigate this genus on a worldwide basis.

Fifty herbaria throughout the world were asked for plant material and 35 of these supplied samples. The specimens had originally been collected from all parts of the tropical and warm-temperate world and represented a sampling density of about one per 2000 km^2. The survey resulted in the reidentification of all previously-known hyperaccumulators of nickel and the discovery of five additional species (all from New Caledonia) with this same character. Fourteen previously unknown strong accumulators (100-1000 $\mu g/g$ nickel in dry material) were also discovered. From the collection localities of the hyperaccumulators and strong accumulators, it was possible to pinpoint many of the world's major nickel-rich ultramafic occurrences in tropical and warm-temperate regions. The principle was obviously applicable to other species.

Some mention has already been made (Chapter 3) of a nickel survey of all but one of the 168 species of *Alyssum* (Brooks *et al.*,1979a) which resulted in the discovery of over 40 previously-unknown hyperaccumulators of nickel. The collection localities of these plants identified virtually every single major ultramafic rock outcrop in southern Europe and Asia Minor. Although this work was not of much value for mineral exploration, because the area is already very well mapped geologically, it did point to the possibility of carrying out a similar survey in areas which had not been well mapped. An obvious place for this type of work was Southeast Asia.

Brooks and Wither (1977) found hyperaccumulation of nickel in *Rinorea bengalensis*. A survey of the nickel and cobalt contents of 89 herbarium specimens of this species, showed that many of the collection localities coincided with a number of important ultramafic outcrops throughout the region. This is demonstrated in Fig.4.4.

The *Rinorea* data are also shown in Table 4.4 where abundances of nickel and cobalt are arranged under the appropriate rock type (if known) of the collection locality. *Rinorea bengalensis* obviously grows over a wide range of substrates and is a hyperaccumulator of nickel when growing over ultramafic rocks. Over such substrates, the Co/Ni ratio (ca. 0.01) is considerably lower than for plants growing over other rock types. This serves as an additional criterion for pinpointing ultramafic occurrences. It was invariably possible to identify these serpentinic substrates by nickel levels exceeding 3000 μg/g in the plant material. Values above 1000 μg/g usually indicated this type of substrate but there was some overlap with the higher values corresponding to sedimentary (excluding limestone) substrates. Whereas plants growing over ultramafic substrates could almost always be differentiated from those growing over other rock types (i.e. by a combination of high nickel concentrations and low Co/Ni ratios), it was not possible to separate other substrates from each other by the nickel content of this species alone. This however is relatively unimportant from the standpoint of mineral exploration because ultramafic rocks are a favoured target for explorationists as they are hosts for many economic minerals such as nickel, chromium, cobalt and the metals of the platinum group.

Fig.4.5 also shows the distribution of *R.bengalensis* in Southeast Asia and includes only specimens whose collection localities could be established accurately. The range extends all the way from Sri Lanka to the Solomons and the ultramafic areas pinpointed (large circles) represent most of the important regions containing these rocks. In all but two cases, ultramafic rocks were known to occur at the collection localities. The exceptions were the substrate of a plant from Dalman, Nabire, which contained 1.75% nickel, and another from the Sorong area (1.2% nickel). Both were from Irian Jaya (Indonesian New Guinea).

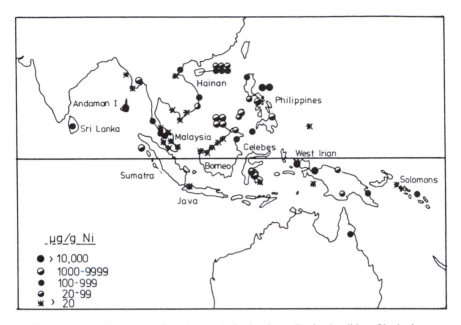

Fig.4.4. Map of Southeast Asia and Australasia showing collection localities of herbarium specimens of *Rinorea bengalensis* and indicating nickel concentrations (µg/g dry weight) in leaf material. Source: Brooks and Wither (1977).

Because of the very high nickel contents and low Co/Ni quotients (0.003 and 0.002 respectively), the above plants were almost certainly growing over ultramafic rocks.

In the Nabire area, the presence of ultramafic rocks can be inferred by extrapolation from partially-surveyed sporadic undifferentiated mafic intrusives forming a long belt passing through Nabire. The location of the specimen from Sorong will probably have to be investigated *in situ* to establish the nature of the substrate.

The presence of previously-unknown ultramafic areas in Irian Jaya was revealed by a chain of events which began with the original collection of material by two Japanese botanists in 1940. The samples were stored for 37 years at the Arnold Arboretum (Harvard University) and then sent to New Zealand for analysis by persons who had never visited Irian Jaya.

Rinorea bengalensis was not selected for analysis in a random manner. Since another hyperaccumulating genus (*Hybanthus*) belonged to the same family (Violaceae), it was logical that this be investigated for other such genera.

It is improbable that biogeochemical prospecting in the herbarium will result in identification of economic minerals *per se* because the method is only likely to reveal potential host rocks rather than the much smaller target of mineralisation within them. The technique still remains an extremely inexpensive means of surveying on a broad scale.

Table 4.4. Nickel and cobalt concentrations (μg/g dry weight) in herbarium specimens of *Rinorea bengalensis*.

Substrate	N	Cobalt		Nickel		
		Mean	Range	Mean	Range	Co/Ni
Ultramafic	21	87	6-545	6860	836-17,500	0.01
Limestone	12	12	1-33	113	2-560	0.11
Other sed.	14	75	3-290	674	3-3000	0.11
Basic rocks	11	27	1-217	103	1-550	0.26
Acid rocks	5	16	5-29	20	2-56	0.80
Unknown	26	38	1-300	177	1-2000	0.21
Overall	89	51	1-545	1810	1-17,500	0.03

Source: Brooks and Wither (1977).

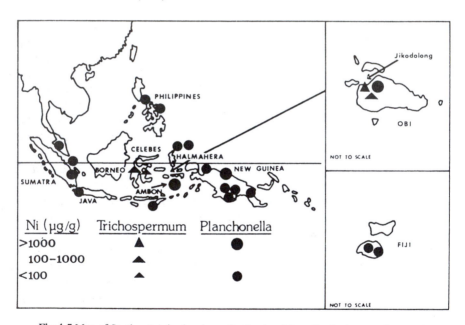

Fig.4.5 Map of Southeast Asia showing collection localities of herbarium specimens of *Planchonella oxyedra* and *Trichospermum kjellbergii*. Nickel concentrations (μg/g dry weight) are also shown. Source: Wither and Brooks (1977).

In another herbarium survey, an entire collection of plants from Obi Island, Indonesia, which had been collected by Dr E de Vogel (Leiden) in 1976, was analysed for nickel. Most of these plants had been growing over a serpentine occurrence on the island. This survey resulted in the identification of two more hyperaccumulators of nickel, *Planchonella oxyedra* and *Trichospermum kjellbergii*.

Once these species had been determined as having a hyperaccumulating capacity, further herbarium specimens of both species were obtained and analysed for nickel (Wither and Brooks, 1977). The data are shown in Fig.4.5 and once again pinpointed ultramafic outcrops throughout Southeast Asia. One of these areas on the island of Ambon had not been known previously to geologists and is deserving of follow-up.

Other herbarium surveys for nickel have involved the identification of seven hyperaccumulators of nickel in New Caledonian species of *Geissois* (Jaffré *et al.*,1979a). All of these species indicated nickeliferous rocks on the island. Similar studies were carried out by Jaffré *et al.* (1979b) on New Caledonian species of *Casearia, Lasiochlamys* and *Xylosma*. As a result of this work, a further 18 "nickel plants" were discovered, all of which indicated ultramafic rocks.

Cobalt

Hyperaccumulators of cobalt are far less numerous than those of nickel. This is not difficult to understand because although nickeliferous ultramafic rocks occupy a significant proportion of the earth's surface, cobalt-rich rocks supporting hyperaccumulators of this element ($>0.1\%$ - dry weight) are, apart from isolated ore occurrences, confined to the copper-cobalt province of Zaïre/Zambia in Central Africa which used to be one of the largest sources of cobalt in the world. These deposits support a highly-unusual and diversified metal-tolerant vegetation that includes the only known hyperaccumulators of cobalt (Brooks and Malaisse, 1985).

The first herbarium study on plants of Shaba Province, Zaïre, was by Brooks (1977) who analysed 19 species of *Haumaniastrum* and found up to 1.02% cobalt in the dried leaves of *H.robertii*. This species had been known as a "copper flower" for many years but because of the co-existence of copper and cobalt in all the Central African deposits, it is not easy to decide which of these elements is really indicated. However pot trials on seedlings of this species (Morrison *et al.*,1979) indicated that *H.robertii* is more tolerant of cobalt than copper and is therefore probably a cobalt indicator. In all, 26 plant species (all from Zaïre) have been recognized as being hyperaccumulators of cobalt (see Chapter 3) as a result of herbarium studies.

The unusual accumulation of cobalt by the Nyssaceae has been reported by Brooks *et al.* (1977b). Herbarium specimens (375) of species of *Nyssa, Camptotheca* and *Davidia* were analysed for cobalt and nickel. All species of *Nyssa* (4 from the United States and 2 from Southeast Asia) possessed a marked ability to accumulate cobalt not only in absolute terms but also relative to nickel. Although the maximum cobalt concentrations (530 μg/g in *N.sylvatica)* were below the threshold defining hyperaccumulation (i.e. 1000 μg/g), these values were orders of magnitude higher than for any other plant species outside the Zaïrean copper/cobalt outcrops.

It was concluded that all species of *Nyssa* could be used for assessing the cobalt status of soils as had already been proposed by Kubota *et al.* (1960) in the case of *N.sylvatica*. *Camptotheca acuminata* showed a much lower uptake of cobalt than *Nyssa* species but this uptake was still considerably higher than for most other plants and strengthened the case for taxonomic inclusion of *Camptotheca* in the Nyssaceae. *Davidia involucrata* showed no capability for accumulating cobalt to a level exceeding normal background for most vegetation and strengthened powerful arguments that this monotypic genus does not belong to the Nyssaceae. This survey therefore, though of marginal significance for mineral prospecting, did have some importance in the field of plant taxonomy.

Copper

There are very few examples of plants which hyperaccumulate copper. All of these are from the copper/cobalt deposits of southeastern Zaïre. Brooks *et al.* (1995) have listed 24 species that could hyperaccumulate copper. In a herbarium survey of *Aeollanthus*, the extraordinary accumulation of copper by *A.biformifolius* was observed (Brooks *et al.*,1978; Malaisse *et al.*,1979). This species is confined to copper-rich substrates and contains the highest copper concentration ever recorded for any phanerogam (flowering plant). The whole plant contained up to 1.37% copper on a dry weight basis. Although there are very few hyperaccumulators of copper, a very much larger number of species is able to indicate copper mineralisation without inordinate uptake of this element. One of these is the Australian "copper flower" *Polycarpaea spirostylis* which like so many of this type of indicator is a herb and belongs to the Caryophyllaceae (pink family). Brooks and Radford (1978) analysed 183 herbarium specimens of 11 out of 12 known species of *Polycarpaea* and found statistically the threshold copper concentrations which appeared to be anomalous. Fourteen specimens were in this category and the collection localities are now being investigated further.

The well-known "kisplanten (pyrite plants) of Fennoscandia include *Lychnis (= Viscaria) alpina* and *Silene (=Melandrium) dioica*. Brooks *et al.* (1979b) analysed over 700 herbarium specimens of these two species for copper, lead and nickel. Anomalous values for copper ($> 20 \mu g/g$), nickel ($> 20 \mu g/g$) and lead ($> 80 \mu g/g$) revealed a number of biogeochemical anomalies, including virtually all of the major copper deposits of Fennoscandia, particularly those of the Røros area of Norway and at Bergslågen, Sweden. The highest value was 250 $\mu g/g$ for a specimen growing over copper slag in Garpenberg at Bergslågen.

Lead "anomalies" should be treated with caution in collection localities close to a main road where automotive emissions can easily pollute vegetation samples. Nickel anomalies invariably coincide with ultramafic rocks.

Lead

There has already been some discussion above of lead values in *Lychnis alpina* and *Silene dioica*. Recently, specimens of *Alyssum wulfenianum* and *Thlaspi rotundifolium* subsp. *cepaeifolium* from the Austria/Italy border area near Raibl have been analysed for lead (Reeves and Brooks, 1983a). Both species are endemic plants confined to a base metal (zinc and lead) mining area. Values of up to 860 $\mu g/g$ in dried leaves of *A.wulfenianum* and 7000 $\mu g/g$ in the *Thlaspi* species were found. No other plant has ever been found with lead levels even approaching the concentrations found in *Thlaspi rotundifolium* subsp. *cepaeifolium*.

Manganese

Very little work has been done on herbarium specimens of manganese-accumulating plants. However, Jaffré (1977, 1979) used herbarium material to identify a number of manganese-accumulating species from New Caledonia. In a later study (Brooks *et al.*,1981), 31 species of *Alyxia* from the same island were analysed for their manganese content. Most of the species showed excessive uptake of manganese with a maximum of 1.15 % in dried leaves of *A.rubricaulis*. The manganese content alone was sufficient to distinguish a number of species from each other.

As was the case with nickel plants, the species with very high concentrations of manganese indicated localities with manganese-rich ultrabasic rocks.

Zinc

One of the best known of all zinc hyperaccumulators is *Thlaspi calaminare* which is confined to the zinc-rich calamine (zinc carbonate) deposits of Western Europe. In a survey of European species of Thlaspi, Reeves and Brooks (1983b) determined that at least nine other species contained over 1 % zinc in their dried leaves. These high values were found not only in zinc-rich substrates but also in a wide range of other rock formations with presumably much lower zinc levels.

A particularly interesting feature of *Thlaspi* species is that they also hyperaccumulate nickel (ten species) and accurately reflect the presence of ultramafic outcrops even when these are quite localized.

References

Bennett,E.M.(1969) *The Genus Hybanthus Jacquin in Australia Including the Cytology and Anatomy of the Western Australian Species*. MSc Thesis, University of Western Australia, Perth.

Brooks,R.R.(1972) *Geobotany and Biogeochemistry in Mineral Exploration*. Harper & Row, New York, 290 pp.

Brooks,R.R.(1977) Copper and cobalt uptake by *Haumaniastrum* species. *Plant and Soil* 48, 541-545.

Brooks,R.R.(1983) *Biological Methods of Prospecting for Minerals*. Wiley, New York, 322 pp.

Brooks,R.R. (1987) *Serpentine and its Vegetation*. Dioscorides Press, Portland, 454 pp.

Brooks,R.R.(ed.)(1992) *Noble Metals and Biological Systems*. CRC Press, Boca Raton, 392 pp.

Brooks,R.R.,Dunn,C.E. and Hall,G.E.M.(1995) *Biological Systems in Mineral Exploration and Processing*. CRC Press, Boca Raton, 538 pp.

Brooks,R.R.,Lee,J.,Reeves,R.D. and Jaffré (1977a) Detection of nickeliferous rocks by analysis of herbarium specimens of indicator plants. *Journal of Geochemical Exploration* 7, 49-57.

Brooks,R.R. and Malaisse,F.(1985) *The Heavy Metal Tolerant Flora of Southcentral Africa*. Balkema, Rotterdam, 199 pp.

Brooks,R.R.,McCleave,J.A. and Schofield,E.K.(1977b) Copper and cobalt uptake by the Nyssaceae. *Taxon* 26, 197-201.

Brooks,R.R.,Morrison,R.S.,Reeves,R.D.,Dudley,T.R. and Akman,Y.(1979a) Hyperaccumulation of nickel by *Alyssum* Linnaeus (Cruciferae). *Proceedings of the Royal Society (London) Section B* 203, 387-403.

Brooks,R.R.,Morrison,R.S.,Reeves,R.D. and Malaisse,F.(1978) Copper and cobalt in African species of *Aeolanthus* Mart. (Plectranthinae, Labiatae). *Plant and Soil* 50, 503-507.

Brooks,R.R. and Radford,C.C.(1978) An evaluation of background and anomalous copper and zinc concentrations in the "copper plant" *Polycarpaea spirostylis* and other Australian species of the genus. *Proceedings of the Australasian Institute of Mining and Metallurgy*. No.268, 33-37.

Brooks,R.R.,Reeves,R.D.,Baker,A.J.M.,Rizzo,J.A. and Diaz Ferreira,H.(1990) The Brazilian Serpentine Plant Expedition (BRASPEX) - April/May 1988. *National Geographic Research* 6, 205-219.

Brooks,R.R.,Trow,J.M. and Bølviken,B.(1979b) Biogeochemical anomalies in Fennoscandia: a study of copper, lead and zinc in *Melandrium dioicum* and *Viscaria alpina. Journal of Geochemical Exploration* 11, 73-87.

Brooks,R.R.,Trow,J.M.,Veillon,J.-M.and Jaffré,T.(1981) Studies on manganese-accumulating *Alyxia* from New Caledonia. *Taxon* 30, 420-423.

Brooks,R.R. and Wither,E.D.(1977) Nickel accumulation by *Rinorea bengalensis* (Wall.)O.K. *Journal of Geochemical Exploration* 7, 295-300.

Brundin,N.(1939) Method of locating metals and minerals in the ground. *United States Patent* No.2, 158,980.

Cannon,H.L.(1957) Description of indicator plants and methods of botanical prospecting for uranium deposits on the Colorado Plateau. *United States Geological Survey Bulletin* 1036-M, 399-516.

Cannon,H.L.(1960) The development of botanical methods of prospecting for uranium on the Colorado Plateau. *United States Geological Survey Bulletin* 1985-A, 1-50.

Cannon,H.L.(1964) Geochemistry of rocks and related soils and vegetation in the Yellow Cat area, Grand County, Utah. *United States Geological Survey Bulletin* 1176, 1-127.

Carlisle,D.,Berry,W.L,Kaplan,I.R. and Watterson,J.R.(eds)(1986) *Mineral Exploration Biological Systems and Organic Matter.* Prentice-Hall, Englewood Cliffs, 465 pp.

Chenery,E.M.(1948) Aluminium in the plant world. *Kew Bulletin* 1948, 173-183.

Chenery,E.M.(1951) Some aspects of the aluminium cycle. *Journal of Soil Science* 2, 97.

Cohen,D.R.,Hoffman,E.L and Nichol,I.(1987) Biogeochemistry: a geochemical method for gold exploration in the Canadian Shield. *Journal of Geochemical Exploration* 29, 49-73.

Cole,M.M.(1971)The importance of environment in biogeographical/geobotanical and biogeochemical investigations. *Canadian Institute of Mining Special Volume* No.11, 414-425.

Cole,M.M.(1973) Geobotanical and biogeochemical investigations in the sclerophyllous woodland and shrub associations of the Eastern Goldfields area of Western Australia with particular reference to the role of *Hybanthus floribundus* (Lindl.) F.Muell. as a nickel indicator and accumulator plant. *J. Applied Ecology* 10, 269-320.

Connor,J.J.and Shacklette,H.T.(1975) Background geochemistry of some rocks, soils, plants, and vegetables in the conterminous United States. *United States Geological Survey Professional Paper* 574-F, 1-168.

Dunn,C.E.(1983a) Uranium biogeochemistry of the NEA/IAEA Athabasca test area. *Geological Survey of Canada Paper* 82-11, 127-132.

Dunn,C.E.(1983b) Detailed biogeochemical studies for uranium in the NEA/IAEA Athabasca test area. *Geological Survey of Canada Paper* 82-11, 259-272.

Dunn,C.E.(1986) Biogeochemistry as an aid to exploration for gold, platinum and palladium in the northern forests of Saskatchewan, Canada. *Journal of Geochemical Exploration* 32, 211-222.

Dunn,C.E.(1989) Reconnaissance-level biogeochemical surveys for gold in Canada. *Transactions of the Institution of Mining and Metallurgy Sec.B* 98, 153-161.

Dunn,C.E.(1995) Intyroduction to biogeochemical prospecting. In: R.R.Brooks, C.E.Dunn and G.E.M.Hall (eds). *Biological Systems in Mineral Exploration and Processing.* Ellis Horwood, Hemel Hempstead, pp.233-242.

Dunn,C.E.,Byman,J. and Ek,J.(1984) Uranium biogeochemistry: A bibliography and report of the state of the art. *International Atomic Energy Agency Technical Document* 327, 1-83.

Dunn,C.E.,Erdman,J.A.,Hall,G.E.M. and Smith,S.C(1992) *Biogeochemical Exploration Simplified (notes for a short course on methods of biogeochemical and geobotanical prospecting, with emphasis on arid terrains).* Association of Exploration Geochemists, Vancouver.

Dunn,C.E.,Hall,G.E.M. and Hoffman,E.L(1989) Platinum group metals in common plants of northern forests: developments in analytical methods, and the application of biogeochemistry to exploration strategies. *Journal of*

Geochemical Exploration. 32, 211-222.

Dunn,C.E.,Hall,G.E.M.   and   Scagel,R.(1993)   *Applied   Biogeochemical Prospecting in Forested Terrain*. Association of Exploration Geochemists, Rexdale, Ontario.

Erdman,J.A and Olsen,J.C.(1985) The use of plants in prospecting for gold: a brief overview with a selected bibliography and topic index. *Journal of Geochemical Exploration* 24, 281-304.

Erdman,J.A.,Leonard,B.F. and McKown,D.M.(1985). A case for plants in exploration -gold in Douglas fir at the Red Mountain stockwork, Yellow Pine District, Idaho. *United States Geological Survey Bulletin* 1658A-S, 141-152.

Fortescue,J.A.C.   and   Hornbrook,E.H.W.(1967)   Progress   report   on biogeochemical research at the Geological Survey of Canada 1963-1966. *Geological Survey of Canada Paper* 67-23, Part I.

Fortescue,J.A.C.   and   Hornbrook,E.H.W.(1969)   Progress   report   on biogeochemical research at the Geological Survey of Canada 1963-1966. *Geological Survey of Canada Paper* 67-23, Part II.

Grabovskaya,L.I.(1965) *Biogeochemical Prospecting Methods* (in Russian) Gosgeolkoma SSSR, Moscow.

Hall,G.E.M.,Pelchat,J.-C. and Dunn,C.E.(1990) The determination of Au,Pd, and Pt in ashed vegetation by ICP-Mass spectrometry and graphite furnace atomic absorption spectrometry. *Journal of Geochemical Exploration* 37, 1-23.

Harbaugh,J.W.(1950) Biogeochemical investigations in the Tri-State district. *Economic Geology* 45, 548-567.

Hoffman,E.L and Booker,E.J.(1986) Biogeochemical prospecting for gold using instrumental neutron activation analysis with reference to some Canadian gold deposits. In: Carlisle,D.,Berry,W.L.,Kaplan,I.R. and Watterson,J.R.(eds) *Mineral Exploration Biological Systems and Organic Matter*. Prentice-Hall, Englewood Cliffs, pp.159-169.

Hornbrook,E.H.W.(1969) Biogeochemical prospecting for molybdenum in west-central British Columbia. *Geological Survey of Canada Paper* 68-56.

Hornbrook,E.H.W.(1970a) Biogeochemical investigations in the Perch Lake area, Chalk River, Ontario. *Geological Survey of Canada Paper*. 70-43.

Hornbrook,E.H.W.(1970b) Biogeochemical prospecting for copper in west-central British Columbia. *Geological Survey of Canada Paper* 69-49.

Jaffré,T.(1977) Accumulation du manganèse par les espèces associeés aux terrains ultrabasiques de Nouvelle Calédonie. *Comptes Rendus de l'Academie des Sciences de Paris Série D*, 284, 1573-1575.

Jaffré,T.(1979) Accumulation du manganèse par les Protéacées de Nouvelle Calédonie. *Comptes Rendus de l'academie des Sciences de Paris Série D* 289, 425-428.

Jaffré,T.,Brooks,R.R. and Trow,J.M.(1979a) Hyperaccumulation of nickel by *Geissois* species. *Plant and Soil* 51, 157-162.

Jaffré,T.,Kersten,W.,Brooks,R.R. and Reeves,R.D.(1979b) Nickel uptake by the Flacourtiaceae of New Caledonia. *Proceedings of the Royal Society of London*

Section B 205, 385-394.

Jaffré,T. and Schmid,M.(1974) Accumulation du nickel par une Rubiacée de Nouvelle Calédonie: *Psychotria douarrei* (G.Beauvisage) Däniker. *Comptes Rendus de l'Academie des Sciences de Paris Série D* 278, 1727-1730.

Kovalevsky,A.L.(1987) *Biogeochemical Exploration for Mineral Deposits.* VNU Science Press, Utrecht, 224 pp.

Kovalevsky,A.L (1991) *Biogeochemistry of Plants* (in Russian) Nauka Press, Novosibirsk, 299 pp.

Kovalevsky,A.L and Kovalevskaya,O.M. (1989) Biogeochemical haloes of gold in various species and parts of plants. *Applied Geochemistry* 4, 369-374.

Kubota,J.,Lazar,V.A. and Beeson,K.C.(1960) The study of cobalt status of soils in Arkansas and Louisiana using the black gum as the indicator plant. *Soil Science Proceedings* 24, 527-528.

Lee,J.,Brooks,R.R.,Reeves,R.D.,Boswell,C.R. and Jaffré,T.(1977) Plant-soil relationships in a New Caledonian serpentine flora. *Plant and Soil* 46, 675-680.

Lounamaa,J.(1956) Trace elements in plants growing wild on different rocks in Finland. *Annales Botanici Societatis Zoologicae Botanicae 'Vanamo'* 29, 1-196.

Lungwitz,E.E.(1900) The lixiviation of gold deposits by vegetation and its geological importance. *Mining Journal of London* 69 500-502.

Malaisse,F.,Grégoire,J.,Brooks,R.R.,Morrison,R.S. and Reeves,R.D.(1979) Copper and cobalt in vegetation of Fungurume, Shaba Province, Zaïre. *Oikos* 33, 472-478.

Malyuga,D.P.(1964) *Biogeochemical Methods of Prospecting.* Consultants Bureau, New York, 205 pp.

Minguzzi,C. and Vergnano,O.(1948) Il contenuto di nichel nelle ceneri di *Alyssum bertolonii* Desv. *Atti della Società Toscana di Scienze Naturali Memorie* 55 1-28.

Morrison,R.S.,Brooks,R.R.,Reeves,R.D. and Malaisse,F.(1979) Copper and cobalt uptake by metallophytes from Zaïre. *Plant and Soil* 53, 535-539.

Parduhn,N.L and Smith, S.C.(1991) Biogeochemistry and geomicrobiology in mineral exploration (short course notes) In: *15th International Geochemical Exploration Symposium, Reno.* Association of Exploration Geochemists.

Perel'man,A.I.(1961) *Geochemistry of Landscapes* (in Russian). Geografichesky Izdatel', Moscow.

Perel'man,A I.(1966) *Landscape Geochemistry* (in Russian) Vysshaya Shkola, Moscow.

Persson,H.(1956) Studies of the so-called "copper mosses". *Journal of the Hattori Botanical Laboratory* 17, 1-18.

Polikarpochkin,V.V. and Polikarpochkina,R.T.(1964) *Biogeochemical Prospecting for Ore Deposits.* Nauka Press, Moscow.

Reeves,R.D.(1992) The hyperaccumulation of nickel by serpentine plants. In: Baker,A.J.M.,Proctor,J. and Reeves,R.D.(eds), *The Vegetation of Ultramafic*

(Serpentine) Soils. Intercept, Andover, pp.253-278.

Reeves,R.D. and Brooks,R.R.(1983a) Hyperaccumulation of lead and zinc by two metallophytes from a mining area in Central Europe. *Environmental Pollution Series A*, 31, 277-287.

Reeves,R.D. and Brooks,R.R.(1983b) European species of *Thlaspi* L. (Cruciferae) as indicators of nickel and zinc. *Journal of Geochemical Exploration* 18, 275-283.

Severne,B.C.(1972) Botanical Methods for Mineral Exploration in Western Australia. PhD Thesis, Massey University, Palmerston North, New Zealand.

Severne,B.C. and Brooks,R.R.(1972) A nickel-accumulating plant from Western Australia.*Planta* 103, 91-94.

Shacklette,H.T. and Connor,J.J.(1973) Airborne chemical elements in Spanish Moss. *United States Geological Survey Professional Paper* 574E, 1-46.

Shacklette,H.T.,Lakin,H.W.,Hubert,A E. and Curtin,G.C.(1970) Absorption of gold by plants. *United States Geological Survey Bulletin* 1314-B, 1-23.

Stafleu,F.A. (ed.)(1974) *Index Herbariorum*. Oostoek, Scheltema and Holkema, Utrecht, 397 pp.

Talipov,R.M.(1966) *Biogeochemical Prospecting for Polymetallic and Copper Deposits in the Conditions of Uzbekistan* (in Russian) FAN Press, Tashkent.

Tkalich,S.M.(1938) Testing plants as indicators in geological prospecting and exploration (in Russian). *Vestnik Dal'nevostochnoi Filiala Akademii Nauka SSSR* 32, 3-25.

Tkalich,S.M.(1959) *Practical Guide to Biogeochemical Prospecting for Ore Deposits* (In Russian). Gosgeolteltekhizdatel', Moscow.

Tkalich,S.M.(1961) *Biogeochemical Anomalies and their Interpretations* (in Russinan). Irkutsk, 41.

Tkalich,S.M.(1970) *Phytogeochemical Prospecting for Ore Deposits*. (in Russian) Nedra Press, Leningrad.

Vernadsky,V.I.(1965) *Chemical Composition of the Biosphere and its Environment* (in Russian) Nauka Press, Moscow.

Vinogradov,A P.(1954) Prospecting for ore deposits by plants and soils (in Russian) *Trudy Biogeokhimichesky Laboratoriya* 10, 3-27.

Warren,H.V. and Delavault,R.E.(1950) Gold and silver content of some trees and horsetails in British Columbia. *Bulletin of the Geological Society of America* 61, 123-128.

Warren,H.V.,Delavault,R.E. and Barakso,J.(1968) The arsenic content of Douglas fir as a guide to some gold, silver, and base metal deposits. *Canadian Mining and Metallurgy Bulletin* 61, 860-867.

Wither,E.D.(1977) Biogeochemical Studies in Southeast Asia by Use of Herbarium Material. MSc Thesis, Massey University, Palmerston North, New Zealand.

Wither,E.D. and Brooks,R.R.(1977) Hyperaccumulation of nickel by some plants of Southeast Asia. *Journal of Geochemical Exploration* 8, 579-583.

Chapter five:

Seaweeds as Hyperaccumulators

C.E. Dunn

Geological Survey of Canada, 601 Booth St, Ottawa, Canada K1A OE8

Introduction

Seaweeds are marine macro algae of which more than 40,000 species are known
(Vinogradov, 1953). They can be classified into three main groups according to
their habitat and colour. The green seaweeds (Class Chlorophyceae) are mostly
from the upper tidal zone; brown seaweeds (Class Phaeophyceae) are mostly in
the mid-tidal zone; and red seaweeds (Class Rhodophyceae) are mostly from the
low tidal zone. In temperate to warm seas the more rigid coralline algae (Class
Corallinaceae) are common. Since most of the metal-accumulating species occur
in the first three classes of seaweed, there will be little reference to the
Corallinaceae in this chapter. In addition to the macro algae there are a few green
rooted aquatic plants that are angiosperms. They commonly have the appearance
of grass but are not related to true grasses. They flower underwater and pollen
is dispersed by water. Examples are eelgrass (*Zostera marina*) and surfgrass
(*Phyllospadix scouleri*), The following discussion will be confined to the macro
algae.

Most seaweeds occur on or close to the shore in waters of normal salinity (i.e.
3.5% NaCl). Some, notably the green seaweeds, can survive in fresh water.
Where there is a significant influx of fresh water into the marine environment,
such as in fjords fed by melting snow and ice, surface waters may be brackish
and relatively few species flourish. In such environments the common rock-weed
Fucus is able to survive. There can be a wide range in chemical composition
among different genera of seaweed, and many can accumulate significant amounts
of certain chemical elements from the ambient seawater.

The basic knowledge base on the elemental composition of seaweed and other
marine organisms was prepared in the 1930s and 1940s by the renowned Russian
scientist A.P.Vinogradov. This fundamental work, which was largely
contemporaneous with, and as profound and original as V.M.Goldschmidt's
classic work on geochemistry, has served as the basis for subsequent studies on

119

the chemistry of marine organisms. Fortunately for English-speaking scientists, Vinogradov's original work in Russian was translated through a grant from the Sears Foundation, and synthesised into one benchmark text (Vinogradov, 1953).

The second chapter of Vinogradov's translated work has 113 pages of information which deals with the chemical composition of non-planktonic marine algae. His estimates were that the world's total quantity of seaweed (excluding the calcareous algae) is of the order of several billion tonnes. Most of this material is distributed in the littoral zone, attached to rocks and sediments, Thus there is a vast amount of plant life in the tidal zones throughout the world which is available to accumulate metals and other elements brought into the marine environment from weathering of the adjacent land masses, and dissolved in the seas. A notable mass of seaweed far from the shore is the huge concentration of *Sargassum* sp. (gulf weed) which extends across the surface of several hundred thousand square kilometres of the Sargasso Sea, and was probably transported there by ocean currents.

Seaweeds are primitive non-vascular plants and so they accumulate nutrients and trace elements by direct absorption from surrounding seawater. Unlike land-based vascular plants which, through their root systems, absorb much of their nutrient requirement from soil, seaweed has no true root system to perform a similar task. The equivalent structure is a *holdfast* which performs purely the physical role of securing seaweed to the underlying rock or sediment. Element uptake occurs by simple ion exchange in which cations in the water are exchanged after membrane transport on to intercellular negatively charged polysaccharides (Skipnes *et al.*,1975). Some elements are transferred to specific cellular sites where they are probably bound to polyphenols (Jensen, 1984).

Brooks *et al.* (1977) coined the term *hyperaccumulator* to describe plants growing on serpentine and which contain > 1000 $\mu g/g$ nickel in dry material. A similar concentration threshold was later adopted for other elements such as cobalt and copper. This amount of nickel represents a concentration about 100 times greater than values to be expected in non-accumulating plants growing on the same substrate, and is therefore a useful yardstick by which to define other hyperaccumulator species. However, seaweeds accumulate little or none of an element from the rock to which they are attached because their composition is determined by the chemistry of the seawater in which they grow. The local rock chemistry provides, therefore, only an indirect (but potentially significant) influence on the elemental composition of seaweed. Throughout this chapter discussion of element concentrations will include *relative enrichment* of elements, with hyperaccumulation being assessed only by comparison with vascular plant species.

An approach to recognising hyperaccumulation with regard to seaweeds is to compare element concentrations in dry seaweed with those in dry land plants. Thus, to follow the intent of Brooks *et al.* (1977), if a seaweed contains more than 100 times the amount of an element found as a common level in a land plant, it can be considered a hyperaccumulator of that element. This concept is pursued later by comparing concentrations in seaweeds with the *standard reference plant* (Markert, 1994).

Structure and Basic Chemical Composition of Seaweeds

The size and shape of seaweeds varies greatly, but most are made up of one or more stems (or *stipes*), laminae (or *blades*), and sometimes bladders (or *sacs*). Attachment to the substrate is by a root-like structure known as a holdfast. One of the most common seaweeds is the brown "common rockweed" or "wrack" of the genus *Fucus*. Like many species it is cartilaginous and flattened. When covered by tidal waters it stands erect and up to 50 cm high. It is regularly and repeatedly dichotomously branched with swollen hollow receptacles toward the ends of its branches. The brown "bull kelp" *Nereocystis luetkeana* has a profusely branched holdfast. At a growth rate of 5-6 cm/day, it is the fastest-growing plant known in either the marine or terrestrial environment. It is an annual with a stipe that may exceed 25 m in length terminating in a float (bladder) from which a cluster of up to 20 blades may grow. Each blade is 6-15 cm broad and up to 4.5 m long. By contrast the "brown gulf weed" *Sargassum muticum*, common along the shores of western Canada, is profusely branched and up to 2 m high. It has a short stipe (3 cm) from which grow flattened primary branches bearing long filiform secondary branches. The latter have further branchlets upon which there are stalked vesicles. Different again is the small (1-10 cm high) yellowish-brown saccate species *Scytosiphon bullosus*. These and other common species of the west coast of North America are succinctly described and clearly illustrated in Scagel (1967).

The water content of green, brown and red seaweeds may range from 55 to 90%. Commonly it is about 80% and quite evenly distributed among the stems, laminae and, where present, the bladder. The ash content of dry matter (after ignition at 450-500°C) is commonly an order of magnitude greater than the 2-5% typical of terrestrial plants. The exceptions in land plants are those with a silica framework, such as the horsetail *Equisetum*, which can yield 20% ash. In the brown seaweeds high ash levels are common for the kelps (*Laminaria* and *Macrocystis* [giant kelp] 30-40%; *Nereocystis* [bull kelp) commonly 50-60%). Some green seaweeds have been reported with 60% or more ash yield (Dunn, 1990). The more common seaweed ash yield of 20-30% represents some 3-5% of the living, wet plant. In the calcareous algae (Corallinaceae) the water content is much lower and the ash content may be up to 90%. Relatively low ash yields of 15-18% are obtained from the brown alga *Ascophyllum nodosum* (knotted wrack) from the coast of Norway (Sharp and Bølviken, 1978), and only 10-15% yield is reported from *Porphyra*, known to the Japanese as *Nori*, and used extensively in Sushi dishes.

The ash of seaweeds is approximately 80% water-soluble. This soluble portion contains chlorides, sulphates, phosphates, carbonates, alkaline and alkaline-earth metals (Vinogradov, 1953). Table 5.1 shows the average composition of several species of common brown seaweed, and shows that oxides of the alkali metals comprise from 30 to 50% of the ash. These elemental concentrations are similar to those found in many land plants, except for the extremely high content of sodium and chlorine in the seaweeds. Such hyperaccumulation of sodium and chlorine is, of course, to be expected in light of the high salt content of seawater. The sulphur data are, as noted by Vinogradov (1953), probably underestimates, because some sulphur volatilises during reduction to ash; indeed, some sulphur is released during air drying of samples. Of the brown seaweeds, the large kelps *Macrocystis* and *Nereocystis* are the genera that concentrate potassium most and commonly have more than twice as much potassium as sodium. The *Laminaria* have less potassium, but still more potassium than sodium. The rockweed *Fucus* has similar amounts of potassium and sodium, or slightly more sodium. Small encrusting forms such as *Codium fragile* have substantially more sodium than potassium.

Table 5.1. Average composition (%) of ash from common brown seaweeds (Vinogradov, 1953).

	Macrocystis (Giant kelp)	*Nereocystis* (Bull kelp)	*Laminaria* (Kelp)	*Fucus* (Rockweed/Kelp)
CaO	6.7	2.8	9.6	15.0
Cl	34.0	39.0	25.0	18.0
Fe_2O_3	0.4	0.2	0.4	0.8
K_2O	34.0	37.0	25.0	8.7
MgO	3.6	2.4	5.8	9.7
Na_2O	14.0	14.0	18.0	23.0
P_2O_5	1.7	1.4	2.7	2.8
SiO_2	-	-	0.7	1.0
SO_3	6.4	3.6	13.0	23.0

Trace Elements in Seaweeds

Introduction

The National Union of Biological Sciences has developed the concept of the *standard reference plant* (Markert, 1994). This is a fictitious plant considered to be of average composition, and therefore element concentrations in specific plants can be measured as enrichments or depletions with respect to the reference plant. Elemental concentrations in this reference plant were established from the analysis of many plant species, excluding those that are typical accumulator or rejector plants. Table 5.2 lists the composition of this reference plant, and also shows a 100-times concentration based on the definition of a hyperaccumulator by Brooks *et al.* (1977). In addition, values are shown of average element concentrations in 21 species of brown, green and red seaweed from nine non-mineralised and non-polluted sites along the west shores of Vancouver Island. Substrates were either beach sands or diorite. As such, these data can be considered as representing natural *background* sites and provide an overview of those elements that are typically enriched in seaweeds.

Table 5.2. Trace element content (µg/g dry weight) of *reference land plant*, and of seaweeds from nine 'background' sites on the west coast of Vancouver Island. Determinations at background sites were by instrumental neutron activation analysis.

Element	Ref.Plant[1]	Hyper. Level[2]	Mean seaweed[3]	Seaweed hyperaccumulators
Ag	0.2[4]	20	<0.8	-
As	0.1	10	8.2	Most: especially *Sargassum, Macrocystis, Nereocystis, Gigartina, Egregia, Hedophyllum*
Au	0.001[4]	0.1	<4	-
Ba	40	4000	<40	-
Br	4	400	643	Most: notably *Prionitis*
Ce	0.5	50	0.94	-
Co	0.2	20	2.5	-
Cr	1.5	150	2.2	-
Cs	0.2	20	0.11	-
Cu	10	1000	12	-
Hf	0.05	5	0.2	-
I	3	300	238	Most: notably some brown seaweeds - *Hedophyllum, Macrocystis, Nereocystis*
La	0.2	20	0.57	-
Mo	0.5[4]	50	<0.8	-
Ni	1.5	150	<20	-
Pb	1	100	-	-
Rb	50	5000	23	-
Sb	0.1[4]	10	0.11	-
Sc	0.02	2	0.49	2 µg/g in *Ectocarpus*
Sr	50	5000	696	Moderately high levels in many species (1500 µg/g in *Sargassum*)
Th	0.005	0.5	0.06	-
U	0.01	1	0.44	*Fucus, Sargassum*; locally in other brown seaweeds
V	0.5	50	5.9	Locally in *Ulva*
Zn	50	5000	37	-

[1] From Markert (1994).

[2] Hyperaccumulation level - 100 times concentration in 'reference' plant (Markert, 1994).

[3] Average content of elements in 101 samples representing 21 common seaweed species from nine sites near the Bamfield Marine Station, west Vancouver Island.

[4] Values for Ag, Au, Mo and Sb (from Markert, 1994) are an order of magnitude higher than commonly found in plants from North America.

Footnotes to Table 5.2 above indicate that the levels of silver, gold, molybdenum and antimony recommended for the reference plant are substantially higher than those typically found in plants from North America.

Iodine

Perhaps the element best known to accumulate in seaweeds in high concentrations
is iodine. Courtois (1813) reported the discovery of iodine in the ash of the
seaweed 'varec' from France. Over the following decade many researchers
reported the presence of iodine in seaweeds, and by the 1840s an iodine industry,
using seaweed, had begun in Scotland and France.

There is some cause for concern in quoting the iodine content of seaweed ash,
because some chemical species of iodine readily volatilise and therefore the data
quoted in the literature commonly only represent a part of the iodine content of
seaweed. Most iodine is organically bound, especially with potassium and
rubidium, but some occurs in elemental form and is lost on reduction of the dry
tissue to ash. Since it is not possible to determine from most published texts
exactly how much iodine is lost during ashing, it should be assumed that the
concentrations quoted are usually underestimates of the true iodine concentrations.
Many of the early studies on iodine content of seaweed were performed on ash,
and the data converted to a dry weight basis.

The iodine content of seaweeds is commonly more than 100 times greater
than in land plants, making seaweed a true hyperaccumulator of iodine. In
general, the brown seaweeds have the highest iodine content, with less in red
seaweeds and the least in the green. However, even the green seaweeds are richer
in iodine than land plants. Of the brown seaweeds, *Laminaria* is reported to have
up to 5.5 % iodine in ash (Marchand, 1866), although there is commonly 0.5-1 %
of this element. The rockweed *Fucus* has a lower iodine content with 1 % as high,
and 100 - 1000 $\mu g/g$ as a common range. In the author's experience, substantially
lower concentrations may occur. Red seaweeds have mostly 400-1200 $\mu g/g$ iodine
in ash, but can range from 20 $\mu g/g$ to more than 1 %. The green seaweeds may
have as little as a few $\mu g/g$ iodine and rarely more than a few hundred $\mu g/g$.

Several factors are reported to control the iodine content of seaweeds. For
brown seaweeds, in general the greater the depth of habitat the higher the iodine
content. Secondly, iodine contents are commonly highest in the summer months.
Thirdly, iodine concentrations are usually higher in young than old growth.

Bromine

Although there is a thousand times more bromine than iodine in seawater, there
are few seaweeds that are known to concentrate bromine to high levels. However,
all seaweeds are richer in bromine than land plants. With regard to analytical
data, the same comments apply as those for iodine. Bromine, unless incorporated
in the structure of a stable compound, is a volatile element of which a
considerable proportion may be released during the ashing process. Data quoted
for bromine should therefore be considered as low estimates of the bromine
content of the fresh seaweed.

From the compilation by Vinogradov (1953) it appears that in general the red and brown seaweeds contain 0.5-1% bromine in ash (approximately 2000-4000 $\mu g/g$ in dry matter), and the green seaweeds are poorest in bromine. There are no true hyperaccumulators of bromine although the range of 2000-7000 $\mu g/g$ bromine in ash which appears from the literature to be typical for *Laminaria* and *Fucus*, represents a significant bromine enrichment over that found in rocks, sediments and most land plants.

Analyses of a range of common seaweeds from the west coast of Vancouver Island have shown that the red seaweed *Prionitis lanceolata* yields consistently higher concentrations of bromine (5000-6000 $\mu g/g$ in ash) than the other species tested. Table 5.3 summarizes the content of several major elements and halogens in the ash of species collected from near the marine station at Bamfield on the west coast of Vancouver Island.

Table 5.3. Concentrations (% except for Fe, Br and I which are in $\mu g/g$) of some major elements and in the ash of seaweeds from west Vancouver Island (near Bamfield Marine Station). All analytical determinations by instrumental neutron activation analysis. Species identification by R.K. Scagel (Pacific Phytometric Consultants, Vancouver).

Species	N	Ash	K	Na	Ca	Fe	Mg	Cl	Br	I
BROWN SEAWEEDS										
Alaria marginata	5	26.2	16.4	18.4	4.4	500	2.8	28.6	1420	715
Costaria costata	1	26.3	24.4	13.3	5.0	700	2.9	25.9	1000	800
Cryptosiphon woodii	1	36.5	20.6	15.5	6.0	8000	-	-	1600	-
Cymathere triplicata	1	39.8	34.7	9.4	2.1	500	1.6	35.0	1400	170
Ectocarpus dimorphus	2	22.7	7.5	14.0	5.2	5250	2.5	8.7	1650	381
Egregia menziesii	7	33.7	26.0	11.5	4.4	3140	2.4	29.0	2186	550
Fucus gardneri	9	25.3	16.1	18.5	4.9	850	3.2	20.7	1411	157
Hedophyllum sessile	4	33.0	20.7	16.1	3.3	1380	2.8	28.4	1950	3600
Leathesia difformis	4	60.6	14.1	14.1	5.6	3800	2.7	32.8	2225	<5
Macrocystis integrifolia	7	35.1	27.4	11.4	3.2	520	2.0	31.9	2343	1700
Nereocystis lueteana										
(bladder)	1	50.3	37.2	9.4	1.4	<500	1.5	38.3	1300	560
(frond)	6	43.0	26.0	13.1	2.9	930	1.9	33.6	2050	1200
(stipe)	6	56.2	35.6	10.2	1.7	<500	1.7	38.0	1550	733
Postelsia palmaeformis (frond)	1	33.8	23.6	15.5	1.8	<500	-	-	1500	-
P.palmaeformis (stipe)	11	43.1	38.6	8.2	1.6	<500	-	-	2200	-
Sargassum muticum	7	30.4	29.2	14.8	5.1	760	3.6	29.2	981	65
RED SEAWEEDS										
Gigartina exasperata	3	25.2	12.3	19.3	2.7	630	3.4	16.8	1627	<5
Halosaccion glandiforme	5	28.5	12.1	12.9	9.6	5540	1.9	22.0	1600	<5
Iridaea splendens	4	26.4	6.2	25.1	2.4	1580	3.5	10.8	1312	<5
Prionitis lanceolata	6	24.0	15.0	14.8	10.3	1700	2.4	10.3	5117	150
GREEN SEAWEEDS										
Codium fragile	1	39.2	7.0	22.7	3.7	4000	3.7	37.6	100	<5
Enteromorpha intestinalis	4	32.7	8.3	14.7	7.6	5250	5.3	10.5	888	<5
Spongomorpha coalita	7	56.8	15.9	7.6	11.7	6560	1.3	15.5	1230	<5
Ulva fenistrata	5	33.1	15.4	15.0	4.2	4700	5.4	19.1	1700	<5

Seaweeds in Mineral Exploration and Environmental Monitoring

Introduction

Building upon the work of Vinogradov (1953), Black and Mitchell (1952), and Bryan (1969), Haug (1972) suggested that seaweed might be a useful sampling medium for detecting influx of metals into the marine environment. A close correspondence between the metal content of seaweed and that of seawater should be expected because algae are non-vascular plants. His studies of areas fed by streams enriched in base metals found that zinc, cadmium and copper reached, respectively, 43, 35 and 20 times normal values. Since that time there have been a number of studies involving the analysis of seaweed that have been directed toward mineral exploration (e.g. Haug *et al.*, 1974; Lysholm, 1972; Bollingberg, 1975; Sharp and Bølviken, 1979; Bollingberg and Cooke, 1985; Dunn, 1990; Dunn *et al.*, 1993), and environmental monitoring (Fuge and James, 1973, 1974; Morris and Bale, 1975; Skipnes *et al.*,1975; Bryan and Hummerstone, 1973; Cullinane and Whelan, 1982; Luoma *et al.*, 1982; Woolston *et al.*, 1982; Forsberg *et al.*, 1988). This newly acquired information has provided improved insight to those species which are able to accumulate significant amounts of metals.

Base metals

From the analysis of two species of rockweed (*Fucus distichus* and *F. vesiculosus*) near a base metal deposit in Maarmorilik fjord of western Greenland, Bollingberg (1975) found an increase in lead content from a background level of 1 $\mu g/g$ Pb to 200 $\mu g/g$, thereby establishing hyperaccumulation of lead in these species. Similarly, zinc increased from 1.6 to 488 $\mu g/g$, copper from 1.9 to 8.9 $\mu g/g$, and cadmium from 0.57 to 8.0 $\mu g/g$. In subsequent studies of that area, carried out after the Black Angel mine had gone into production, increased levels of metals were recorded in the same species of seaweed, reaching maxima in dry weight of 611 $\mu g/g$ Pb and 2060 $\mu g/g$ Zn (Bollingberg and Cooke, 1985).

Locally, increased concentrations of nickel and manganese have been found associated with the substrate, whereas increases in zinc, cadmium, iron, cobalt and vanadium have been attributed to influx of fresh water (Sharp and Bølviken, 1979).

Downslope from the carbonate-hosted copper-gold deposit of the abandoned Little Billie mine on Texada Island, near Vancouver, nine species of seaweed were collected for analysis (Dunn, 1990). A 33-element instrumental neutron activation analysis on ashed specimens provided baseline data for elements not commonly reported in seaweeds. No species showed a clear tendency to accumulate gold or copper. Of note, were moderate levels of arsenic, uranium and strontium in the brown seaweeds, and relative enrichment of iron, aluminium and the rare-earth elements in green seaweeds. One sample of the sea lettuce *Ulva* yielded 15 μg/g molybdenum, perhaps reflecting the known association of molybdenum with the mineral deposit.

In southwestern British Columbia acid rock drainage seeps down a mountainside from an abandoned copper mine (Britannia) into Howe Sound, a few kilometres north of Vancouver. This fjord-like inlet has prolific growth of the rockweed species *Fucus gardneri* lining its shores. For a distance of more than 1 km on either side of the drainage from Britannia into Howe Sound there is no seaweed growth. Where it first appears it is stunted and samples have yielded up to 3200 μg/g copper in ash (1000 μg/g copper dry weight, or 200 μg/g wet weight - Dunn *et al.*, 1993), attesting to the high degree of metal accumulation that the seaweed can withstand before concentrations become detrimental to its growth.

From the data in Table 5.2, this confirms that *Fucus gardneri* can be a hyperaccumulator of copper. Background concentrations in ash of rockweed from the southern part of Howe Sound are 60-70 μg/g copper. The contrast between zinc concentrations is less pronounced, with 1540 μg/g zinc in the ash of rockweed from near Britannia Beach, and 300 μg/g zinc near the entrance to Howe Sound. Farther out to sea, on the shores of Texada Island in the Strait of Georgia, and on the west side of Vancouver Island, levels of zinc in rockweed ranged from 40 to 160 μg/g in ash (12-50 μg/g dry weight, and 3-10 μg/g wet weight).

Arsenic

Data on common seaweeds from southwestern British Columbia indicate that the brown seaweeds can accumulate arsenic to a greater extent than the green or red (Dunn, 1990). *Sargassum* appears capable of scavenging arsenic to a higher degree than other genera. Table 5.4 lists arsenic data in the ash of species from sites bear the Bamfield Marine Station. The data show that the brown seaweeds *Sargassum* (gulf weed), *Egregia* (feather boa kelp), *Macrocystis* (giant kelp), and *Nereocystis* (bull kelp, especially the fronds) are those which accumulate arsenic to the highest levels.

Of the red seaweeds *Gigartina* and *Prionitis* are capable of concentrating arsenic, whereas none of the four green seaweeds that were tested contained significant concentrations of arsenic. By comparison with most species of land plants, *Sargassum* is a true hyperaccumulator of arsenic, since in the Bamfield area it contains a maximum of 230 μg/g arsenic in ash (70 μg/g in dry matter), whereas on the adjacent shore tissues of red alder (*Alnus rubrum*), Sitka spruce (*Picea sitchensis*), and western red cedar (*Thuja plicata*) yielded a maximum of 3.5 μg/g arsenic in ash (0.07 μg/g in dry tissue). Near a zone of undisturbed arsenic-rich mineralization on the east shore of Bowen Island (Howe Sound, near Vancouver), *Fucus gardneri* yielded 7-12 μg/g arsenic and a single sample of

Sargassum contained 36 μg/g arsenic in dry tissue.

Uranium

Uranium in land plants is commonly present in concentrations of only 10-20 ng/g (ppb) in dry tissue, although there are reports of substantially higher levels near zones of uranium mineralization. Studies of seaweeds show that in many species concentrations in excess of 1 μg/g uranium are common background levels.

Table 5.4. Arsenic concentrations (μg/g in ash) in seaweeds from west Vancouver Island (near Bamfield Marine Station, sites 1-9) and Texada Island (Strait of Georgia, sites 10,11). All analytical determinations by instrumental neutron activation analysis. Species identification by R.K. Scagel (Pacific Phytometric Consultants, Vancouver).

	1	2	3	4	5	6	7	8	9	10	11
BROWN SEAWEEDS											
Alaria marginata	4	24	26	28	-	-	17	-	-	-	-
Costaria costata	-	-	27	-	-	-	-	-	-	-	-
Cymathere triplicata	-	13	-	-	-	-	-	-	-	-	-
Ectocarpus dimorphus	25	15	-	-	-	-	-	-	-	-	-
Egregia menziesii	-	-	49	19	29	36	33	69	24	-	-
Fucus gardneri	<2	23	21	11	8	15	18	35	2	24	12
Hedophyllum sessile	21	23	40	-	-	-	-	-	34	-	-
Leathesia difformis	-	-	5	-	-	24	8	<2	-	-	-
Macrocystis integrifolia	-	-	40	54	50	71	33	56	-	-	-
Nereocystis lueteana (all)	-	-	-	-	-	-	-	-	-	25	-
" " (frond)	-	10	98	14	6	26	40	-	-	-	-
" " (stipe)	-	28	9	25	27	<2	42	-	-	-	-
Postelsia palmaeformis	-	-	-	-	-	-	-	-	10	-	-
Sargassum muticum	-	-	82	45	63	170	82	130	230	140	60
RED SEAWEEDS											
Calliarthron tuberculosum	3.7	3.9	3.6	-	-	4	-	-	-	-	-
Gigartina exasperata	-	-	51	-	-	59	-	-	12	-	-
Halosaccion glandiforme	11	12	12	-	-	8	<3	-	-	-	-
Iridaea splendens	-	-	19	-	-	10	9	-	7	-	-
Prionitis lanceolata	-	25	27	9	-	23	36	66	-	-	-
GREEN SEAWEEDS											
Enteromorpha intestinalis	-	4.4	-	-	-	-	<2	16	-	-	-
Codium fragile	-	-	<2	-	-	-	-	-	-	-	-
Spongomorpha coalita	6.3	-	8.4	1.7	5	13	9	23	-	-	-
Ulva fenistrata	2.4	7.8	-	-	10	<2	21	<2	1.5	13	2.7

Data on the composition of brown seaweeds show that, with respect to the average concentrations in land plants, they are capable of hyperaccumulating uranium. In the Bamfield area, concentrations in *Fucus gardneri* ranged from 0.75 to 1.9 μg/g uranium (dry weight). The same species collected at 36 sites around the shores of Howe Sound, north of Vancouver, had a range of 1 to 4 $\mu g/g$ of this element.

Summary

Although there are many large books that discuss the classification, morphology and occurrence of the many thousands of seaweed species, there are as yet relatively few data on their major and trace element contents. There are only a few studies that have examined the composition of seaweeds in areas of natural metal enrichment, and a few more studies which have monitored pollution from anthropogenic activities. From the available information, however, it appears that, relative to land plant compositions, there are several species which yield a natural hyperaccumulation of a few elements. Of note are:

1 - high levels of iodine, especially in brown seaweeds (e.g. *Laminaria*);

2 - moderate levels of bromine in brown seaweeds (e.g. *Laminaria* and *Fucus*), and in the red seaweed *Prionitis*;

3 - enrichment of arsenic in brown seaweeds, notably *Sargassum*;

4 - enrichment of uranium in brown seaweeds, notably *Fucus*;

5 - moderate enrichment of strontium in some brown seaweeds (*Fucus* and *Sargassum*);

6 - moderate enrichment of vanadium in the green sea lettuce *Ulva*.

Studies of metal uptake in areas of elevated levels of metal flux into the marine environment have shown that lead, zinc, cadmium, and copper can all reach hyperaccumulation status before toxicity inhibits their growth. It appears probable that many other elements could reach a similar status before growth is impaired. Consequently, there is considerable potential for applying seaweed chemistry to the environmental monitoring of coastal areas, and to the exploration for minerals along the enormous extent of the world's coastlines.

References

Black,W.A.P. and Mitchell,R.L.(1952) Trace elements in the common brown algae and in sea water. *Journal of the Marine Biologists Association of the UK* 30, 575-584.

Bollingberg, H.J.(1975) Geochemical prospecting using seaweed, shellfish and fish. *Geochimica et Cosmochimica Acta* 39, 1567-1570.

Bollingberg, H.J. and Cooke Jr., H.R.(1985) Use of seaweed and slope sediment in fjord prospecting for lead-zinc deposits near Maarmorilik, west Greenland. *Journal of Geochemical Exploration* 23, 253-263.

Brooks,R.R., Lee,J., Reeves,R.D. and Jaffré,T.(1977) Detection of nickeliferous rocks by analysis of herbarium specimens of indicator plants. *Journal of Geochemical Exploration* 7, 49-57.

Bryan,G.W.(1969) The absorption of zinc and other metals by the brown seaweed

Laminaria digitata. Journal of the Marine Biology Association UK 49, 225-243.

Bryan,G.W. and Hummerstone,L.G.(1973) Brown seaweeds as an indicator of heavy metals in estuaries in south-west England. *Journal of the Marine Biology Association UK* 53, 705-721.

Courtois,B.(1813) Découverte d'une substance nouvelle dans le vareck. *Analytica Chimica (Phys.)* 1, 88, 304.

Cullinane,J.P. and Whelan,P.M.(1982) Copper, cadmium and zinc in seaweeds from the south coast of Ireland. *Marine Pollution Bulletin* 13, 205-208.

Dunn,C.E.(1990) Results of a biogeochemical orientation study on seaweed in the Strait of Georgia, British Columbia. In: Current Research, Part E, *Geological Survey of Canada*, Paper 90-1E, 347-350.

Dunn,C.E., Percival,J.B., Hall,G.E.M. and Mudroch,A.(1993) Reconnaissance geochemical studies in the Howe Sound drainage basin In: Levings, C.D., Turner, R.B. and Ricketts, B.(eds) *Proceedings of Howe Sound Environmental Science Workshop. Canadian Technical Report, Fisheries and Aquatic Science* 189, 89-95.

Forsberg.Aa, Söderlund,S., Frank,A., Petersson,L.R. and Pedersen,M.(1988) Studies on metal content in the brown seaweed *Fucus vesiculosus* from the archipelago of Stockholm. *Environmental Pollution* 49, 245-263.

Fuge,R. and James,K.H.(1973) Trace element concentrations in brown seaweeds, Cardigan Bay, Wales. *Marine Chemistry* 1:4, 281-293.

Fuge,R. and James,K.H.(1974) Trace metal concentrations in *Fucus* from the Bristol Channel. *Marine Pollution Bulletin* 5, 9-12.

Haug,A.(1972) Akkumulering av tungmetaller i marine alger. *In: Symposium om Tungmetallforurensinger.* NTNF, Oslo, 198-206.

Haug,A., Melsom,S. and Omang,S.(1974) Estimation of heavy metal pollution in two Norwegian fjord areas by analysis of the brown alga *Ascophyllum nodosum. Environmental Pollution* 7, 179-192.

Jensen,A.(1984) Marine ecotoxicological tests with seaweeds. *In* G. Persoone, E. Jaspers, and C. Claus (eds), *Ecotoxicological Testing for the Marine Environment, State University of Ghent and Institute of Marine Science Resources*, Bredene, Belgium, vol.1, 181-193.

Luoma,S.N., Bryan,G.W. and Langston,W.J.(1982) Scavenging of heavy metals from particulates by brown seaweed. *Marine Pollution Bulletin* 13, 394-396.

Lysholm,C.(1972) Tungmetallinnhold i tang som en mulig geokemisk prospekteringsmethode. (Unpublished thesis). Norges Tekniske Högskole, Trondheim, 78 pp.

Marchand, E.(1866) Composition des cendres végétales. *Analytica Chimica (Phys.)* 4,8, 320.

Markert,B.(1994) *Progress Report on the Element Concentration Cadaster Project (ECCE) of INTERCOL/IUBS.* International Union of Biological Sciences, 25th General Assembly, Paris, 54 pp.

Morris,A.W. and Bale,A.J.(1975) The accumulation of   cadmium, copper, manganese and zinc by *Fucus vesiculosus* in the Bristol Channel. *Estuarine Coastal Marine Science* 3, 153-165.

Scagel,R.F.(1967) *Guide to Common Seaweeds of British Columbia.* British Columbia Provincial Museum, Dept. of Recreation and Conservation, Handbook No. 27.

Sharp,W.E. and Bølviken,B.(1979)   Brown algae: a sampling medium for prospecting fjords. In: J.R. Watterson and P.K. Theobald (eds). *Geochemical Exploration 1978,* Association of Exploration Geochemists, Rexdale, Ontario, 347-356.

Skipnes,O., Roald,T., and Haug,A.(1975) Uptake of zinc and strontium by brown algae. *Physiologia Plantarum* 34, 314-320.

Vinogradov,A.P.(1953) *The Elementary Chemical Composition of Marine Organisms.* Sears Foundation for Marine Research, Yale University, New Haven, Memoir II, 647 pp.

Woolston,M.E., Breck,W.E. and Van Loon, G.W.(1982) A sampling study of the brown seaweed *Ascophyllum nodosum* as a marine monitor for trace metals. *Water Resources* 16, 687-691.

Chapter six:

Hyperaccumulation of Metals by Prokaryotic Microorganisms Including the Blue-green Algae (Cyanobacteria)

T.J. Beveridge

Department of Microbiology, College of Biological Science, University of Guelph, Guelph, Ontario, Canada N1G 2W1

Prokaryotic Microorganisms

Antiquity

Prokaryotes are the Earth's oldest lifeform. Their remnants have been discovered as *microfossils* in several ancient cherts and shales from exposed regions of the continental crust which did not undergo dramatic metamorphic events throughout geological time. Some of these have been dated as being approximately 3.0-3.6 billion years old (e.g., the Archaean stromatolites of the Warrawoona group in Australia and the Gunflint Chert of Canada: Walter, 1983, Fig.6.1a).

The shape and form of the cells preserved in rock as microfossils have remarkable similarity to those of present-day bacteria (Figs 6.1b and c). Filaments, cocci and chains can be distinguished within the mineral matrix of the microfossiliferous rock once it has been polished-down to a thinness that will allow imaging with a light microscope. Prokaryotic life must have flourished in Archaean times; its general cellular format seems to have already been well established and has endured through time.

Our ability to distinguish prokaryotic microorganisms in such ancient geological horizons is truly remarkable. Present-day bacteria are composed entirely of "soft tissue". They have no hard endo- or exo-skeleton made of inorganic minerals to ensure that their cellular shape is retained. They consist only of organic matter and the obvious question, then, is how did they manage to have such "soft tissue" preserved in rock for these extremely long periods of time? As we will see in this chapter, the answer is obvious - many bacteria have the innate ability to interact with, and concentrate, environmental inorganic ions which, eventually form minerals Beveridge, 1989a: Beveridge and Doyle, 1989).

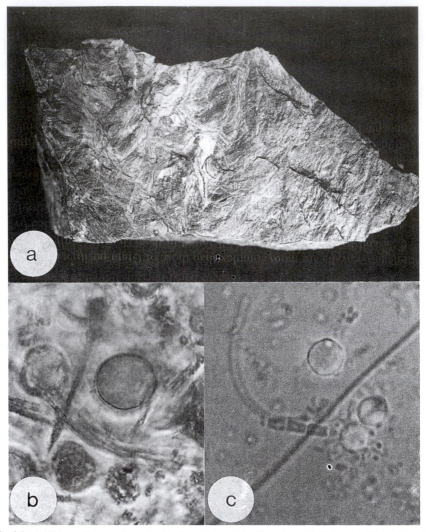

Fig.6.1.(a) Picture of a rock from the Gunflint Chert of the Canadian Shield (north of Lake Superior in Canada) which has been dated to be approximately 2 billion years old. The laminations of an ancient stromatolite can be seen in the centre of the rock. The long axis of the rock is ca. 20-cm and the height is ca. 10-cm.

(b) A bright field photomicrograph of the laminated region of the rock in (a) which has been polished and thinned to ca. 1-2-mm so that light can be passed through it. Microfossils of prokaryotic cells can be seen which are either coccoid or filamentous in shape. The coccoid microfossil in the centre is ca. 20- μm in diameter. This piece of Gunflint Chert was collected by F.G.Ferris during his tenure in the author's laboratory.

(c) A phase-contrast photomicrograph of a prokaryotic community found in a present-day biofilm for comparison the microfossils seen in (b). The coccoid cells are ca. 10-μm.

As they become enshrouded in mineral, their cell shape is preserved even after the organic soft tissue has departed (Ferris *et al.*,1988).

Large numbers of these mineralised cells can actually form tangible deposits. For example, ancient Archaean stromatolites are the layered mineral remains of complex microbial communities which, during growth, arranged themselves into so-called "biofilms" in a stratified manner which optimised their metabolism. As older cells died and became mineralised, newer cells continued to overgrow them in their search for outside nutrients and sunlight. Mineralised layer after mineralised layer was put down one after another and visible, tangible microbial reefs were established which eventually became the microfossiliferous rock we see today (Fig.6.1a and b). The same process is still occurring in certain modern-day environments (e.g. Thompson *et al.*,1990).

General design

All prokaryotic life shares fundamental cellular design features. They are extremely small cells, usually occupying a volume of no more than 1.5-2.0 μm^3 (Beveridge, 1988). Unlike *eukaryotic* cells, they do not have room for a nucleus, mitochondria, lysosomes, endoplasmic reticulum, etc. (i.e., the compartmentalisation of distinct organelles). Instead, they have a rather featureless cytoplasm unless internal membranes for photosynthesis, nitrification or methanotrophy are required, or storage granules are produced (e.g., polyphosphate, glycogen, polyhydroxbutyrate, etc. - Beveridge, 1989b). Cell shape is not produced by an internal scaffolding of microfilaments and microtubules as in eukaryotic cells, but is somehow integrated into the development of the natural contours within the cell wall which lies outside the plasma (cytoplasmic) membrane (Beveridge, 1988, 1989b).

Bacteria rely entirely on diffusion for their nourishment and exchange of waste products; this is one of the prime reasons why they are so small. They must optimise their exchange capacity with the external environment and, by being small, they increase their surface area-to-volume ratio to such an extent that diffusion is maximised (i.e., the interfacial contact of the bacterium to its external fluid phase is increased to the best possible extent - Beveridge, 1988). Simple modifications in shape to the cell (e.g., changing a coccus to a rod) can increase the surface area-to-volume ratio even further. These are extremely small cells that have maximum exposure and interaction with their external surroundings.

Two separate prokaryotic lineages have recently been discovered, the *Bacteria* (eubacteria) and the *Archaea* (archaeobacteria or archaebacteria - Woese and Fox. 1977; Woese *et al.*,1990). Although these are fundamentally different biological domains and the exact cell structure and composition of their cellular parts may differ (Woese and Wolfe, 1985), they still share many of the common general features which I have outlined above. Almost all eubacteria and archaeobacteria possess cell walls that dictate their shape and all are at approximately the same level of size. The major general difference is with the ecological niche that the two domains inhabit. Many archaeobacteria reside in extreme environments (e.g., high salt (*Halobacterium*), strict absence of oxygen (*Methanospirillum*), extreme high temperature (*Pyrodictium*) or low pH and high temperature (*Sulfolobus*) that would be lethal to eubacteria. For this reason, unique and unusual compounds can be found in their membranes (e.g., tetraether, diether, hydroxydiether and macrocyclic diether lipids) and cell walls (e.g., methanochondroitin and pseudomurein) (König, 1988; Sprott *et al.*,1991). Yet, even with these compositional differences in their surface structures, the interfacial physicochemistry of archaeobacteria is similar to that of eubacteria. The remainder of this chapter will emphasize the eubacteria since this is the better studied system.

Structural format and chemical make-up of eubacterial surfaces

These bacteria are divided into two basic varieties depending on how they react to the Gram stain which is used for light microscopy. Gram-positive bacteria are stained purple and have relatively thick (ca. 25 nm) cell walls, whereas Gram-negative bacteria are stained red and have more complex cell walls (Fig.6.2). For these, a thin peptidoglycan layer followed by a bilayered outer membrane is found above the plasma membrane (Fig.6.2b). It is both the structural format and the compositional make-up of the cell walls of these two bacterial varieties which dictates their response to the Gram stain (Beveridge and Davies, 1983).

Gram-positive eubacteria

The surfaces of these bacteria consist of peptidoglycan to which secondary polymers are attached (Beveridge, 1981). Although there are subtle chemical differences to this format in a number of Gram-positive bacteria (e.g., *Mycobacterium*), this general definition will suffice and is represented by *Bacillus subtilis*. The *B. subtilis* wall contains approximately 25 layers of peptidoglycan, all piled on top of one another. Peptidoglycan is a linear polymer consisting of approximately 50 N-acetyl glucosamyl-N-acetyl muramyl dimers per peptidoglycan strand. A short pentapeptide (L-ala-D-glu-meso-dpm-D-ala-D-ala) is connected to each N-acetyl muramyl residue and these serve to cross-link adjacent glycan strands together whether they be laterally opposite one another, or directly above or below. Usually, about 50% of all muramyl peptides are covalently cross-linked in this manner. The interconnection of glycan strands along the x, y and z axes (in space) produces an extremely resilient sacculus which completely surrounds the cell.

Secondary polymers are also frequently attached to the muramyl moiety of the peptidoglycan. In *B. subtilis*, these consist of either teichoic acid or teichuronic acid and the levels of each in the wall are determined by the amount of phosphate in the growth medium (Beveridge, 1981). In the presence of phosphate, teichoic acid is synthesized and it is a glyerol- or ribitol-based linear polymer through which the organic moieties are connected by phosphodiester linkages. When phosphate is limiting, teichuronic acid is produced and this resembles the former polymer except the phosphates are now replaced by uronic acids. In both cases, the teichoic and teichuronic polymers are relatively long (ca. 30-50 residues), flexible molecules which wind through the inter-peptidoglycan spaces to form an

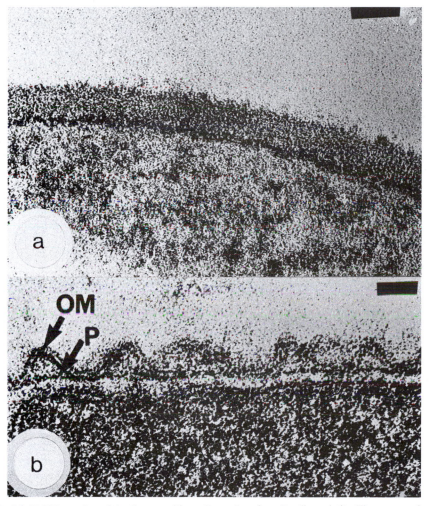

Fig.6.2.(a) Thin section of the Gram-positive cell envelope from *Bacillus subtilis*. The arrow points to the 25 nm thick cell wall. Bar = 25-nm.

(b) Thin section of the Gram-negative cell envelope of *Aquaspirillum serpens* strain VHL. The peptidoglycan layer (P) can be seen immediately above the plasma membrane. Above the peptidoglycan layer is the outer membrane (OM). The region between the plasma and outer membranes is called the periplasmic space. Bar = 25-nm.

amorphous matrix with the peptidoglycan (Fig.6.3a). Because peptidoglycan and teichuronic acid each possess many carboxylate groups and teichoic acids contain ionisable phosphates (Beveridge and Murray, 1980; Doyle *et al.*,1980) at neutral pH, these cell walls possess an overall electronegative charge (Beveridge, 1989a).

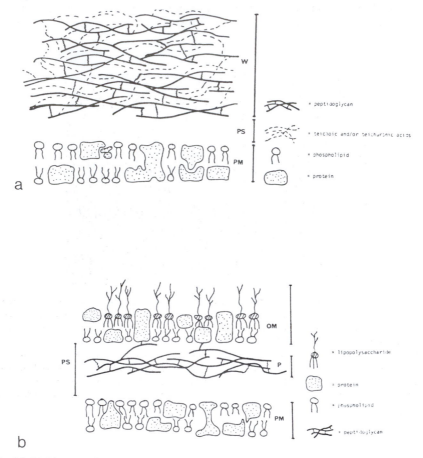

Fig.6.3.(a) Diagram of the arrangement of molecules within a typical Gram-positive cell envelope. Above the lipid-protein bilayer of the plasma membrane (PM) lies a thin periplasmic space (PS) and a thick cell wall (W). The key to the major molecules within the diagram is seen on the right.

(b) Diagram of the arrangement of molecules within a typical Gram-negative cell envelope. Above the PM is a thin peptidoglycan layer (P) and the outer membrane (OM) which consists of protein, phospholipid and lipopolysaccharide (see key at right). The region between the PM and OM is called the periplasmic space (PS). The long O-side chains of the LPS extend far into the external environment and shield the other outer membrane constituents.

Bacteria must insert new material into their pre-existing walls in order to grow in size and there is a complex set of events which discards older material from the uppermost wall layers while, at the same time, new material is bonded into the lower regions (Doyle and Koch, 1987; Koch, 1995). Specialised enzymes named peptidoglycan hydrolases (or autolysins) are used to clip the covalent bonds of the outermost wall fabric so that a "picket-fence" network is seen (Fig.6.4a and b; Graham and Beveridge, 1994; Beveridge and Graham, 1991). This serves two important physicochemical functions; it drastically increases the overall surface area-to-volume ratio that would be expected of a more intact surface and it produces many more electronegative sites (i.e., due to the bonds broken by the hydrolases).

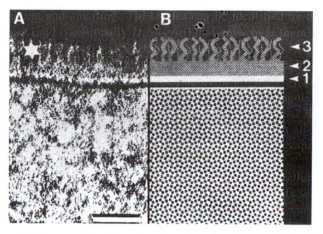

Fig.6.4. (a) Dark-field image of a thin-sectioned portion of a freeze-substituted *B. subtilis* cell envelope. The "star" shows the picket-fence arrangement of the outer region of the cell wall, produced by the peptidoglycan hydrolases (autolysins), which tremendously increases the surface area of the bacterium (cf. this Figure with Fig.6.2a). Bar = 30-nm.

(b) A diagram of the image seen in (a) which shows the three distinct regions of the cell wall that are produced by a process called "cell wall turnover" (see Beveridge and Graham, 1991 and Graham and Beveridge, 1994 for more details). For this chapter, region No.3 is important because of the tremendous increase in surface area for the cell. (A similar figure was previously published in Graham and Beveridge, 1994 and this is reproduced by the authors' permission.)

Gram-negative eubacteria

Gram-negative walls are more complex than those of Gram-positive bacteria (cf. Figs 6.2a and b). Peptidoglycan is still found in them, but it is only 1-3 layers thick and is designated to the periplasmic space (i.e., that region bounded by the plasma and outer membranes, Beveridge, 1995; Fig.6.3b). This peptidoglycan layer shares the same chemistry of that found in *B. subtilis* cell walls and is, therefore, electronegative at neutral pH. The outer membrane is found above this layer and contains a unique lipid, lipopolysaccharide (LPS), on its outer face (Beveridge, 1981). LPS is a three-component molecule. One component, lipid A, is a phosphorylated molecule with extensive fatty acid chains which sink the molecule deeply into the hydrophobic domain of the bilayer. Immediately above lipid A, the core polysaccharide is attached and is composed of heptose, phosphate, 2-keto-3-deoxyoctonate (KDO) and other sugars.

The carboxylate groups of the KDOs and the phosphates are ionised at neutral

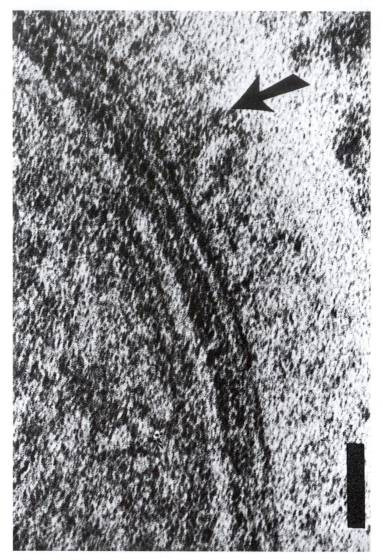

Fig.6.5. Thin section of the cell envelope of a freeze-substituted *Pseudomonas aeruginosa* strain PA01 cell. The arrow points to the long O-side chains of the lipopolysaccharide (LPS) which extend from the bilayered surface of the outer membrane. Bar = 30-nm. (Previously published in Lam *et al.*,1992 and reprinted with the authors' permission.)

pH. Above this core region, O-side chains extend outwards into the external milieu; these are frequently composed of 3-4 sugar repeats that can contain ionisable charge groups. Sometimes a single cell can possess more than one type of LPS. For example, *Pseudomonas aeruginosa* (a common bacterium in the environment which can be an opportunistic pathogen) has two separate LPSs on its outer membrane, A- and B-band LPSs. The former is a relatively neutral LPS because it contains few phosphates and the O-side chain consists of polyrhamnose. B-band LPS on the other hand contains 12 phosphates, 3 KDOs and an O-side chain comprised of a trimer of two uronic acid derivatives plus an N-acetyl fucosamine. This trimer is repeated along the chain many times to give B-band LPSs several different lengths, the longest being about 50 repeats which extends almost 40-nm above the outer face of the outer membrane (Fig.6.5 - Lam *et al.*,1992). Clearly, these long, highly charged LPSs shield the other outer membrane components (e.g., proteins and phospholipids) from the environment (Fig.6.3b). Like Gram-positive surfaces, these Gram-negative cell walls because of their topography have a greater surface area than was originally suspected and they are highly charged so that they interact strongly with external inorganic ions.

Biofilm and planktonic forms

Bacteria can be free-living (planktonic) and are often highly motile in this form. Since the turn of the century, researchers have often chosen to grow bacteria in highly nutritious media in the laboratory. These conditions encourage planktonic growth. Yet, this form of growth is rarely seen under nutrient-limiting conditions in natural habitats; the biofilm mode of growth is preferred (Lappin-Scott and Costerton, 1995). In this type of growth, bacteria take advantage of the ability of a natural large surface (e.g., a rock or plant) to attract and concentrate organic nutrients from the more dilute fluid phase. Therefore, planktonic bacteria will search out and attach to natural interfaces in order to increase their nutrient abundance. Initial attachment to the solid substratum is weak and easily reversible but within hours these bacteria are difficult to dislodge (Marshall, 1992). Attachment depends on the character of both the substratum and bacterial surfaces. For example, Gram-negative bacteria often rely on the charge of the LPS. The cell can frequently modulate this charge to "fit" the solid interface to which the bacterium must adhere. When *P.aeruginosa* adheres to a hydrophilic surface (e.g., glass), B-band LPS is best (Makin and Beveridge, 1996). When it encounters a hydrophobic surface, A-band LPS suffices (Makin and Beveridge, 1996).

As these bacteria grow and divide on the substratum, they exude a matrix of extracellular polymers (most frequently, exopolysaccharide) that encase the entire colony in a highly porous charged material. For example, *P.aeruginosa* secretes alginate which is a viscous linear exopolysaccharide consisting of a copolymer of two uronic acids (i.e., D-mannuronate and its C-5 epimer, L-gluturonate; Evans and Linker, 1973). The "slipperiness" encountered with such solid natural interfaces as rocks in tidal pools and freshwater streams is due to these viscous microbial biofilms.

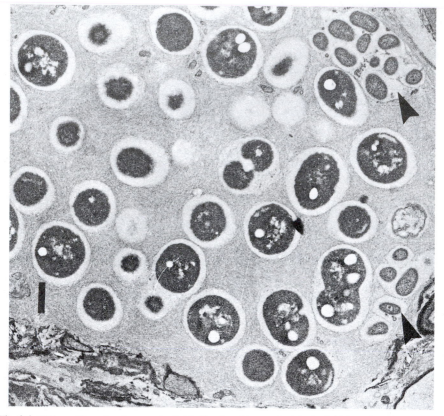

Fig.6.6. Thin section of a photosynthetic biofilm found in a soda lake in British Columbia, Canada. Most of this region of the biofilm consists of large filamentous cyanobacteria which are surrounded by copious extracellular polymer. The arrows point to smaller heterotrophs (bacteria that are nourished by organics given-off by surrounding bacteria and decomposing matter) that are also part of this community. At the bottom right of the figure, some bacteria are beginning to mineralise. This figure can be compared to the microfossils and bacteria seen in Fig.6.1b and c. Bar = 1-μm. (This figure was kindly supplied by S. Schultze-Lam of my laboratory.)

Natural biofilms are rarely monocultures of a single bacterial species. Instead, many different microorganisms seek-out solid interfaces on which to grow so that complex microbial consortia are formed within biofilms (Fig.6.6). Microbial hierarchies are established over time and the biofilms become stratified. This stratification is frequently seen in habitats which are exposed to sunlight. Biofilms become so extensive and thick that "microbial mats" are produced; the upper layers are green because of the chlorophyll contained by cyanobacteria and other algae. Underlying layers are yellow or orange and are due to the pigments found in the anaerobic phototrophs. No light penetrates to the lower regions which are brown to deep black because sulphate-reducing bacteria emit H_2S during their metabolism which forms black precipitates with surrounding metals. Gradients of many varieties are formed in these and thinner microbial amalgams; for example, redox, pH, chemical and nutrient gradients are often seen between the top and bottom of a biofilm. In fact, it is not unusual for strict metabolic interdependencies to be established between different microbial partners (e.g., methane-utilising methylotrophs will often colonise near methane-producing methanogens).

Because natural microbial biofilms consist of bacteria and their exopolymers, they present a highly interactive matrix to inorganic ions within the environment.

Surface Interaction with Environmental Inorganic Ions

So far this chapter has discussed the design features of bacteria, the make-up of their surfaces, and their growth patterns within natural habitats. It has also emphasized their potential reactivity with electrolytes within their environment. All of this was necessary to establish a working knowledge to understand why bacteria have such an innate large capacity to concentrate and mineralise metals within their local environments. It is not unusual for a bacterium to precipitate an amount of metal from solution under simple laboratory conditions that is equal to its organic mass (Fig.6.7a). And, similar phenomena can be seen in natural environments (Fig.6.7b). It is the high surface area-to-volume ratio of bacteria combined with their highly reactive cell walls which account for the hyperaccumulation of metals.

The interaction of metal ions with bacterial surfaces is at least a two-step phenomenon (Beveridge and Murray, 1976). The first event is a stoichiometric interaction between metal ion and reactive surface groups on the bacterium. This event lowers the energy barrier for additional metal complexation and nucleates the precipitation of additional metal. As metal ions are added to the nucleation foci, counter ions from the external milieu are also added. In this way, counter ions help dictate the chemical composition of the precipitate so that metal hydroxides, carbonates, sulphates, sulphides, phosphates, etc. can be formed; the eventual mineral phase is determined by the availability of metal and counter ion, and the chemical environment (pH, Eh, etc.) of the local fluid phase. The bacterial surface can be viewed as a convenient, but rather passive, interface on which metal precipitates can accumulate. Over time, these hydrous precipitates dehydrate and bona fide crystalline mineral phases are developed (Beveridge, 1989a).

For Gram-positive bacteria, such as *B.subtilis*, it is the carboxylate groups of the peptidoglycan which are most important for metal accumulation (Beveridge and Murray, 1980; Doyle *et al.*,1980). For those bacillae that have cell walls in which the secondary polymers are the major component (e.g., *B.licheniformis*), teichoic acid is the most important (Beveridge *et al.*,1982). LPS is the critical component for Gram-negative surfaces; here, the phosphoryl groups of the lipid A and core polysaccharide regions are highly reactive in laboratory strains of *Escherichia coli* which lack O-side chains (Ferris and Beveridge, 1984). Carboxyl groups of the KDO residues in the core region of LPS also contribute but to a lesser extent (Ferris and Beveridge, 1986).

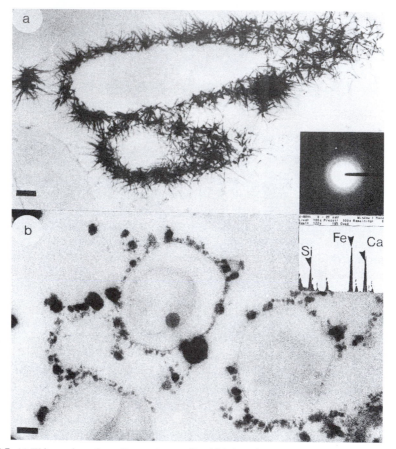

Fig.6.7. (a) Thin section of two *P.aeruginosa* cells which have been suspended in a solution of 1 mM lanthanum nitrate for 10 minutes at 20°C and washed clean of soluble salts after the reaction. The cells have not been contrasted by the normal stains used for electron microscopy so that all of the electron density surrounding each cell is from lanthanum precipitates. Bar = 250-nm. The bottom right shows a selected area electron diffraction (SAED) transform and the powder diagram suggests that the lanthanum precipitate is crystalline. (Previously published in Mullen *et al.*,1989 and reprinted with the authors' permission.)

(b) Thin section of a cell in a biofilm that rested on a granitic rock found in a freshwater stream through which leachates percolate from the pyritic Kam Kotia mine tailings near Timmins, Ontario, Canada. Like (a), this specimen has not been contrasted by electron microscopical stains. Bar = 250-nm. The energy-dispersive X-ray spectroscopic (EDS) analysis at bottom right reveals that these fine-grained minerals are rich in iron and phosphate.

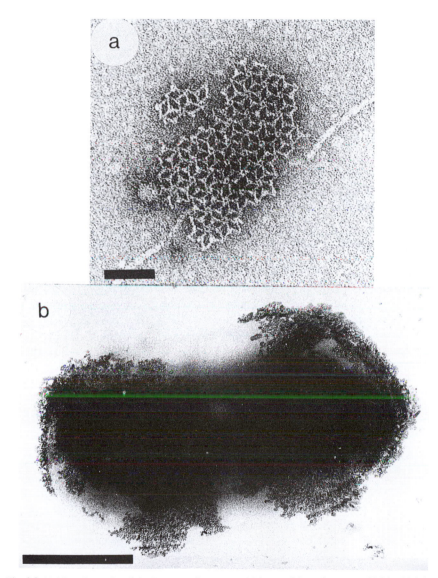

Fig.6.8 (a) Negative stain of the hexagonally-arranged S-layer of *Synechococcus* GL24 which has a centre-to-centre spacing of 22-nm. Bar = 50-nm.

(b) Unstained whole-mount of a *Synechococcus* GL24 cell growing in Fayetteville Green Lake water. Since no stains have been used the electron density is entirely due to the mineralisation of the S-layer. In this case, since the bacterium was grown in the dark so that photosynthesis was absent, the mineral phase is $CaSO_4 \cdot 2H_2O$ (gypsum). Bar = 250-nm. (These images were kindly provided by S. Schultze-Lam of the author's laboratory.)

Those bacteria with highly charged, long O-side chains on their LPS (e.g., B-band LPS on *P.aeruginosa*) can immobilise even more metal (Mullen *et al.*,1989). In this instance, metal precipitates extend up to 250 nm away from the bacterium (Fig.6.7a). Although it is impossible to identify the exact type of bacteria present when looking at natural samples (especially biofilms) by electron microscopy, these cells also possess copious amount of metal which mimic the size of those seen in the laboratory experiments (cf. Figs 6.7a and b). Energy dispersive X-ray spectroscopy (EDS) and selected area electron diffraction (SAED), techniques which require the use of a transmission electron microscope, can give compositional and crystallinity analyses of these metallic deposits (for more information see Beveridge *et al.*,1997). SAED and EDS analyses of the minerals in Fig.6.7a and b are shown in the corner of each figure.

Ironically, living and respiring bacteria do not always have as high a metal deposition capacity as dead or metabolically-inactive bacteria. Respiration energises the plasma membrane and activates a proton-motive force across the bilayer so that protons (H^+) are pumped from the cytoplasm through the membrane into the fabric of the wall. Here, the protons compete with electropositive metal ions and reduce metal adsorption to the bacteria (Urrutia *et al.*,1992).

The above phenomenon is not seen with dead bacteria, which accumulate large quantities of metal, and this emphasizes that the initial nucleation product can be laid-down without an "active" contribution from the bacterium. Once initiated, the precipitate will continue to grow given enough time and external complexing ions. In fact, in freshwater and marine systems, we envision a gentle rain of live and dead bacteria, their particulate residues, and their cellular aggregates constantly cleansing the water column of heavy metals as they settle and concentrate to the underlying sediment.

Fine-grain Mineral Formation by a Cyanobacterium

The predominant cyanobacterium in the mixolimnion of Fayetteville Green Lake, New York is a relatively small (ca. 1.5 μm^3), Gram-negative eubacterium named *Synechococcus* GL24. It possesses an extra layer on its outer membrane that consists of a paracrystalline array of protein, a so-called S-layer (Sleytr *et al.*,1996). These are self-assembly (entropy-driven) systems and, as the proteins fold and arrange themselves on the outer membrane, the polar amino acids of each proteinaceous subunit often become internalised leaving the non-polar residues exposed (Beveridge, 1994). The S-layer of *Synechococcus* GL24 makes this bacterium more hydrophobic than its more "naked" counterparts such as *E. coli* or *B. subtilis* (Schultze-Lam and Beveridge, 1994a). This S-layer has hexagonal symmetry with a centre-to-centre spacing of 22 nm (Fig.6.8a) and the S-protein has an $M_r = 104$-109 kDa; mass estimations combining the M_r of the S-protein to the surface array suggest that 6 proteins combine to form one hexagon (i.e., a unit cell of the lattice).

Even though this S-layer makes the surface of the cyanobacterium relatively unwettable, labelling studies with an electropositive molecule (cytochrome C) have shown that electronegative sites (presumably COO⁻) are periodically situated along the slender "arms" of the protein in the lattice (Schultze-Lam and Beveridge, 1994a). These sites are highly interactive with metal ions. Indeed, because the water of Green Lake contains Ca^{2+} and SO_4^{2-} in relatively high abundance (Thompson et al.,1990), Ca^{2+} complexes to the S-protein and SO_4^{2-} can act as a counter ion, and the mineral gypsum ($CaSO_4.2H_2O$) is formed.

This S-layer on *Synechococcus* GL24 is an exquisite interface for the precipitation of calcium and the ensuing development of calcium minerals (Schultze-Lam, 1992). In fact, the cyanobacterium controls the entire "solid mineral field" of Fayetteville Green Lake. Over time, it has formed both a calcium carbonaceous "reef" or bioherm (ca. 10-m high) and a thick marl sediment on the lake's floor. The pH of the lake is neutral and at this pH, gypsum is the preferred mineral phase. Yet, the reef and marl sediment consist of calcite ($CaCO_3$). Laboratory experiments simulating lake waters solved this enigma.

Synechococcus GL24 is a photosynthetic bacterium that is most metabolically active in the presence of sunlight. During metabolism, hydroxyl ions (OH⁻) are excreted from the bacteria and raise the pH of the local environment surrounding each cell; as the pH approaches 8.0, calcite becomes the preferred mineral phase. Under artificial conditions, if the bacteria are kept in the dark, gypsum is formed and, if they are grown in the presence of light, calcite is produced (Schultze-Lam et al.,1992). These laboratory results fit in well with our data from the field; maximum calcite production is during the sunny summer months and this drops off during the winter. Carbon isotopic data also confirm our OH⁻ hypothesis since the calcite is enriched in ^{13}C whereas the bacteria are enriched in ^{12}C (Thompson et al.,1997). In view of these results, it is possible that bacterial nucleation contributes significantly to so-called "whiting events" in marine and fresh waters on a global scale.

It is uncertain exactly why bacteria such as *Synechococcus* GL24 would choose to enshroud themselves in a solid, dense mineral phase, but preliminary experiments suggest that these bacteria are not particularly enticing to nor are they easily digested by predaceous protozoa (Koval, 1993). These minerals may ensure that the cells are not a food source for higher microbiota. Yet, there is an additional problem of having a mineralised encasement surrounding each cell....how can the cells expand their size during growth and how can they divide? As *Synechococcus* GL24 grows, we have noticed that it sloughs-off cell wall material, especially those regions that are particularly well-coated with precipitates (Fig.6.8b). Presumably, this provides enough breaks in the mineral shroud to allow growth and expansion. Because the liberated cell wall fragments already possess minerals, they continue to promote additional precipitation while floating free from the bacterium and could add significantly to the mineral budget. Surprisingly, the sloughing-off phenomenon is calcium-mineral dependent. Strontium (a calcium analogue) can also be precipitated by *Synechococcus* GL24, but these mineral phases do not slough from the bacteria and the cells eventually die (Schultze-Lam and Beveridge, 1994b).

Silicates

Silica is the most abundant mineral in the Earth's crust and, so far, this chapter has emphasized metals. This is because the overall charge on bacterial surfaces is usually electronegative. Yet, scattered throughout this matrix of anionic sites are occasional amine groups. These can interact with and bind SiO_3^{2-} (Urrutia and Beveridge, 1993a).

Multivalent metal cations also can contribute to silicate development via their ability to salt-bridge SiO_3^{2-} to either phosphate or carboxylate groups on the cell wall and, in this way, metal silicates are produced (Urrutia and Beveridge, 1993a and 1994). Eventually, short-range ordered silicates are developed, even when cells are in the presence of organic ligands possessing metal-complexing ability (e.g., fulvic acids, Urrutia and Beveridge, 1995). Once formed, these fine-grained silicates (in themselves) make good adsorption surfaces for additional environmental metals (Walker *et al.*, 1989). Eventually, large silicaceous bacterial aggregates or 'flocs' can be formed that are visible to the eye. Their reactive surfaces make them highly interactive to a variety of organic and inorganic pollutants (Rao and Dutka, 1993). Surprisingly, studies to determine the ease with which metals can be leached from such complexes show that metals are more tenaciously bound to the bacterial fraction than to the silicates (Flemming *et al.*, 1990; Urrutia and Beveridge, 1993b).

Concluding Remarks

Bacteria are not only efficient at immobilising metals from solution, they are also efficient templates for the precipitation of silicaceous materials. To see these processes in action, a researcher needs only to visit and sample those natural environments in which prokaryotic life resides. The development of fine-grain minerals on their cell surfaces is part of these cells' innate character. Because prokaryotes can exist under almost any imaginable earthly setting, including those involving extremes of temperature, pressure, pH, anoxicity, salt, etc., their production of minerals must be an on-going phenomenon that exists almost everywhere. Given the ancient lineage of these lifeforms, the amount of mineral mass produced since the Archaean is almost unimaginable!

Acknowledgements

The work from the author's laboratory reported in this chapter has been supported by an on-going Natural Sciences and Engineering Research Council of Canada grant. In this research, I am grateful for the hard work and friendship of a large number of students and research associates, and their names will often been seen associated with mine in the Reference section of this chapter. One of these, S. Schultze-Lam, has also aided me by supplying Figs 6.6 and 6.8. I thank them all.

References

Beveridge,T.J.(1981) Ultrastructure, chemistry, and function of the bacterial wall. *International Review of Cytology* 12, 229-317.

Beveridge,T.J.(1988) The bacterial surface: General considerations towards design and function. *Canadian Journal of Microbiology* 34, 363-372.

Beveridge,T.J.(1989a) The role of cellular design in bacterial metal accumulation and mineralization. *Annual Review of Microbiology* 43, 147-171.

Beveridge,T.J.(1989b) The structure of bacteria. In: Leadbetter,E.R. and Poindexter,J.S.(eds). *Bacteria in Nature: A Treatise on the Interaction of Bacteria and their Habitats v.3.* Plenum Publishing Co., New York, pp.1-65.

Beveridge,T.J.(1994) Bacterial S-layers. *Current Opinions in Structural Biology* 4, 204-212.

Beveridge,T.J.(1995) The periplasmic space and periplasm in Gram-positive and Gram-negative bacteria. *ASM News* 61, 125-130.

Beveridge,T.J. and Davies,J.A.(1983) Cellular response of *Bacillus subtilis* and *Escherichia coli* to the Gram stain. *Journal of Bacteriology* 156, 846-858.

Beveridge,T.J. and Doyle,R.J.(eds)(1989) *Metal Ions and Bacteria*. Wiley, New York, pp.1-461.

Beveridge,T.J.,Forsberg,C.W. and Doyle,R.J.(1982) Major sites of metal binding in *Bacillus licheniformis* walls. *Journal of Bacteriology* 150, 1438-1448.

Beveridge,T.J. and Graham,L.L.(1991) Surface layers of bacteria. *Microbiological Reviews* 55, 684-705.

Beveridge,T.J.,Hughes,M.N.,Lee,H.,Leung,K.T.,Poole,R.K.,Savvaidis,I., Silver, S. and Trevors,J.T.(1997) Metal-microbe interactions: contemporary approaches. *Advances in Microbial Physiology* 38, 177-243.

Beveridge,T.J. and Murray,R.G.E.(1976) Uptake and retention of metals by cell walls of *Bacillus subtilis*. *Journal of Bacteriology* 127, 1502-1518.

Beveridge,T.J. and Murray,R.G.E.(1980) Sites of metal deposition in the cell wall of *Bacillus subtilis*. *Journal of Bacteriology* 141, 876-887.

Doyle,R.J.,Matthews,T.H. and Streips,U.N.(1980) Chemical basis for selectivity of metal ions by the *Bacillus subtilis* cell wall. *Journal of Bacteriology* 143, 471-480.

Doyle,R.J. and Koch,A.L.(1987) The functions of autolysins in the growth and division of *Bacillus subtilis*. *Critical Reviews in Microbiology* 15, 169-222.

Evans,L.R. and Linker A.(1973) Production and characterization of the slime polysaccharide of *Pseudomonas aeruginosa*. *Journal of Bacteriology* 170, 1452-1460.

Ferris,F.G. and Beveridge,T.J.(1984) Binding of a paramagnetic metal cation to *Escherichia coli* K-12 outer membrane vesicles. *FEMS Microbiological Letters* 24, 43-46.

Ferris,F.G. and Beveridge,T.J.(1986) Site specificity of metallic ion binding in *Escherichia coli* K-12 lipopolysaccharide. *Canadian Journal of Microbiology* 32, 52-55.

Ferris,F.G.,Fyfe, W.S. and Beveridge,T.J.(1988) Metallic ion binding by *Bacillus subtilis*: implications for the fossilization of microorganisms. *Geology* 16, 149-152.

Flemming,C.A.,Ferris,F.G.,Beveridge,T.J. and Bailey,G.W.(1990) Remobilization of toxic heavy metals adsorbed to bacterial wall-clay composites. *Applied and Environmental Microbiology* 56, 3191-3203.

Graham,L.L. and Beveridge,T.J.(1994) Structural differentiation of the *Bacillus subtilis* cell wall. *Journal of Bacteriology* 176, 1413-1421.

Koch,A.L.(1995) *Bacterial Growth and Form*. Chapman & Hall, New York, pp.1-423.

König,H.(1988) Archaeobacterial cell envelopes. *Canadian Journal of Microbiology* 34, 395-406.

Koval,S.F.(1993) Predation on bacteria possessing S-layers. In: Beveridge,T.J. and Koval,S.F.(eds), *Advances in Bacterial Paracrystalline Surface Layers*. Plenum Press, New York, pp.85-92.

Lam,J.S.,Graham,L.L.,Lightfoot,J.,Dasgupta,T. and Beveridge,T.J.(1992) Ultrastructural examination of the lipopolysaccharides of *Pseudomonas aeruginosa* strains and their isogenic rough mutants by freeze-substitution. *Journal of Bacteriology* 174, 7159-7167.

Lappin-Scott,H.M. and Costerton,J.W.(eds)(1995) *Microbial Biofilms*. Cambridge University Press, Cambridge, pp.1-310.

Makin,S.A. and Beveridge,T.J.(1996) The influence of A-band and B-band lipopolysaccharide on the surface characteristics and adhesion of *Pseudomonas aeruginosa* to surfaces. *Microbiology* 142, 299-307.

Marshall,K.C.(1992) Biofilms: an overview of bacterial adhesion, activity, and control at surfaces. *ASM News* 58, 200-207.

Mullen,M.D.,Wolf,D.C.,Ferris,F.G.,Beveridge,T.J.,Flemming,C.A. and Bailey,G.W.(1989) Bacterial sorption of heavy metals. *Applied and Environmental Microbiology* 55, 3143-3149.

Rao,S. and Dutka,B.(eds)(1993) *Particulate Matter and Aquatic Environments*. Lewis Publishers, Chicago, pp.1-205.

Schultze-Lam,S. and Beveridge,T.J.(1994a) Physicochemical characteristics of the mineral-forming S-layer from the *Synechococcus* GL24. *Canadian Journal of Microbiology* 40, 216-223.

Schultze-Lam,S. and Beveridge,T.J.(1994b) Nucleation of celestite and strontianite on a cyanobacterial S-layer. *Applied and Environmental Microbiology* 60, 447-453.

Schultze-Lam,S.,Harauz,G. and Beveridge,T.J.(1992) Participation of a cyanobacterial S layer in fine-grain mineral formation. *Journal of Bacteriology* 174, 7971-7981.

Sleytr,U.B.,Messner,P.,Pum,D. and Sára,M.(eds)(1996) *Crystalline Bacterial Cell Surface Proteins*. R.G. Landes Co./Academic Press, Austin, pp.1-230.

Sprott,G.D.,Meloche,M. and Richards,J.C.(1991) Proportions of diether, macro-cyclic diether and tetraether lipids in *Methanococcus jannaschii* grown at different temperatures. *Journal of Bacteriology* 173, 3907-3910.

Thompson,J.B.,Ferris,F.G. and Smith,D.A.(1990) Geomicrobiology and sedimentology of the mixolimnion and chemocline in Fayetteville Green Lake, New York. *Palaois* 5, 52-75.

Thompson,J.B.,Schultze-Lam,S.,Beveridge,T.J. and DesMarais,D.J.(1997) Whiting events: biogenic origin due to the photosynthetic activity of cyanobacterial picoplankton. *Limnology and Oceanography* 42, 133-141.

Urrutia,M.M. and Beveridge,T.J.(1993a) Mechanism of silicate binding to the bacterial cell wall in *Bacillus subtilis*. *Journal of Bacteriology* 175, 1936-1945.

Urrutia,M.M. and Beveridge,T.J.(1993b) Remobilization of heavy metals retained as oxyhydroxides or silicates by *Bacillus subtilis* cells. *Applied and Environmental Microbiology* 59, 4323-4329.

Urrutia,M.M. and Beveridge,T.J.(1994) Formation of fine-grained metal and silicate precipitates on a bacterial surface (*Bacillus subtilis*). *Chemical Geology* 116, 261-280.

Urrutia, M.M. and Beveridge,T.J.(1995) Formation of short-range ordered aluminosilicates in the presence of a bacterial surface (*Bacillus subtilis*) and organic ligands. *Geoderma* 65, 149-165.

Urrutia,M.M.,Kemper,M. Doyle,R.J. and Beveridge,T.J.(1992) The membrane-induced proton motive force influences the metal binding ability of *Bacillus subtilis* cell walls. *Applied and Environmental Microbiology* 58, 3837-3844.

Walker,S.G.,Flemming,C.A.,Ferris,F.G.,Bailey,G.W. and Beveridge,T.J. (1989) Physicochemical interaction of *Escherichia coli* envelopes and *Bacillus subtilis* walls with two clays and ability of the composite to immobilize heavy metals from solution. *Applied and Environmental Microbiology* 55, 2976-2984.

Walter,M.R.(1983) Archean Stromatolites: Evidence of the Earth's earliest benthos. In: Schopf,J.W.(ed.), *Earth's Earliest Biosphere: Its Origins and Evolution*. Princeton University Press, Princeton, pp.214-239.

Woese,C.R. and Fox,G.E.(1977) Phylogenetic structure of the prokaryotic domain: the primary kingdoms. *Proceedings of the National Academy of Sciences USA* 74, 5088-5090.

Woese,C.R. and Wolfe,R.S.(eds)(1985) *The Bacteria: A Treatise on Structure and Function. V.8. Archaebacteria*. Academic Press, New York, pp.1-582.

Woese,C.R.,Kandler,O. and Wheelis,M.L.(1990) Towards a natural system of organisms: proposal for the domains *Archaea, Bacteria,* and *Eucarya*. *Proceedings of the National Academy of Sciences USA* 87, 4576-4579.

Chapter seven:

Phytoarchaeology and Hyperaccumulators

R.R. Brooks
Department of Soil Science, Massey University, Palmerston North, New Zealand

Introduction

In Chapters 3 and 4 of this book, it has been shown that geobotany and biogeochemistry can be applied in using plants (including hyperaccumulators) to detect mineralisation in the substrate in which these species grow. It is but a small step to use plants to predict the presence of ancient mine sites in the neighbourhood of these mineralised areas.

In a recent book, Brooks and Johannes (1990) have coined the term *phytoarchaeology* to describe the relationship between vegetation and archaeological remains. They have divided the discipline into a number of subheadings involving the use of geobotany and biogeochemistry to detect these ancient sites, and have also discussed the influence of ancient trade routes on anthropogenic modification of the vegetation. Mediaeval mines in Europe and smelter sites in Central Africa have been identified by the use of vegetation. These topics will all be discussed below with information largely derived from the above book.

Geobotany and Exploitation of Ancient Mines

Introduction

The subject of geobotany in mineral exploration has already been fully discussed in Chapter 3 and the full details will not need to be repeated here. It is sufficient to say that certain plant species and plant communities that include various hyperaccumulator species will indicate potential mining sites that may well contain ancient artefacts or mine workings.

Indicator plants are species found over mineralised ground and are of two main types: *local* and *universal*. Local indicators have only local indicating properties and at other sites they may also grow over non-mineralised ground. By contrast, universal indicators indicate minerals at all sites where they occur and therefore are of great use in exploration. Indicator plants have been known for many years, the 17th century copper miners of Scandinavia were guided to their targets by the local indicator *Viscaria (Lychnis) alpina,* which was known as the *kisplante* (pyrite plant) and which even today is found over the ancient copper mines of Røros in central Norway (Vogt, 1942).

Apart from species changes in vegetation over mining sites compared to plants growing over the surrounding background, certain types of mineralisation have a characteristic flora that can be identified readily without mapping procedures. These include basiphilous (limestone), halophyte, selenium, serpentine, cupricolous, and zinc (Galmei) floras (Brooks, 1983, 1987). Not all of these are really relevant to the search for ancient mining activities but some, particularly zinc and cupricolous floras, that contain a fair number of highly specific hyperaccumulators, are very useful and will be discussed below with specific case histories.

Ancient mine sites

Introduction

The study of indicator plants has been hindered by the existence of a persistent folklore on the alleged indicating ability of some species. A myth once established, is often perpetuated by a string of review articles which merely quote previous reviewers. A good example of this is furnished by the work of Nemec *et al.* (1936) who alleged that the horsetail, *Equisetum palustre, from* Czechoslovakia contained 610 μg/g (ppm) gold in its ash. This would have corresponded to about 120 μg/g in dry matter (the ash of *Equisetum* is around 20% of the dry weight) and would certainly have qualified this species as a hyperaccumulator of gold.

Subsequent work by Cannon *et al.* (1968) indicated that the ash never contained more than 1 μg/g of gold. It is not suggested that Nemec *et al.* falsified their data, but they do appear to have been mistaken in that they precipitated sulphides from solutions of plant material, weighed the precipitate and assumed it was all gold. In fact, as we now know, copper, arsenic, and lead, all of which form insoluble sulphides, are much more common in horsetails than is gold. Elemental abundances in horsetails are shown in Table 7.1 and one of the species (*E. arvense*) is illustrated in Fig.7.1.

In a later paper, Brussell (1978) claimed to have discovered gold in the emission spectrum of the ash of *E. hyemale* from the United States, but the spectral lines used were weak lines of gold subject to massive interference from other elements in the plant material.

There is some evidence, however, that horsetails frequently do grow over ancient gold mines since these plants appear to favour disturbed ground and are tolerant of arsenic (frequent companion of gold), an element normally very toxic to plants.

The question of the association of horsetails with gold has been investigated by Brooks *et al.* (1981) who determined gold, arsenic and antimony (all these elements are associated together geochemically) in horsetails growing over abandoned gold mining sites in Nova Scotia, eastern Canada. Six of the 25 species of horsetail were analysed: *E.arvense*, *E.fluviatale*, *E.hyemale*, *E.palustre*, *E.scirpioides*, and *E.sylvaticum*.

In no case did the gold content of the horsetails exceed 1 µg/g (ppm) in sharp contrast to the very high values obtained by Nemec *et al.* (1936). Of particular interest, however, was the very high arsenic content of the samples. Concentrations of this element in the phyllodes ranged from 41 to 738 µg/g, accompanying antimony contents of 4-77 µg/g.

Background arsenic levels in horsetails in Nova Scotia were of the order of

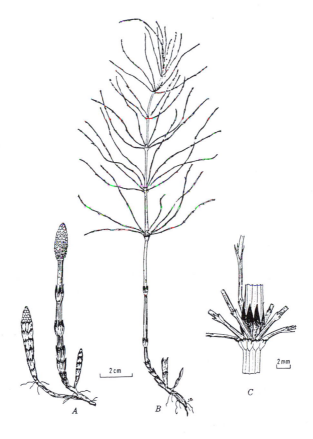

Fig.7.1. Sketch of *Equisetum arvense* L. A - fertile stems with spore-producing cones, B - sterile stems with part of rhizome, C - node of sterile stem. Source: Cannon *et al.* (1968).

3 µg/g, a value far higher than for vegetation in general (0.2 µg/g - Brooks, 1983). The high arsenic levels in Nova Scotian horsetails result from the highly arseniferous nature of the Meguma Series rocks covering most of the southern part of Nova Scotia. The rocks are so arseniferous that in many places, such as Montague Mines near Halifax, mining activities have resulted in extensive contamination of water supplies by this element.

Table 7.1. Mean and maximum elemental concentrations (µg/g dry weight) found in *Equisetum* species compared with mean for all types of vegetation.

Element	A	B	C	D
Barium	77	86	900	Hg
Boron	41	114	320	-
Chromium	0.9	<1	3.6	-
Cobalt	0.05	0.4	6.0	Pb/Zn
Iron	100	236	3000	Pb/Zn
Gold	0.001	0.01	0.10	Hg
Lead	5.0	2.4	84	Zn/Pb
Manganese	230	50	5600	Mo
Molybdenum	0.90	<1	30	Mo
Nickel	4.6	1.4	56	Mo
Silver	0.10	<0.10	0.40	Au
Strontium	120	60	300	-
Titanium	32	56	240	-
Vanadium	1.4	<3	10	Mo
Zinc	85	<40	1800	Zn
Zirconium	1.4	<10	46	-

A - average metal content of all types of vegetation in non-mineralised soil, B - average metal content of specimens of *Equisetum* not growing in mineralised soils, C - maximum metal content of all *Equisetum* analysed, D - metals produced in region where maximum metal content was found in *Equisetum*. Source Cannon *et al.* (1968).

The tolerance of horsetails to arsenic when combined with the frequency of their occurrence over auriferous mine tailings, leads to the conclusion that the genus *Equisetum* contains species that are in fact local indicators of the gold with which the arsenic is associated. The original mythology may therefore have some basis of fact, albeit based on an indirect local association with gold rather than accumulation of this element by the plants.

The gold mines of Nova Scotia date back only to the 19th century and these sites are hardly archaeological in the true sense of the word. However, the research carried out at these sites has served to shed light on the role of horsetails as indirect indicators of gold mineralisation.

A species of onion (*Allium* sp.) also appears to have the reputation of being an indicator of gold and silver mineralisation in China. According to Boyle (1979), early Chinese records of the 9th century AD describe the association of *A.fistulosum* with silver and *A.bakeri* with gold. This association may be indirect rather than direct and could reflect the presence of sulphur which is found in both

onions and sulphide minerals (hosts for silver and gold). *Allium* also has the
reputation of hyperaccumulating arsenic which is often associated with gold.

Cave del Predil lead/zinc mines

The Cave del Predil (formerly Raibl) region of Northern Italy (Fig.7.2) is the site
of a complex of large and small mines that have been operating since mediaeval
times. The first evidence for lead mining in the region dates back to 1506 AD,
although a manuscript from 1456 AD mentions sulphide mining near Raibl. The
area near this mine, and the river terraces downstream, are populated by an
extraordinary plant with an unusual capability to tolerate and hyperaccumulate
both lead and zinc. This plant is *Thlaspi rotundifolium* subsp. *cepaeifolium*. Its
ecology and biogeochemistry have been investigated by Reeves and Brooks (1983)
and Ernst (1974).

The river Silizza (Schlitza) flows from Lago del Predil adjacent to the mines
and descends via Tarvisio (Tarvis) into the Gailtal and then via Villach into the
Drau River. On the west bank of the Silizza near Cave del Predil and extending
into the lower slopes of Monte Re (Königsberg) are numerous mine workings

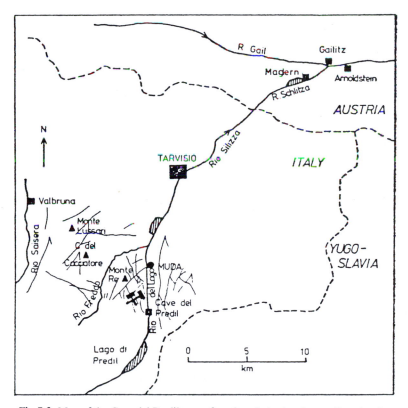

Fig.7.2. Map of the Cave del Predil area of northern Italy showing mediaeval and
modern lead/zinc mine workings. Source: Reeves and Brooks (1983).

which are found as far north as the confluence of the Freddo (Kaltwasser) and Silizza rivers (Fig.7.2). The tailings from the workings have been transported into the river where the fluvial sediments and gravels have become stained with dark ore-rich material. The tailings have been transported some distance by the river and have accumulated at various sites downstream such as at Arnoldstein in Austria near the confluence of the Gail and Schlitza rivers. The metallophytes are often found in these accumulations. Their local distribution is therefore from Lago del Predil to Arnoldstein with intermediate occurrences at Muda (Mauth) on the west bank of the Silizza between the Freddo River and Tarvisio, and at Maglern a short distance upstream from Arnoldstein. At the last locality there is also a zinc mine whose tailings make some contribution to the contaminated gravels at this site.

Fig.7.3. *Thlaspi rotundifolium* subsp. *cepaeifolium*. This plant typically hyperaccumulates up to 8200 μg/g (dry weight) lead. Photo by R.R.Brooks.

Thlaspi rotundifolium subsp. *cepaeifolium* (Fig.7.3) is one of the dominant members of a heavy metal plant community described by Ernst (1974) that includes a metal-resistant ecotype of *Minuartia verna*. It can hyperaccumulate up to 8200 μg/g (0.7%) lead and 1.73% zinc in its dry tissue (see also Chapter 3). If these values are not inflated by wind-borne dust contamination, this species is the only known hyperaccumulator of lead. Another species, *Alyssum wulfenianum,* is an acccumulator of this element and can contain up to 860 μg/g lead and 2500 μg/g zinc (Reeves and Brooks, 1983). The *Thlaspi* metallophyte is found at scattered mineralised and contaminated localities from Jauken in Austria some 55 km to the northwest of Cave del Predil to Pecnik-Peca in Croatia some 105 km to the east.

According to Ernst (1974), the soils supporting this *Thlaspi* subspecies have lead contents ranging from 0.0617% (Bleiberg ob Villach) to 0.417% (Jauken), and zinc contents ranging from 0.7% (Crna) to 14.6% (Jauken), together with small amounts of copper (0-0117%).

The heavy-metal community associated with the Raibl Mine is known as the *Thlaspietum cepaeifolii* association (Ernst, 1964, 1974 - see Chapter 3 for a discussion of phytosociological communities). The physiognomy of this species-poor community is characterised by dark-green cushions of *Thlaspi rotundifolium* subsp. *cepaeifolium* and light-green cushions of an ecotype of *Minuartia verna* that is particularly resistant to heavy metals.

The Derbyshire lead mines of England

The Pennine Range of northern England, the Mendips of southwest England, and the mountains of North Wales contain a large number of base metal (lead/zinc/copper) mining sites which in many cases, have been worked since Roman times.

The spoil heaps and mine tailings of these workings are still in evidence today and provide an environment extremely hostile to both plants and animals. For example, Chisnall and Markland (1971) found that cattle and horses were still dying from lead poisoning by contaminated water (2-5 μg/mL lead) adjacent to old Roman mine tips near Matlock in the Pennines.

The location of abandoned lead mines in the British Isles is given in Fig.7.4. The metal-tolerant flora of lead/zinc mineralisation has been classified by Ernst (1976) within the phytosociological class *Violetea calaminariae*. Within the Pennines, Mendips and North Wales is an association known as the *Minuartio-Thlaspietum alpestris*. Compared with the heavy-metal communities of western Europe, the *Minuatio-Thlaspietum* association is impoverished in numbers of species. The most noteworthy feature is the complete absence of *Armeria maritima* and a somewhat rare appearance of *Silene cucubalus*, both very common species of the Western European lead/zinc deposits. The most obvious of the members of the British communities is *Thlaspi alpestre*, a well known hyperaccumulator of zinc (Baumann, 1887). Although a true *T.alpestre* does indeed exist (it is found mainly in alpine regions of Europe), the species found in northern England is more properly known as *Thlaspi caerulescens* but the earlier name will be retained in this chapter to avoid confusion as most of the literature for this region refers to *T.alpestre*.

The plant communities in Britain have furthermore been greatly disturbed by active mining subsequent to the original Roman exploitation.

The lead mines of the Pennine Range of northern England have been worked since before Roman times (Ford and Rieuwerts, 1983). Their distribution in Derbyshire can be seen from Fig.7.4. The workings are virtually confined to metalliferous veins in the Carboniferous limestone of the Peak District in Derbyshire. The mineralisation is largely galena (lead sulphide) with associated sphalerite (zinc sulphide). The average lead content is about 5% in the veins capable of being worked. It is estimated that some 3-6 million tonnes of lead ores have been recovered since mining began over 2000 years ago. Zinc production is estimated at 250,000-500,000 tonnes. The mines ceased operation by the early 20th century and have been replaced by other mines producing large amounts of barite (barium sulphate), calcite (calcium carbonate), and fluorite (calcium fluoride). All of these minerals had been discarded as waste during the lead mining period.

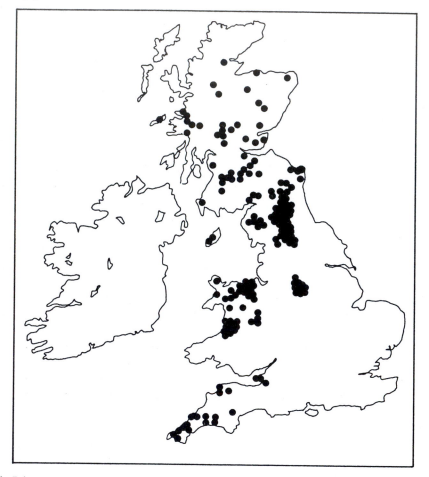

Fig.7.4. Distribution of ancient lead mines in Britain. Sources: various, reported by Hajar (1987).

Lead ore was probably exploited in the Peak District even in the pre-Roman period, though there is no definite proof of this. It is known however, that there is evidence of Roman exploitation in the shape of lead "pigs" (ingots). At least 27 of these have been recovered and were distinguished by the letters "LVT", "LVTVD", or "LVTVDARES". These refer to the settlement of Lutudarum where the smelting was carried out at a site near Matlock in the southern Pennines.

Very few Roman artefacts have been recovered from the mine sites because of subsequent disturbance, although some objects associated with lead mining operations have been discovered at Elton, Crich, and Longstone Edge. More recently, an opencast lead vein bridged by a Roman wall has been discovered at Roystone Grange near Ballidon (Ford and Rieuwerts, 1983). In the Matlock Museum there is a Roman smelting hearth (L. Willies pers. comm.) operated with foot bellows, and is one of only three or four so far discovered in Britain.

After the withdrawal of the Romans, the Saxons and Danes continued mining on a small scale in the 600 years before the Norman Conquest. The Domesday Survey undertaken in 1086, some 20 years after the Norman Conquest, lists 7 lead smelters at Bakewell, Ashford, Crich and Matlock. The lead was used mainly for castles and numerous religious houses in the 11th, 12th and 13th Centuries.

The period from 1700 to 1750 was one of particularly active mining. Thereafter, production remained relatively constant until the middle of the 19th century when a steady decline began. The last operation to close was the Mill Close Mine in Darley Dale. This ceased operations just before World War II.

The vegetation of spoil heaps and outcrops of the Pennine lead mines has been studied by several workers including Clarke and Clark (1981), Shaw (1984), Shimwell and Laurie (1972), Baker (1987), and Hajar (1987). The vegetation is influenced by a number of unfavourable edaphic factors in addition to the toxic effects of heavy metals such as lead, cadmium and excess zinc. These factors include surface instability, low organic status of the soil, and severe deficiency of plant nutrients such as nitrogen and potassium. The process of colonisation is slow and is governed initially by species capable of giving rise to specialised races or ecotypes capable of successful growth and reproduction on otherwise phytotoxic substrates.

Among the most important of the plants that grow on Pennine mine sites are *Minuartia verna* and *Thlaspi alpestre*, now known as *Thlaspi caerulescens* (Fig.7.5). These are normally considered to be alpine taxa in many parts of Europe, but in Britain they are strongly associated with soils contaminated by heavy metals. This association was noted as early as in the 16th century in the case of *M. verna*, and in the 19th century for *T. alpestre*. The distribution of these two metallophytes in the British Isles in relation to the location of lead mines is shown in Fig.7.6. Within the southern Pennines the distribution of both species is strongly correlated with the location of local lead mines.

From experiments carried out by Hajar (1987), it appears that *T. alpestre* tends to be more commonly associated with bare soil than does *M. verna*, but the latter is found over a wider range of soil pH values. Both species occupy similar ecological niches (Hajar, 1987); i.e. infertile but somewhat disturbed sites with broken vegetation on metalliferous mining spoil heaps.

In making a general comparison of the above species, Hajar (1987) came to the following conclusions:

   1 - *T. alpestre* has a more restricted distribution in Europe and the British Isles than does *M. verna*;

   2 - in Derbyshire, *M. verna* is found over a wider range of soil pH values;

   3 - *T. alpestre* is particularly associated with high exposures of bare soil;

   4 - *T. alpestre* may be less restricted to north-facing slopes.

Fig.7.5. *Thlaspi caerulescens* growing over lead/zinc mine waste
in New Zealand. Photo by R.R.Brooks.

   Many other plant species are found over lead-contaminated soils in the Pennine Range. These include *Viola lutea*. This species is related to the classical zinc hyperaccumulator *Viola calaminaria* (Baumann, 1885) and is known to accumulate zinc, though not to the same extent as the latter. It is frequently associated with zinc deposits and ancient mine sites. *Viola lutea* is usually yellow in colour but some populations (e.g. at Nenthead in Cumberland) are mauve.

Exactly the same situations pertains to *V. calaminaria* that is yellow in the *locus classicus* near Aachen and has a mauve counterpart in Westphalia (Ernst (1974).

In the Derbyshire lead mining region, the positions of most of the main veins have been known for several hundred years. There may be cases however, where the distribution of *Minuartia verna* and *Thlaspi alpestre* might indicate the presence of hitherto unrecorded lead veins or outcrops together with associated mediaeval or even Roman artefacts. Even if this turns out not to be the case, the story of the relationship between vegetation and mine sites in this interesting region is still well worth recording.

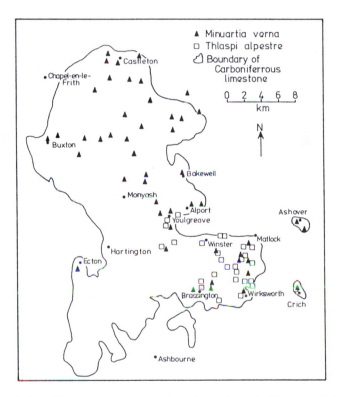

Fig.7.6. Distribution of *Minuartia verna* and *Thlaspi alpestre* in Derbyshire, UK. Cf. Fig.7.4 that shows the location of Ancient lead mines. Source: various, reported by Hajar (1987).

The Mendip lead mining district

Another lead mining district in which *Thlaspi alpestre* is to be found, is located in the Mendip Hills some 20 km south of the city of Bristol in southwest England. The mines were known to be in operation in 49 AD (6 years after the Roman conquest) and there is some evidence that mining activities began as far back as the 3rd century BC (Anon.,1970).

Evidence for mining during Roman times has been afforded by the discovery of 23 dated "pigs" of lead as far afield as St. Valery-sur-Somme in France.

Thereafter, the record became obscure until Richard I gave permission for lead mining in 1189 AD.

The lead mining industry gradually declined due to flooding in the lower levels of the mines, exhaustion of surface workings, and increased competition from foreign ores. All mining operations had ended by 1850. The slags and wastes however, were so rich in lead, and were of such a vast extent, that companies sprang up to reprocess waste material. Silver was also mined and some 3000 ounces had been produced by the end of the 19th century. The reprocessing plants gradually shut down as the price of lead fell: the Waldegrave works in 1876, and the St. Cuthberts works in 1910. These works were once combined under the name of "Priddy Minery". They offer a rich harvest of interesting remains including smoke flues, ruined buildings, and spoil heaps. Roman remains are also to be found at the site.

The area around the old Priddy smelter is so heavily contaminated that conifers planted in recent years have failed to flourish over the site and its immediate environs.

The vegetation of the Mendip mining area has not been studied to the same degree as that of the southern Pennines. However, Baker (1974) has studied the occurrence of heavy metal-tolerant plants in the area. Among these is *Silene maritima* which Baker found at five stations in the area, usually along with the ubiquitous hyperaccumulator *Thlaspi alpestre*. His observations were as follows:

...Between Shipham and Rowberrow. *Silene maritima* was abundant on old mining ground. There is now very little evidence of the calamine workings in this area. One site was inspected by the author in 1971 just north of Shipham. The workings are largely covered over with a grassy turf. Rocky outcrops where visible, probably mark the location of shallow calamine pits. A few plants of *S. maritima* were recorded at the periphery of these outcrops along with *Thlaspi alpestre*.

Between Star and Shipham. *S. maritima* was found on ground left hummocky by former calamine workings, now calcareous pasture. The species was found in small quantity here and was associated with *Thlaspi alpestre, Genista tinctoria* and *Anthyllis vulneraria*. Plants were found along the periphery of calamine pits as above, and also on more rocky limestone cliffs.

The zinc deposits of Western Germany and Eastern Belgium

If one were to be asked to name the *locus classicus* where the concept of hyperaccumulation of metals by plants originated, there is little doubt that the zinc deposits of western Germany and eastern Belgium would spring to mind. It was here, near Aachen, that Baumann (1887) reported the unusual accumulation (ca. 1% in dry matter) of zinc, a finding whose significance lay unrealised until the 1990s when the idea of *phytoremediation* (see Chapter 12) was born. The zinc-tolerant vegetation community became known as a *Galmei flora* and its present distribution is shown in Fig.7.7.

The geology of the calamine deposits has been described by Schwickerath (1931) and by Gussone (1961). The zinc-rich soils extend in a wide southwest to northeast-trending band from just east of Aachen to southeast of Le Rocheux in eastern Belgium. The deposits are found within strips of limestone and dolomite and are characterised by numerous prospecting shafts and adits, spoil heaps, and

ruined buildings from earlier mining activities. These give a desolate appearance
enhanced by the surrounding limestone hills which are almost completely denuded
of trees and shrubs.

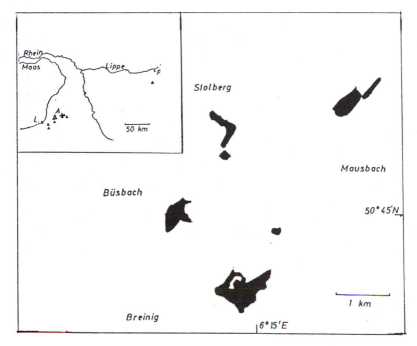

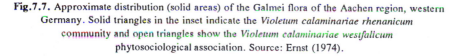

Fig.7.7. Approximate distribution (solid areas) of the Galmei flora of the Aachen region, western
Germany. Solid triangles in the inset indicate the *Violetum calaminariae rhenanicum*
community and open triangles show the *Violetum calaminariae westfalicum*
phytosociological association. Source: Ernst (1974).

Mining activity at Aachen dates back to Roman times, because Roman coins
(lst century AD) and implements have been found at mine sites and spoil heaps
near Gressenich and Breinigerberg. Plinius (77 AD) reports mining activity in
"Germania" where an ore known as *cadmia* was extracted. There is no evidence
however, for a continuous period of ore extraction after withdrawal of the
Romans up to the first record of Middle Ages activity in 1344 AD - a gap of
about 1000 years.

One of the most important of the early mining sites was in the area bounded
by Maubach, Stolberg, Busbach and Breinig (Fig.7.7). The last of the mines to
operate was at Diepenlinchen near Maubach where working ceased at about the
time of World War I. The number of previously-operated mines is very large and
some of these were of great importance. In more recent times, some of the mine
tips were worked to remove more of their remaining lead and zinc.

The lead/zinc ore deposits of the region are associated with limestones of the
Devonian and Lower Carboniferous periods. These ores are often accompanied

by sulphides and more frequently by marcasite (white iron pyrites). Copper mineralisation is scarce.

Fig.7.8. *Viola calaminaria* growing over zinc-rich soils near Aachen, Germany. The hyperaccumulation of zinc (up to 1% dry weight) was first reported in 1885. Photo by Dieter Johannes.

One of the earliest mines to be developed was on the Brockenberg where Galmei, a collective name for carbonates and silicates of zinc (sometimes erroneously called calamine), was mined. Evidence is found in numerous ancient mine workings. Mining activity at the Breinigerberg is likewise very ancient and has been traced back to Roman times. Operations here ceased in the 1870s due to the high cost of draining the mines.

The Galmei flora of the ore-rich soils of the region has been described by Schwickerath (1931), Savelsbergh (1976) and Kremer (1982). The community includes such famous metallophytes as *Thlaspi calaminare*, which is the first to appear in the spring. Other members of the community include *Viola calaminaria* (Fig.7.8), *Minuartia verna* var. *caespitosa*, *Armeria maritima* subsp. *elongata*, and *Silene vulgaris* subsp. *humilis*. The last particularly favours collapsed mine shafts or other depressions known locally as *Pingen*. The mixture of colours afforded by the rose-red *Armeria* and yellow *Viola* gives the tone to this community which is known as the *Violetum calaminariae rhenanicum* phytosociological association (Ernst, 1969). In Table 7.2 there is a summary of the European plant associations within the three alliances of the order *Violetalia calaminariae* and Class *Violetea calaminariae*. The reader is referred to Chapter 3 for an explanation of the nomenclature of phytosociology.

Table 7.2 Phytosociological classification of metal-tolerant plants found over Central European Roman and mediaeval mining sites.

Category	Character species	Occurrence
Class - *Violetea calaminariae*	*Silene vulgaris*	
Order - *Violetalia calaminariae*	*Minuartia verna*	
Alliance 1 - *Galio anisophylli-Minuartion vernae*	-	Alpine distribution
Alliance 2 - *Thlaspion-calaminaris*	*Thlaspi calaminare** *Cardaminopsis halleri**	
Association 1 - *Violetum calaminariae rhenanicum*	*Viola calaminaria** (yellow form)	Zinc ores west of the Rhine
Association 2 - *Violetum calaminariae westfalicum*	*Viola calaminaria** (violet form)	Zinc ores east of the Rhine
Alliance 3 - *Armerion halleri*		
Association 1 - *Armerietum halleri halleri*	-	Harz and northern foothills

*Hyperaccumulator of zinc.
Source: Brooks (1987).

The range of Galmei plants is being reduced by industrial development in the region and is now only a fraction of its original extent. There are very few undisturbed sites supporting this community except in eastern Belgium near Le Rocheux. Ernst (1974) has described another Galmei community situated in Westphalia. Here the *Violetum calaminariae rhenanicum* association is replaced by the *Violetum calaminariae westfalicum*. *Viola calaminaria westfalicum* with its reddish violet flowers dominates the community from May to October.

The origin of the Galmei flora is of some interest. The flora is of Quaternary origin and is probably a relic of the Ice Ages, certainly in regard to *Viola calaminaria, Minuartia verna* and *Thlaspi alpestre*. Most of their companions are found elsewhere only in alpine regions.

Ancient copper mines of Daye, China

The Daye region of Hubei Province in central China contains large copper deposits that have been exploited for nearly 3000 years. Unlike many other mining provinces where later mining has destroyed earlier artefacts, the region still retains evidence of early mining activities.

After rainfall, the local mountains are often stained with *tonglü* (verdigris) due to leaching of secondary copper minerals, and are named Tonglüshan (shan = mountain). Early miners were attracted to the copper deposits not only because of the verdigris but also because of the ubiquitous presence of the "copper plant" *Elsholtzia haichowensis* (Fig.7.9) that colonises the phytotoxic soil. A description of this plant has been given by Se Sjue-Tszin and Sjui Ban Lian (1953).

This plant has been analysed by Dr. Xiaoe Yang (pers. comm. 1996) and was found to contain up to 1000 μg/g copper in dry material. It is the only hyperaccumulator of copper ever found outside Zaïre.

Fig.7.9. *Elsholtzia haichowensis* a hyperaccumulator of copper growing over ancient copper mines and slag heaps in the Daye copper province of central China. Photo by D.Ager.

A fascinating description of the ancient Daye mining province and of the "copper flower" as well as of the numerous ancient mining artefacts discovered in the region, has been given by Huangshi Museum *et al.* (1980).

Found on Tonglüshan are no less than 400,000 tonnes of ancient slag, ancient shafts and drifts densely distributed underground, and remains of copper smelting furnaces built in various dynasties, all of which bear witness to the scale and skill of mining and smelting in those days.

By C-14 dating of the mine posts and wooden handles of tools, as well as from pottery styles, it was found that some of the mines belong to the Spring and Autumn Period (770-479 BC) or earlier. Bronze and wooden implements were used in these mines. Iron and steel tools replaced bronze in the pits of the period from the Warring States to the Han Dynasty (5th century BC to 2nd century AD). Smelting furnaces and potteries of later periods have been unearthed, and are mainly of the Song Dynasty (960-1279 AD).

Among the tools found in the ancient drifts prior to the 6th century BC were wooden shovels, rakes and hammers, bronze adzes, chisels and pickaxes. Cast iron was discovered in China by the end of the 6th century BC and in the pits of the Han Dynasty, bronze tools were replaced by iron and included axes, sledgehammers, rakes, hoes, and chisels. With these simple tools, the miners excavated shafts and drifts as deep as 50 m. The excavation depth was between 20 and 30 m in the earlier pits. The shafts and drills were supported by square morticed frames. Some timber props were cut into spears and wedged into the walls to keep them in place. Most of the walls and ceilings of the shafts and drifts were protected by panelling.

Pits of the Warring States and Han Period were generally 40-50 m deep. The shafts were propped up by stacking square frames one on top of another. Instead of panels, fine wooden sticks and bamboo mats were used for lining the walls of shafts and drifts.

The Tonglüshan area of Hubei province remains one of the best preserved sites of ancient mining activities anywhere on earth and is particularly noteworthy in that it carried the faithful copper indicator and hyperaccumulator *Elsholtzia haichowensis*. It is a remarkable illustration of the concept of phytoarchaeology.

Indication of Ancient Trade Routes by a Hyperaccumulator of nickel

A remarkable phytoarchaeological example of how the presence of a nickel hyperaccumulator has evidenced ancient trading routes is to be found on the island of Corsica.

There used to be a stand of *Alyssum corsicum* (Fig.7.10) in the suburbs of the port city of Bastia in the northeast of that island. The site consists of a few hectares of ultramafic (serpentine) soil, a substrate extremely unfavourable for most plants because of its high content of nickel and magnesium and its low nutrient status. These ultramafic soils usually carry a highly specific and unusual flora (Brooks, 1987) characterised by a paucity of species, a high degree of endemism, and the appearance of abnormal and stunted forms.

Alyssum corsicum was discovered early in the 19th century (Duby, 1828) and at that time was thought to be endemic to Corsica. It is endemic to ultramafic rocks and requires this type of substrate for growth and seed germination. However, in the 1950s, D.Huber-Morath discovered that this plant was widespread in western Anatolia, Turkey in an area that had been an important source of grain in antiquity.

It appears that *A.corsicum* was brought to Corsica inadvertently as weed seed in grain shipments from Anatolia. Fortuitously, there was a small outcrop of ultramafic rocks near the port where the grain and its "weeds" fell literally upon "stony ground" so that a small colony of the plants became established and remained almost to this day.

Fig.7.10. Stand of *Alyssum corsicum* on ultramafic soils in the suburbs of Bastia, Corsica.
The colony probably became established when Venetian traders brought the *Alyssum*
as weed seeds in grain imported from Anatolia. Photo by R.D.Reeves.

The traders who brought *Alyssum corsicum* to Corsica have been thought to have been either the Phoenecians or the Venetians. It is true that Phoenecians visited Corsica in the 6th century BC from their strongholds in Sicily and Sardinia, but the only port in Corsica at that time was Alalia (Aleria), some 60 km south of the present site of Bastia. Corsica was at that time controlled by the Greek Phocaeans who were conquered later in the same century by the Etruscans. It is therefore unlikely that the Phoenecians brought *A. corsicum* to Corsica. It is probable that the Venetians brought the grain and its "weeds" from Anatolia because they had important trading links between western Anatolia (a source of grain) and Corsica (a source of timber).

The fate of the only colony of *Alyssum corsicum* in the western Mediterranean now hangs in the balance since developers have already built over this unique site to extend the city (which can be clearly seen in the background of Fig.7.10). Conservationists led by the late Madame Marcelle Conrad of Miomo near Bastia tried hard to halt the march of "progress" but apparently in vain. I was unable to find any specimens of this plant when I last visited the island in 1995.

Hyperaccumulators as Indicators of Precolonial Copper Workings in Zaïre

The copper/cobalt deposits of Zaïre and Zambia are the largest in the world with some 100 outcrops spread over an area of about 25,000 km². One of the best known of the "copper flowers" is *Haumaniastrum robertii*, a member of the mint (Lamiaceae) family, found to the west of the Shaban Copper Arc (see Fig.7.11).

In Zaïre, the plant is found primarily in the region between Kolwezi and Likasi. To the east of the Shaban Copper Arc, near Lubumbashi, this species is replaced by its close relative *H.katangense* which is an indicator of copper and grows exclusively on copper-rich soils (Brooks *et al.*,1980, 1987). It can contain a maximum of 0.20% cobalt and 0.84% copper (Brooks *et al.*,1987). The distribution of *H.katangense* has led to important and exciting discoveries in the field of archaeology (Plaen *et al.*,1982).

Haumaniastrum katangense has escaped from its original habitat and is found on man-made substrates which are lightly or heavily mineralised in copper. Such substrates include sites of furnaces, traditionally used in precolonial days for the production of small copper crosses; sites of exploitation of copper mines during the early decades of colonisation; and the verges of dirt roads which have been dressed with gangue from mining activities (the copper content of this gangue can sometimes approach 3%).

The full distribution of *H.katangense* has been described by Malaisse and Brooks (1982). We will here only discuss its occurrence over sites of ancient precolonial mining and smelting. These sites can be identified by dense carpets of *H.katangense* (and occasionally *H.robertii*) that grow over them. Only at one locality (Luisha) are the two species found together.

During the precolonial period, the 14th century Kabambian culture developed in Zaïre (Maret, 1979). One of the most important artefacts of this culture is copper in the form of *croisettes* (crosses) used originally as ingots and later as currency (Fig.7.12). In the course of the Kabambian, the crosses diminished in size and became more uniform. They gradually evolved into an all-purpose unit of value. These copper crosses, which are well dated, provide a valuable clue for dating other sites in Zaire.

The artisans used copper smelters which were located at rather specialised sites. The furnaces were backed on to abandoned termite hills which thereby provided a gradient suitable for the flow of molten copper. They were surrounded by earth walls about a metre high which must have contaminated the moulds and smelting conduits. The furnaces were situated near dambos (periodically-inundated savanna) or near rivers which provided the necessary water. The sites were also near stands of *Pterocarpus tinctorius* which were used to produce charcoal. A reconstructed smelter is shown in Fig.7.13.

The old sites discovered along the Luano, Ruashi, Kilobelobe, or Kafubu rivers (Fig.7.14) have the same common characteristics: the adjacent open forest has disappeared; there are few species growing on the summits of the termite mounds; *H.katangense* grows in a dense circle along the base of the mounds (Fig.7.15). These sites are sometimes arranged in groups of 30 along 5 km of river bank, as is the case at Ruashi. Each occurrence of *H.katangense* faithfully delineates a subsoil containing an archaeological horizon linked to a smelter.

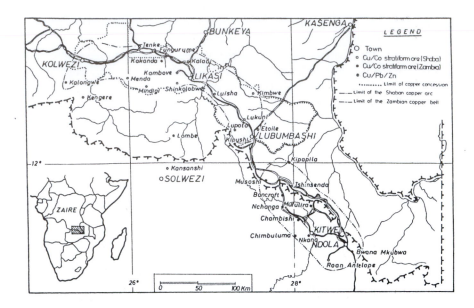

Fig.7.11. Map of Central Africa showing the Shaban Copper Arc and Zambian Copper Belt.

Fig.7.12. Copper "croisettes" used as monetary units by the 14th century Kabambian culture of Zaïre.

Within a radius of 3 or 4 m, this horizon furnishes remains of conduits, moulds and rejected copper crosses, together with abundant smelter waste. This waste is composed of partially-melted malachite, or copper amalgamated with the earth of the retaining walls, as well as scoria and charcoal.

Fig.7.13. Reconstruction of a Kabambian copper smelter formed from an abandoned termite hill in Zaïre.

Typical artefacts from one of the sites (near Kilobelobe) have been described. Beneath an overburden of humic soil not exceeding 10 cm in depth, layers of 5-9 cm thickness contained metallurgical residues. One typical excavation of 2 m² in area and 10-20 cm in depth, revealed 200 g of charcoal and copper slag, 350 g of fragments of burnt clay from furnaces and conduits, 160 g of malachite and 150 g of copper.

A phytosociological study of the Kilobelobe site showed a carpet dominated by *Haumaniastrum katangense* and *Bulbostylis mucronata*, both hyper-accumulators (see Chapter 3). Along the periphery, *Nephrolepis undulata* and *Athraxon quartinianus* were well developed. Several clumps of *Alectra sessiliflora* var. *senegalensis, Antherotoma naudinii* and *Polygala* sp. were observed.

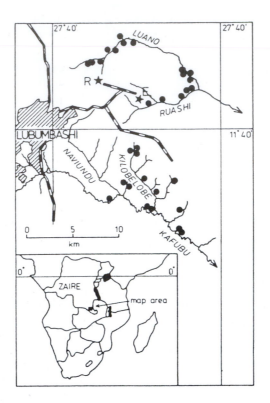

Fig.7.14. Sites of ancient copper smelters near Lubumbashi, Zaïre.
(R = Ruashi, E = Etoile Mine). Source: Plaen *et al.* (1982).

Study of the above stratigraphy and the quantity of waste, showed that activity had been brief and had been confined to a single season. About 200 of these precolonial sites have now been discovered and have led to excavations resulting in the unearthing of many copper crosses ranging in weight from a few grammes up to 36 kg. *Haumaniastrum katangense* does not indicate the presence of ancient smelting activities when the remains are buried beneath a thick non-metalliferous surface layer. It should also be noted that not all archaeological sites have been revealed by the presence of *H.katangense* alone. Nevertheless, without the aid of this "copper flower", the number of discoveries would have been only about one third of those so far recorded.

Biogeochemistry and Ancient Mining Activities

Introduction

Biogeochemical methods of prospecting (see Chapter 4) involve the chemical analysis of plant material to identify mineralisation in the substrate. It is but a short step to the concept of using the same technique to identify ancient mine workings and their archaeological artefacts. Unlike geobotanical methods such as the discovery of smelter sites in Zaïre, biogeochemistry is less useful because the mine sites are very obvious to the naked eye because of the disturbance of the land. There are nevertheless cases where this technique can be useful.

Fig.7.15. View of an abandoned copper smelter in Zaïre. The structure is surrounded by *Haumaniastrum katangense*.

The famous Mansfeld copper deposits of eastern Germany provide a good example of the successful phytoarchaeological use of biogeochemistry. These deposits have been in use for 1000 years and the older workings have now weathered to ground level so that only analysis of soils and vegetation can indicate where the original workings were located (Brooks and Johannes, 1990). As none of the plant species analysed were hyperaccumulators, the topic will not be discussed further.

There have been a few biogeochemical studies of ancient British mine sites as well as on mosses and lichens growing over them. These topics will be discussed below.

Biogeochemical investigations of ancient British mine sites

The lead and zinc content of two typical metal-tolerant plant species growing over ancient and mediaeval mine sites in Britain have been determined by Hajar

(1987). The species are *Minuartia verna* and *Thlaspi alpestre*. The data are shown below in Table 7.3.

Table 7.3. Maximum lead and zinc concentrations (% dry weight) in *Minuartia verna* (MV), *Thlaspi alpestre* (TA), and soils over ancient and mediaeval mining sites in Britain.

Location	Lead			Zinc		
	MV	TA	Soils	MV	TA	Soils
S.Pennines	0.05	0.27	9.50	0.75	2.50	18.10
Tideslow Rake	0.50	-	5.33	0.07	-	0.46
Slaley	0.002	0.009	0.03	0.22	1.90	3.60
Grinton	-	-	-	0.17	1.13	-
Grisedale	0.04	0.06	0.13	0.16	0.84	1.47
Copperthwaite	-	-	-	0.13	1.13	-
Langthwaite	0.004	0.004	0.03	0.08	0.81	0.44
Grassington	0.02	0.008	7.81	0.32	2.32	0.89
Mendips	-	-	0.85	0.08	0.57	4.00
Trelogan	0.16	-	0.05	0.32	-	0.22
Black Rocks	0.26	0.06	8.49	0.36	1.91	1.62
Bonsall Moor	0.07	0.02	1.74	0.05	1.30	2.01
Bradford Dale	0.14	0.02	5.92	0.13	2.11	1.07
Dirtlow Rake	0.10	-	5.55	0.04	-	0.36
Whites Rake	0.17	0.02	2.80	0.16	1.75	0.59
Grattondale	0.02	-	0.77	0.07	-	2.57
Wensleydale	0.01	0.006	0.83	0.11	1.62	1.47
Mean	0.11	0.048	3.20	0.19	1.53	2.59

Source: Brooks and Johannes (1990).

It is clear that both of these species restrict lead uptake, probably at their root systems, and that *T. alpestre* is clearly a hyperaccumulator of zinc. The maximum zinc concentration in *M. verna* was 0.75% (7500 $\mu g/g$) but does not quite qualify this species to have hyperaccumulator status.

Biogeochemical studies of mosses and lichens over mediaeval mine sites

Mosses and lichens are not generally supposed to be hyperaccumulators of heavy metals even though they are frequently associated with the mineralisation that often delineates ancient and mediaeval mine sites. The reason is that it is extraordinarily difficult to separate these low-growing primitive plants from contamination by the soil surface or from wind-borne dust.

There are indeed "copper mosses" (Persson, 1956) such as *Mielichhoferia mielichhoferi* that are virtually confined to sulphide deposits containing copper and other heavy metals.

The western Harz region of Germany is one of the greatest of the European mineral provinces. Operations have been in progress for about 1000 years and only ceased a decade ago. Bode (1928) and Bornhardt (1943) have recorded the presence of 191 mediaeval mine sites, smelters and slag heaps. Mining was carried out principally over the Rammelsberg shown in the background behind the city of Goslar in Fig.7.16.

Rumour has it that the name "Rammelsberg" is derived from a certain Herr Ramm, who when told by his master Otto I of Saxony to kill a boar for the royal

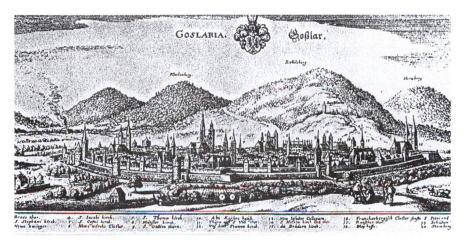

Fig.7.16. 17th century engraving of the city of Goslar (western Harz) showing the Rammelsberg in the background. Engraving by M.Merian courtesy of the Goslar Museum (City Archives of Goslar).

Table 7.4. Mean iron and copper contents (μg/g dry weight) in rocks and lichens growing over mediaeval slag heaps on the Rammelsberg, western Harz, Germany.

Species or substrate	Iron		Copper	
	Slags	Rock	Slags	Rock
Acarospora montana	15,400	1770	89	29
A.sinopica	44,330	62,500	1100	940
A.smaragdula var.	15,100	-	60	-
Cladonia arbuscula	1070	558	80	15
Cornicularia aculeata	424	270	37	25
Lecidea macrocarpa	14,580	-	160	-
Leconora epanora	5832	3164	220	27
Rhizocarpon oederi	5833	-	1670	-
Stereocaulon dactylophyllum	6000	250	80	35
S.vesuvianum	8300	450	100	15
Slag 1	30,000	-	9800	-
Slag 2	34,000	-	5400	-

Source: Lange and Ziegler (1963).

table tethered his horse for a while and when he returned found that the impatient animal had uncovered the soil with his hooves and revealed golden streaks of sulphide ore in the rock beneath.

The vegetation cover of the Rammelsberg is typical of that of sulphide minerals. The plant association is dominated by *Armeria maritima* subsp. *halleri* and includes the well known zinc hyperaccumulator *Cardaminopsis halleri*. Where there is little vegetation cover, there is a community of mosses and lichens known as the *Acarosporetum sinopicae* association (Hilitzer, 1923). The copper and iron content of 10 species of this association are shown below in Table 7.4.

Noeske *et al.* (1970) determined that five lichen species (*Acarospora smaragdula, Lecanora hercynica, Lecidea macrocarpa, Stereocaulon denudatum* and *S. nanodes*) growing over 14th century slag heaps had a high elemental content compared with a control species (*Cornicularia aculeata*) growing over Middle Triassic carbonate rocks. The iron content of these five species ranged from 4875 to 14,409 μg/g compared with 563 μg/g for the control.

References

Anon.(1970) The Mendip lead industry. *Fieldworker* 1, 132-137.

Baker,A.J.M.(1974) Heavy Metal Tolerance and Population Differentiation in *Silene Maritima* With. PhD Thesis, University of London.

Baker,A.J.M.(1987) Metal tolerance. *New Phytologist* 106, 93-111.

Baumann,A. (1885) Das Verhalten von Zinksalzen gegen Pflanzen und im Boden. *Landwirtschaftliche Versuchsstation* 31, 1-53.

Bode,A.(1928) Reste über Hüttenbetriebe im West- und Mittelharz. *Jahrbuch der Deutschen Geographischen Gesellschaft* 1928, 141-197.

Bornhardt,W.(1943) Der Oberharzer Bergbau im Mittelalter. *Archiv der Landes-Volkskunde Niedersachsens* 1943, 449-502.

Boyle,R.W.(1979) The geochemistry of gold and its deposits. *Geological Survey of Canada Bulletin* 280, 580 pp.

Brooks,R.R.(1983) *Biological Methods of Prospecting for Minerals*. Wiley, New York, 322 pp.

Brooks,R.R.(1987) *Serpentine and its Vegetation*. Dioscorides Press, Portland, 454 pp.

Brooks,R.R. and Johannes,D.(1990) *Phytoarchaeology*. Dioscorides Press, Portland, 224 pp.

Brooks,R.R.,Holzbecher,J. and Ryan,D.E.(1981) Horsetails as indirect indicators of gold mines. *Journal of Geochemical Exploration* 16, 21-26.

Brooks,R.R.,Naidu,S.D.,Malaisse,F. and Lee,J.(1987) The elemental content of metallophytes from the copper/cobalt deposits of Central Africa. *Bulletin de la Société Royal Botanique de Belgique* 119, 179-191.

Brooks,R.R.,Reeves,R.D.,Morrison,R.S. and Malaisse,F.(1980) Hyper-accumulation of copper and cobalt: a review. *Bulletin de la Société Royale Botanique de Belgique* 113, 166-172.

Brussell,D.(1978) *Equisetum* stores gold. *Phytologia* 38, 469-473.

Cannon,H.L.,Shacklette,H.T. and Bastron,H.(1968) Metal absorption by *Equisetum* (horsetail). *United States Geological Survey Bulletin 1278A, 1-21.*

Chisnall,K.T. and Markland,J.(1971) Contamination of pasture land by lead. *Journal of the Association of Public Analysts* 9, 116-118.

Clarke,R.K. and Clark,S.C.(1981) Floristic diversity in relation to soil characteristics in a lead mining complex in the Pennines, England. *New Phytologist* 87, 799-815.

Duby,J.E.(1828) *Alyssum corsicum.* In: De Candolle (ed.) *Botanica Gallica* 1, 34.

Ernst,W.(1964) Ökologisch-soziologische Untersuchungen in den Schwermetall-gesellschaften Mitteleuropas unter Einschluss der Alpen. *Abhandlungen des Landesmuseums für Naturkunde* 27, 1-54.

Ernst,W.(1969)*Die Schwermetallvegetation Europas.* Doctoral Thesis, University of Münster, 184 pp.

Ernst,W.(1974) *Schwermetallvegetation der Erde.* Fischer, Stuttgart, 194 pp.

Ernst,W.(1976) Violetia calaminaria. In: R.Tüxen (ed.). *Prodrome of the European Plant Communities V.3.* Cramer, Vaduz,

Ford,T.D. and Rieuwerts,J.H.(1983) *Lead Mining in the Peak District, 3rd edn,* Peak Park Joint Planning Board, Bakewell.

Gussone,R.(1961) Die Blei-Zinkerzlagerstätten der Gegend von Aachen. In: *Mineralogische und Geologische Streifzüge durch die Nördliche Eifel.* VFMG, Heidelberg, pp.19-25.

Hajar,A.S.M.(1987) The Comparative Ecology of Minuartia verna (L.) and Thlaspi alpestre L. in the Southern Pennines with Particular Reference to Heavy Metal Tolerance. PhD Thesis, University of Sheffield.

Hilitzer,A.(1923) Les lichens des roches amphiboliques aux environs de Vseruby. *Cas. Narodny Museum* 1923, 1-14.

Huangshi Museum *et al.*(1980) *Tonglüshan (Mt. Verdigris), Daye - a Pearl among Ancient Mines.* Cultural Relics Publishing House, Beijing.

Kremer,B.(1982) Schwermetallpflanzen und Galmeiflur. *Rheinische Heimatspflege* 19, 34-38.

Lange,O. and Ziegler,H.(1963) Der Schwermetallgehalt von Flechten aus dem Acarosporetum sinopicae auf Erzschlackenhalden des Harzes. *Mitteilungen der Floristischen-Soziologischen Arbeitsgemeinschaft* 10, 156-177.

Malaisse,F. and Brooks,R.R.(1982) Colonisation of modified metalliferous environments in Zaïre by the copper plant *Haumaniastrum katangense. Plant and Soil* 64, 289-293.

Maret,P.de (1979) Luba roots: the first complete Iron Age sequence in Zaïre. *Current Anthropology* 20, 233-234.

Nemec,B.,Babicka,A. and Oborsky,A.(1936) The occurrence of gold in horsetails (in Ger.) *Bulletin Internationale de l'Academie des Sciences de Bohême* 1-7, 1-13.

Noeske,O.,Lauchlii,A.,Lange,O.L.,Vieweg,G.H. and Ziegler,H.(1970)

Konzentration und Lokalisierung von Schwermetallen in Flechten der Ezschlackenhalden des Harzes. *Flechtensymposium 1969 Vorträge des Gesamtgebietes der Botanik* 4, 67-69.

Persson,H.(1956) Studies of the so-called "copper mosses". *Journal of the Hattori Botanical Laboratory* 17, 1-19.

Plaen,G.de,Malaisse,F. and Brooks,R.R.(1982) The copper flowers of Central Africa and their significance for archaeological and mineral prospecting. *Endeavour* 6, 72-77.

Reeves,R.D. and Brooks,R.R.(1983) Hyperaccumulation of lead and zinc by two metallophytes from a mining area in Central Europe. *Environmental Pollution* 31, 277-287.

Savelsbergh,E.(1976) Die Vegetationskundliche Bedeutung und Schutzwürdigkeit des Breinigerberges bei Stolberg unter Berucksichtigung geologischer und geschichtlicher Aspekte. *Göttinger Floristischer Rundbericht* 9, 127-133.

Schwickerath,M.(1931) Das Violetum calaminariae der Zinkböden in der Umgebung Aachens. Eine pflanzensoziologische Studie. *Beitrag der Naturdenkmalpflege.* 14, 463-502.

Se Sjue Tszin and Sjui Ban Lian (1953) *Eschscholtzia haichowensis* Sun - a plant that can reveal the presence of copper-bearing strata. *Dichzi Sjuozheo* 32, 360-368.

Shaw,S.C.(1984) Ecophysiological Studies on Heavy Metal Tolerance in Plants Colonising Tideswell Rake, Derbyshire. PhD Thesis, Sheffield University.

Shimwell,D. and Laurie,A.E.(1972) Lead and zinc contamination of vegetation in the Southern Pennines. *Environmental Pollution* 3, 291-301.

Vogt,T.(1942) Geokjemisk og geobotanisk malmeleting. II *Viscaria alpina* (L.)G.Don som "kisplante". *Kongelige Norske Videnskabers Selskab Forhandlinger* 15, 5-8.

Chapter eight:

Hyperaccumulation as a Plant Defensive Strategy

R.S. Boyd

Department of Botany and Microbiology, and Alabama Agricultural Experiment Station, Auburn University, AL 36849-5407, USA

Why Hyperaccumulate?

Heavy metals have an environmental reputation of being toxic pollutants (Chapter 12). The phenomenon of metal hyperaccumulation raises many questions in the minds of scientists (Baker and Brooks, 1989), not the least of which is what value this behaviour has for a plant that is hyperaccumulating metal in its tissues. It initially seems odd that an organism would accumulate a toxic substance in its tissues. Furthermore, the uptake of metals from the soil by hyperaccumulators does not appear to be a passive phenomenon. Metal contents of hyper-accumulators do not increase linearly as the metal concentration in the substrate increases. This is unlike the behaviour of other types of plant (Baker,1981; Brooks, 1987). Instead, Morrison *et al.* (1980) showed that metal hyperaccumulators will even concentrate metals in their tissues when grown on relatively low-metal soils, indicating an active uptake and sequestration process.

Explanations for the phenomenon of metal hyperaccumulation have been suggested by many authors. Boyd and Martens (1992) summarised the literature in this regard, and placed these explanations into six general categories. These explanations are presented in Table 8.1, along with some of the references that have suggested each explanation. Most of these explanations have a functional basis that includes some positive selective value to the plant. The exception, the inadvertent uptake explanation, suggests that metal intake in hyperaccumulators is a consequence of an enhanced ion uptake system. This uptake system inadvertently results in the concentrating of metal ions in plants growing in metalliferous soils. This explanation is a particular scientific challenge because it is difficult to design experiments that would refute or support it.

However, a similar 'inadvertent' explanation has been advanced to explain the evolution of arsenate tolerance in *Holcus lanatus* (Meharg *et al.*,1993). Arsenate tolerant plants of this species can be found at both high- and low-arsenate sites. Thus, some plants in populations in low-arsenate sites are preadapted to be arsenate-tolerant. Meharg *et al.* (1993) explained this phenomenon as follows: Arsenate and phosphorus share a common high affinity uptake system in this species, and arsenate tolerance stems from suppression of this system. The high affinity uptake system has a low selective value in nutrient-poor habitats because phosphorus uptake in such habitats is diffusion-limited rather than uptake-limited. Therefore, maintenance of the presumably costly high affinity uptake system is selected against in these habitats. As a result, populations in nutrient-poor habitats have suppressed high affinity uptake systems. The inadvertent result was that, when faced with high levels of arsenate in the soil, plants in nutrient-poor habitats take up little arsenate. These plants are therefore preadapted to tolerate arsenate and thus can survive in habitats with high levels of arsenate (Meharg *et al.*,1993). Boyd and Martens (1998) have found that populations of the nickel hyper-accumulator *Thlaspi montanum* var. *montanum* growing on non-metalliferous soils possess nickel hyperaccumulation ability. They suggested that one possible explanation for this ability might be that the mechanisms that underlie nickel hyperaccumulation may have other physiological functions (as with phosphorus uptake in *Holcus*) that are as yet unexplored. Obviously, further research will be needed to address this 'inadvertent uptake' hypothesis with regard to metal hyperaccumulation.

Table 8.1. Explanations advanced for metal (mainly nickel) hyperaccumulation in plants.

Explanation	Papers suggesting explanation
Inadvertent uptake	Baker and Walker (1989), Severne and Brooks (1972)
Metal tolerance (sequestration)	Antonovics *et al.* (1971), Baker (1981,1987) Kruckeberg *et al.* (1993)
Disposal from plant body	Baker (1981), Ernst (1972), Farago and Cole (1988), Wild (1978)
Drought resistance	Baker and Walker (1989), Robertson (1992), Severne (1974)
Interference with other plants	Baker and Brooks (1989), Gabrielli *et al.* (1991)
Pathogen/herbivore defence	Ernst (1987), Ernst *et al.* (1990), Reeves *et al.* (1981)

The other explanations for metal hyperaccumulation in Table 8.1 have selective benefits accruing to the plant. Postulated benefits include enhanced tolerance of metal (tolerance hypothesis), in which the metal is envisioned as being sequestered in physiologically inert areas of the plant body. This would then allow a plant's critical biological machinery to operate in relative isolation from the metals. Another benefit may be the removal of relatively toxic metals from the plant body. It has been suggested that plants can rid themselves of metals by placing them in tissues that can be abscised from the plant (disposal hypothesis). Some authors have speculated that hyperaccumulated nickel may allow plants to better withstand drought stress (drought resistance hypothesis), although the mechanism for this function is relatively unclear.

Two hypotheses in Table 8.1 involve interactions between hyperaccumulators and other organisms in their environment. One, the interference hypothesis, suggests that metal hyperaccumulators enrich the surface soil beneath them with metals when they shed high-metal plant parts (leaves, etc.). Decomposition of this plant material releases metals into the surface layer of the soil. The elevated surface soil metal content then inhibits growth of other (less metal-tolerant) species and provides a competitive advantage to the hyperaccumulator. This hypothesis has been termed *elemental allelopathy* by Boyd and Martens (unpub.) because of the parallels between it and classical allelopathy (e.g. Rice, 1974). The difference is that classical allelopathy depends upon secondary chemicals produced by a plant's biochemical machinery, whereas elemental allelopathy uses element cycling to produce the allelopathic effect. The other organismal interaction hypothesis is the defence hypothesis. This hypothesis suggests that elevated tissue levels of metals function to defend plants from herbivores and/or pathogens. This hypothesis follows from the general toxicity of many metals and the high concentrations of metals present in hyperaccumulator plants.

It should be noted that metal hyperaccumulation may well have more than one function in a given metal hyperaccumulator species. Thus the explanations offered in Table 8.1 should not be considered as mutually exclusive. In fact, given the apparent multiple evolution of metal hyperaccumulation in unrelated groups of plants (Brooks, 1987), metal hyperaccumulation may fulfil multiple functions both within and between evolutionary lines of metal hyperaccumulators. This may be especially true considering the differing properties of the elements hyper-accumulated by plants. It is quite conceivable that these metals will differ in their functions. For example, one element (e.g. nickel) may have a defensive role whereas another (e.g. zinc) might be involved in producing increased drought tolerance. It will take a concerted effort by a number of researchers before these questions are answered satisfactorily.

In their review of these ecological explanations for metal hyperaccumulation, Boyd and Martens (1992) pointed out that there was little experimental evidence to support any of these explanations for metal hyperaccumulation in plants. This is still true today. However, recent efforts have been made to explore the defence hypothesis for metal hyperaccumulation. As a result, it is currently the best-explored explanation and is supported by several lines of evidence. This chapter summarises this research and presents an updated view of the evidence regarding the defence hypothesis. It also points to some of the exciting questions that remain to be answered regarding the ecological function(s) of metal hyperaccumulation in plants.

Metal Toxicity

The biological basis for the defence hypothesis of metal hyperaccumulation is the general toxicity of metals in relatively high doses to organisms. Metals have been used by humans as poisons for some time. For example, the use of copper in Bordeaux mixture (invented in 1885) to prevent fungal attack upon grapevines (Agrios, 1988), symbolises their ability to disrupt biological systems. It has even been suggested that nickel poisoning, stemming from the Cretaceous impact of a nickel-containing asteroid, was ultimately responsible for the extinction of dinosaurs (Beard, 1990). Borovik (1990) points out that metals can wreak biological havoc in several ways. These include displacing essential metal ions from biomolecules, modifying the conformation of active biomolecules such as enzymes, blocking essential functional groups, and disrupting biomolecule integrity. These effects are due to the attraction of metal ions to oxygen, nitrogen, and sulphur atoms: atoms that play important roles in the function of biological systems (Borovik, 1990).

An emphasis on dosage when discussing metal toxicity is important, as several metals are required by plants in at least some small quantity. These metals: Zn, Mg, Mn, and Cu, are generally accepted as essential plant nutrients (Marschner, 1995) and therefore are required for normal plant growth. One of the more interesting aspects of the hyperaccumulation of nickel, the element hyperaccumulated by the greatest number of plant species (see Chapter 1, Table 1.1), was that nickel was not an essential nutrient. Thus, its uptake and hyperaccumulation by plants was even more remarkable. However, there is some evidence that nickel now should be added to the list of essential plant micronutrients (Brown et al.,1987, Dalton et al.,1988). Co, Cd, Se, and Tl are other metals that are hyperaccumulated by at least a few plant species (see Chapter 1, Table 1.1). Currently, none of these is considered an essential plant nutrient (Marschner, 1995).

Focus on Defence

A current list of studies pertinent to the defence hypothesis is summarised in Table 8.2. Most of this research has involved nickel hyperaccumulators, but some zinc and copper hyperaccumulators have been examined as well. The earliest data regarding a possible defensive function of metals in hyperaccumulators were provided by Ernst (1987). He observed adjacent European populations of *Silene vulgaris* (=*Silene cucubalus*). This species can accumulate large quantities of copper (up to 1400 μg/g dry weight [dw] in its leaf tissues). Ernst (1987) observed populations growing on copper-rich or chalk grassland soils, and noted that the seed capsules of the chalk grassland population were often destroyed by moth larvae. In contrast, the plants growing on high-copper soils did not suffer attacks by this moth. Surprisingly, copper concentrations of the capsules and seeds were similar.

Table 8.2. Summary of experimental evidence regarding the plant defence hypothesis of metal hyperaccumulation. Organisms are characterised by their ecological relationship with the hyperaccumulator species, the metal hyperaccumulated, and whether the plant was judged to be successfully defended against attack.

Plant species	Herbivore/pathogen	Metal/ Effective?	Reference
	FOLIVORES		
Silene vulgaris (S.cucubalus)	Hadena cucubalis (Lepidoptera:Noctuidae)	Cu/yes	Ernst (1987)
Streptanthus polygaloides	Pieris rapae (Lepidoptera:Pieridae)	Ni/yes	Martens & Boyd (1994)
S.polygaloides	Euchloe hyantis hyantis (Lepidoptera:Pieridae)	Ni/yes	Martens & Boyd (1994)
S.polygaloides	Unknown grasshopper (Orthoptera:Acrididae)	Ni/yes	Martens & Boyd (1994)
Thlaspi montanum var. montanum	Pieris rapae (Lepidoptera:Pieridae)	Ni/yes	Boyd & Martens (1994)
S.polygaloides	Unknown vertebrate	Ni/no	Martens & Boyd (unpub)
Cardaminopsis halleri	Plutella maculipennis (Lepidoptera:Plutellidae)	Zn/no	Ernst et al. (1990)
Thlaspi caerulescens	Schistocerca gregaria (Orthoptera:Acrididae)	Zn/yes	Pollard & Baker (1997)
T.caerulescens	Deroceras carvanae (Pulmonata:Limacidae)	Zn/yes	Pollard & Baker (1997)
T.caerulescens	Pieris brassicae (Lepidoptera:Pieridae)	Zn/yes	Pollard & Baker (1997)
	PHLOEM PARASITES		
C.halleri	Brachycaudis lychnidis (Homoptera:Aphidae)	Zn/no	Ernst et al. (1990)
S.polygaloides	Acyrthosiphon pisum (Homoptera:Aphidae)	Ni/no	Boyd et al. (unpub.)
S.polygaloides	Cuscuta californica var. breviflora	Ni/no	Boyd et al. (1994)
	PATHOGENIC FUNGUS		
S.polygaloides	Erisyphe polygoni (powdery mildew)	Ni/yes	Boyd et al. (1994)
	PATHOGENIC BACTERIUM		
S.polygaloides	Xanthomonas campestris pv. campestris (bacterium)	Ni/yes	Boyd et al. (1994)
	NECROTROPHIC FUNGUS		
S.polygaloides	Alternaria brassicicola (Imperfect fungus)	Ni/yes	Boyd et al. (1994)

When Ernst (1987) transferred caterpillars to plants on metalliferous soils, they died. However, the explanation was not a simple one: the capsules of those plants were not sufficiently large to support the growth of caterpillars to pupation. Thus, after consuming the available seed capsules, the caterpillars were forced to eat the leaves of the plants, which did have greatly elevated copper contents relative to the leaves of chalk grassland plants. It was consumption of the leaves that poisoned the larvae and prevented the insect from effectively attacking plants growing on metalliferous soil.

The first explicit effort to test the defence hypothesis were experiments by Martens and Boyd (1994) and Boyd and Martens (1994), using the North American nickel hyperaccumulators *Streptanthus polygaloides* and *Thlaspi montanum* var. *montanum* (Table 8.2). Their initial work focused on insect folivores, and involved force-feeding experiments using leaf material harvested from plants growing on greenhouse soils that were either enriched with nickel or not. Response of the insects to the resulting high- and low-metal tissue demonstrated acute toxicity in almost every case, and all cases showed decreased growth in insects fed high-nickel leaves. They thus concluded that high-nickel leaves defended plants against herbivores by killing them once an effective dose was ingested.

These studies documented a very great difference in nickel content between leaves of plants grown on high- and low-nickel soils. However, leaves were shown to vary in other parameters (e.g. zinc content, moisture content), so that the role of nickel in the toxic response was not absolutely clear cut. To focus on the potential role of nickel as the operational toxin, they used artificial insect diet amended with nickel. This would allow them to demonstrate that nickel concentrations in the range found in nickel hyperaccumulating plants caused acute toxicity in the experimental insects. Using larvae of the folivore *Pieris rapae*, Boyd and Martens (1994) and Martens and Boyd (1994) documented decreased survival of larvae when diet nickel concentrations were >1000 $\mu g/g$. They therefore concluded that the nickel concentration of high-nickel leaves was alone adequate to explain the toxic effect of the high-nickel leaves.

The most recent demonstration of the defensive function of hyperaccumulated metals is work by Pollard and Baker (1997) with the zinc hyperaccumulator *Thlaspi caerulescens* (Table 8.2). In experiments in which folivores were allowed to choose between high- and low-zinc plants, three species of organisms (a locust, a slug, and a caterpillar) all preferred to consume low-zinc leaves. Toxicity of high-zinc leaves was not demonstrated, but these experiments did allow Pollard and Baker (1997) to conclude that zinc hyperaccumulation provided a selective advantage to hyperaccumulating plants.

The studies discussed above from Table 8.2 have demonstrated a defensive effect of hyperaccumulated metal, but this table summarises other studies in which no defensive effect was detected. These studies show that high levels of plant metals may not deter some organisms from damaging plants. This is, at least in part, due to the types of ecological interactions that have been studied. For example, most of the studies in Table 8.2 that deal with folivores have demonstrated that metals do limit damage. These studies either have documented greater mortality for insects eating high-metal leaves (Boyd and Martens, 1994; Martens and Boyd, 1994) or, in choice experiments, have showed lessened herbivory on high-metal leaves (Pollard and Baker, 1997). The reason these studies have shown metals to be effective is probably two-fold: first, many of the folivores used were bioassay herbivores. These animals were able to feed and grow on low-metal leaves, but were not native to habitats where the hyperaccumulators were found. They were used as experimental subjects due to their availability for laboratory-based research. For example, Boyd and Martens (1994) used larvae of the widespread introduced butterfly *Pieris rapae*, and Pollard and Baker (1997) used locusts, slugs, and *Pieris brassicae* larvae (Table 8.2). The only exception was the study of Martens and Boyd (1994), which used *Pieris rapae* as a 'bioassay' herbivore but also included two field-collected herbivores (a grasshopper species and caterpillars of a butterfly species). In those cases, the field-collected herbivores both were negatively affected when forced to consume high-metal leaves (Table 8.2). The second reason that studies of folivores have demonstrated a defensive effect of metal is that the folivores used were not selective feeders. They consumed entire leaves and hence received the entire dose of metal available from those leaves.

Selective tissue feeders may be able to avoid receiving the entire dose of metal contained in hyperaccumulator leaves. For example, there have been three studies dealing with phloem parasites (Table 8.2). Two of these studies involved aphids (Ernst *et al.*, 1990; Boyd *et al.*, unpub.), whereas the third examined a parasitic flowering plant (Boyd and Martens, unpub.). All showed that high metal levels did not prevent plants from being attacked. Two of these studies (Boyd and Martens, unpub. and Ernst *et al.*, 1990) documented attack in the field by naturally-occurring parasites, whereas the third (Boyd *et al.*, unpub.) was a manipulative greenhouse study using an aphid species not native to the hyperaccumulator's habitat. A few other studies found in Table 8.2 also have documented successful attack of some organisms on hyperaccumulators. These will be discussed in some detail below, in describing circumvention of metal-based defences. It is important to realise that these studies do not refute the defensive role of hyperaccumulated metals. Plant defences may be breached by some organisms, but that does not obviate their status as defences (e.g. Belovsky and Schmitz, 1994). Instead, failures of a defence simply help to describe the limitations of that defence.

A defence is most effective if it deters attack rather than acting on an attacking organism after damage has occurred. Martens and Boyd (1994) tested the oviposition preference of female *Pieris rapae* and found no difference in egg load between high- and low-metal plants. Thus, high-metal plants were not spared the arrival of caterpillars. However, several weeks after eggs hatched, high-metal plants were significantly less damaged by larvae. The few studies that have tested feeding preferences of folivores (e.g. Martens and Boyd, 1994 with larvae of *Pieris rapae*, and Pollard and Baker, 1997 with a slug and a locust species) generally have found that herbivores do damage high-metal leaves. They simply damage them to a lesser extent than low-metal leaves. An intriguing exception was reported by Pollard and Baker (1997) with larvae of *Pieris brassicae* offered high- and low-zinc leaves of *Thlaspi caerulescens* (Table 8.2). In this case, low-zinc leaves were partially consumed (25-90% eaten), but high-zinc leaves showed no evidence of feeding, even when examined with a microscope. It would be interesting to know what cue prevented larvae from feeding on the high-zinc leaves.

The studies listed in Table 8.2 also vary in the experimental approach used. Some are simply field observations indicating whether a hyperaccumulator was successfully defended from attack (Ernst *et al.*,1990; Boyd and Martens, unpub.; Boyd *et al.*, unpub.). Others have taken a more rigorous experimental approach aimed at isolating the impact of metals. These experiments have taken advantage of the fact that metal hyperaccumulating plant species can be raised on both metalliferous and non-metalliferous substrata in a greenhouse setting (e.g. Boyd and Martens, 1994; Martens and Boyd, 1994; Boyd *et al.*,1994; Pollard and Baker, 1997). The resultant plants either hyperaccumulate metal or have much lower concentrations. Experiments comparing mortality rates or ingestion rates of herbivores/pathogens attacking these plants can then more reliably isolate the effect of metal content.

Laboratory-based experiments should be supplemented with field investigations, but little field evidence currently is available. One of the first field-based reports regarding the defence hypothesis was from a survey of herbivore damage to leaves from plants growing on a serpentine site in the Philippines (Proctor *et al.*,1989). Damage to leaves of the nickel-hyperaccumulator species *Shorea tenuiramulosa* was not markedly lower than that documented for leaves of other co-occurring plants, and the authors concluded that this evidence did not support the defence hypothesis. Such surveys are of limited value as tests of the defence hypothesis, however, because of the difficulty of interpreting differences in amounts of leaf damage between plant species. Similarly, observations that certain insects can successfully attack hyperaccumulators (e.g. Ernst *et al.*,1990) do not necessarily demonstrate that a defensive toll is not being taken upon the attacking organism.

To date, little experimental fieldwork to test the defence hypothesis has been done. Perhaps the first experiment regarding the defence hypothesis was the transfer of caterpillars between low-metal and copper-hyperaccumulating plants in the field performed by Ernst (1987). Although there was no control group, all the caterpillars placed upon high-copper plants died. The only other field experiment is that of Martens and Boyd (unpub.) using the nickel hyperaccumulator *Streptanthus polygaloides* in California. They used high- and low-metal plants, and caging and herbicide treatments, in an attempt to determine the defensive effectiveness of nickel as well as determine what groups of herbivores had the greatest impacts on the plants. Initial observations showed that low-metal plants were damaged by insects more than high-metal plants. However, by the end of the experiment, there was no difference in total damage to high-

and low-metal plants. This was due to an unknown (presumably) vertebrate herbivore that was excluded by cages, but was capable of eating large amounts of plant material. This animal fed on plants unprotected by cages regardless of plant nickel content. The activities of this animal swamped out the relatively minor defensive effects of metals against insect herbivores that had been observed in the early part of this experiment.

The experiment of Martens and Boyd (unpub.) illustrates a potential problem of field-based tests of the defence hypothesis. Basically, failure of an experiment to demonstrate a defensive function of metals has two interpretations. One is that metals in plants do not have a defensive function. The second is that the experiment was not conducted at the appropriate ecological scale to detect a genuinely-extant defensive function. Defences may be important only against certain organisms at particular sites or points in time. Field experiments must therefore rely on detailed knowledge of the natural history of the organisms involved, along with an appropriate experimental design and a certain amount of fortuitous timing, to accurately reflect the biological situation. Despite this level of difficulty, field-based tests are sorely needed to extend our understanding of the ecological functions of metal hyperaccumulation.

A final consideration regarding the defensive function of hyperaccumulated metals is the exact location of those metals at the tissue level. An early study of the western Australian nickel hyperaccumulator *Hybanthus floribundus* (Farago and Cole, 1988) showed high levels of nickel in leaf epidermal cells. Several recent studies of other species show that the highest concentrations of metals are in outer layers of plant leaves and roots. Vázquez *et al.* (1994) studied the zinc hyperaccumulator *Thlaspi caerulescens*, and found that most zinc was located in epidermis and subepidermal parenchyma cells of both leaves and roots. Mesjasz-Przybylowicz *et al.* (1996) examined nickel distribution in leaves of the African nickel hyperaccumulators *Berkheya zeyheri* and *Senecio coronatus*. They found the highest nickel concentrations in epidermis, and adjacent parenchyma tissue also had high nickel levels in some cases. Localisation of metals at the periphery of the plant body is consistent with a defensive function, as outer tissues must be attacked first by herbivores/pathogens as they seek access to deeper layers of plant cells.

An important question that has rarely been addressed is the threshold concentration of a metal that might create an effective defence. There are two issues pertinent to this question: minimum metal content to create a negative effect on an attacking organism and the role of synergism between defences of differing types in creating a harmful outcome. It appears that the minimal metal concentrations that cause acute toxicity may be relatively high. Experiments with artificial diets amended with nickel showed mortality of *Pieris rapae* caterpillars to be increased by >1000 $\mu g/g$ nickel (Martens and Boyd, 1994; Boyd and Martens, 1994). Culture of the plant pathogenic bacterium *Xanthomonas campestris* pv. *campestris* showed marked inhibition of growth at a lower nickel concentration (400 $\mu g/g$). Artificial diet experiments with an aphid species showed no demonstrable increase in mortality until the nickel content reached 2500 μg/g (Boyd et al., unpub.).

However, defences need not be lethal to have a beneficial effect for the plant. For herbivores, reductions in larval growth rate or decreases in adult size and fecundity may be adequately harmful to be advantageous from the plant perspective. In this light, threshold concentrations of metals are probably lower than those required for acute toxicity. The interesting possibility here is that metal concentrations found in the more numerous metal-accumulating serpentine plant species (Brooks, 1987) also may have a defensive role. Unfortunately, experimental investigation to answer this question is lacking. Previous work (e.g. Martens and Boyd, 1994; Boyd and Martens, 1994) has focused upon acute toxicity rather than sublethal effects of metals.

The second issue, that of synergy of defences, may be particularly important regarding sublethal quantities of defending metals. Many plants, including hyperaccumulators, have secondary defensive compounds. If there is synergy between elemental and secondary defences, relatively low doses of each may result in a level of toxicity beyond additive expectations. The main implication of this line of reasoning is that the defensive role of metals, already documented for some hyperaccumulators, may be extended to plants other than hyperaccumulators. Many plants growing on metalliferous soils have elevated metal contents (Brooks, 1987). It may be that their tissue metal levels, although lower than those of hyperaccumulators, still afford them some protection against attack by their enemies.

Elemental Defence: the Advantages of Using Metals

Plants defend themselves against attack in a number of ways, and chemical defences are one common approach used by plants (Berenbaum, 1995). Chemical defences may be divided into secondary chemicals, which are organic compounds manufactured by a plant's biochemical machinery, and elements taken up from the soil (elemental defences). Martens and Boyd (1994) proposed that elemental defences may have some advantages over secondary chemical defences. First, elements cannot be degraded by counterdefences of herbivores. Many herbivores have circumvented secondary chemical defences of plants by using enzymes (e.g. mixed function oxidases) that modify a defensive compound and make it less toxic or more easily excreted (Howe and Westley, 1988). Elements, because of their elemental nature, cannot be modified by herbivores in a similar manner. Second, elemental defences may be relatively inexpensive in metabolic terms. There are no synthetic costs associated with elements, which are extracted from the soil solution. The primary metabolic costs associated with elemental defences would likely be those associated with the construction of the complexing molecules involved in transport and sequestration of metal ions. There is some evidence that these complexing molecules are relatively small molecular weight molecules, such as organic acids (Lee et al.,1978; Brooks et al.,1981) or amino acids (Krämer et *al.*,1996). Because of the small sizes of these molecules, their construction may incur a relatively small metabolic cost to the plant. Third, elemental defences may be relatively novel defences in plants. Certainly, the percentage of metal hyperaccumulators among the Earth's flora is exceedingly small: only a few hundred species known to date (see Chapter 1) out of hundreds of thousands of species. Even on metalliferous sites, often only one or two hyperaccumulator species might be present in a particular habitat (Brooks, 1987). The novelty of an elemental defence increases its defensive value, as it is less likely that herbivore/pathogen species will have faced the defence previously and will have had an opportunity to evolve a counterdefence.

Finally, it should be noted that plant elemental defences have an advantage as a model system for studies of plant defence. Plant metal content can be manipulated by controlling the metal content of the growth medium. Thus, experiments can compare the response of attacking organisms to plants that are either defended or not, but otherwise are similar. While it has been noted (e.g. Martens and Boyd, 1994) that high- and low-metal plants are not identical in all respects excepting metal content, this approach does allow an investigator to manipulate metal content in an experimental setting. In combination with studies that use artificial diet or growth-media to further isolate the effects of a metal on an herbivore/pathogen, studies of elemental defences have a high degree of scientific rigour. This degree of scientific rigour is similar to that of studies that use defence-deficient mutants to study other types of plant defences (e.g. Ausubel *et al.*,1995).

Circumventing Metal-based Defences

Plants defend themselves against attack in many ways, including chemical defences. However, every plant defence known can be at least partially circumvented (Grubb, 1992). There is no reason to believe that elemental defences are an exception to this principle. There are at least three ways that herbivores/pathogens can circumvent metal-based defences: 1) through avoidance of the metal, 2) by dilution of the metal in the diet, and 3) by possession of physiological metal tolerance.

Metal avoidance

The few studies that have examined the variability of metal contents of tissues/organs within hyperaccumulators usually have shown great variability. For example, the nickel content of tissues of the nickel hyperaccumulator *Psychotria douarrei* varied from 2100 μg/g in trunk wood to 92,500 μg/g in leaves (Jaffré and Schmid, 1974). Many other studies of metal contents in hyperaccumulators (e.g. Ernst *et al.*,1990 for zinc, Jaffré, 1977 for manganese, Ernst *et al.*,1990 for copper, Reeves *et al.*,1981 for nickel) have shown that plant tissues can vary markedly in metal content within a single species. The influence of other factors (plant age, leaf age, etc.) is relatively unknown. A recent study by Boyd and Jaffré (unpub.) of *Psychotria douarrei* demonstrated insignificant variation between plants due to age, but great variation in nickel content between young and old leaves. Thus, young leaves were less well-defended by nickel than older ones. Studies of the nickel-hyperaccumulating annual *Streptanthus polygaloides* showed metal contents averaged about 5300 μg/g before flowering but decreased to only 2200 μg/g by late flowering (Nicks and Chambers, 1995). Variation of metal content allows herbivores or pathogens to avoid the high-metal tissues by limiting their attack to those low in metal.

Some of the studies reported in Table 8.2 suggest that the phloem of hyperaccumulators is not well-defended by metals. Boyd *et al.* (unpub.) found that an aphid species, a pest they encountered as they tried to raise hyperaccumulator plants under greenhouse conditions, was unaffected by feeding on nickel-hyperaccumulating plants of *Streptanthus polygaloides*. Boyd and Martens (unpub.) also documented the phloem parasite *Cuscuta* growing on *S. polygaloides* in the field. Ernst *et al.* (1990) reported that an aphid (*Brachycaudis lychnis*) fed on high-zinc plants of *Silene vulgaris* but apparently was not negatively affected.

Diet dilution

A second way to circumvent an elemental defence is to mix high- and low-metal foods so as to achieve a non-toxic dose. This strategy is probably exclusively restricted to herbivores due to their ability to forage selectively. Diet dilution may occur when a taxonomic specialist consumes both high- and low- tissues of a single hyperaccumulator species, but is more likely to be found involving a generalist herbivore that grazes on both hyperaccumulator and non-hyperaccumulator species. By mixing high- and low-metal foods, the total metal load of the diet can be brought below the threshold for negative effects.

Metal tolerance (leading to sequestration)

The third mechanism that could allow an herbivore/pathogen to attack metal-defended tissue is for the herbivore/pathogen to be genuinely metal-tolerant. Here, 'metal tolerance' means the ability to consume relatively high-metal food with no apparent ill effect. This could be due to physiological mechanisms that limit the availability of metal from the food, mechanisms that allow metal uptake from the food but subsequently immobilise or sequester it, or possession of metal-indifferent enzymes etc. that allow the herbivore/pathogen metal content to be high without triggering toxicity. Work with arthropods from metal-polluted areas indicates that metal tolerance can be evolved (Posthuma and Straalen, 1993; Dallinger, 1993). However, studies of insects associated with hyperaccumulator plant species are very few. The report by Ernst *et al.* (1990) of caterpillars of the moth species *Plutella maculipennis* feeding on high-zinc leaves of *Cardaminopsis halleri* is relatively unique in this regard.

The ultimate outcome of metal tolerance on the part of an herbivore/pathogen might be for that organism to, in turn, use metals in its own defence. For example, a metal-tolerant herbivore could evolve to sequester metal in its tissues. This metal might function to deter attack by the herbivore's predators. This would then become a tritrophic level interaction, similar to that found for some secondary plant chemicals (Price *et al.*,1980).

Two cases documenting elevated metal contents of herbivores or pathogens that are associated with metal hyperaccumulators are currently known. One is the report of Ernst *et al.* (1990) of an aphid found on zinc-hyperaccumulating *Silene vulgaris*. Analysis of the aphids revealed 9000 $\mu g/g$ zinc in their bodies. Whether these high-metal aphids were protected from their predators is unknown. The only example that can be cited at this time regarding metals as an antipredator defence deals with a marine worm that accumulates copper. It has been reported that the high-copper body parts of this worm were distasteful to certain fish (Gibbs *et al.*,1981).

The second case of elevated herbivore/pathogen metal content, in this case in the tissues of a parasitic plant, is the report of Boyd and Martens (unpub.) of *Cuscuta* attacking nickel-hyperaccumulating *Streptanthus polygaloides* (Table 8.2). Analysis of the *Cuscuta* stems revealed 800 $\mu g/g$ nickel. Other *Cuscuta* samples taken from the same field site but which were growing upon a non-hyperaccumulator species contained only 11 $\mu g/g$ nickel. Apparently, *Cuscuta* attacking *Streptanthus* was obtaining nickel from its host and incorporating it into its stems. Whether this in turn protected the *Cuscuta* from herbivore attack remains to be discovered.

The evolution of both hyperaccumulation in plants and resistance to elemental defences by herbivores/pathogens requires that there be heritable variation in these traits as the raw material on which natural selection can work. Very little is known about the heritability of hyperaccumulation. Pollard and Baker (1996) documented at least some heritable variation of zinc-hyperaccumulation ability in *Thlaspi caerulescens*. Martens and Boyd (1994) showed that nickel hyperaccumulation was of great selective value when a mixed array of hyperaccumulating and non-hyperaccumulating *Streptanthus polygaloides* plants was subjected to attack by folivorous caterpillars.

Some information also exists regarding metal tolerance in animals (Klerks, 1990; Hopkin, 1989; Heliovaara and Vaisanen, 1993), fungi (Brown and Hall, 1990) and bacteria (Schlegel *et al.*,1991, 1992). Because metal tolerance has been documented in these cases, it is logical to predict that metal-tolerant herbivores/pathogens have also evolved when faced with metal-based plant defences in the field. However, besides the *Cuscuta* example (described above) and our as-yet unfinished research regarding herbivores from serpentines of California and New Caledonia (described below), we know of no information yet regarding the metal contents of herbivores/pathogens collected from metal hyperaccumulators.

Implications for Phytoextraction/Phytomining

One of the promising uses of metal hyperaccumulating plants is to clean metal-contaminated soils or to mine metals from soils via plants (see Chapters 11-15). It would be useful to consider the implications of the points raised in this chapter on these activities. First, it is likely that hyperaccumulating plants will be (initially) relatively free of pathogens and herbivores. The elemental defences of these plants will probably be relatively new phenomena for the herbivores and pathogens native to phytoextraction sites and thus they probably will lack adequate metal tolerance for the dosages contained in hyperaccumulator plant tissue. This assumes that such sites are polluted sites used for phytoremediation, rather than naturally-occurring metalliferous sites that are being used for phytomining.

Phytomining sites may contain naturally-resistant native herbivores and pathogens that essentially might be preadapted to be able to attack a hyperaccumulating crop plant. However, generalist herbivores also may be able to cause damage to a hyperaccumulator crop via diet dilution. Similarly, other herbivores that avoid the metal by specialising on relatively undefended tissue (e.g. aphids feeding on phloem) may also be able to successfully attack a hyperaccumulating plant species. Second, it seems likely that tolerance may evolve in local populations of some herbivores/pathogens, so that new problems will eventually arise even if initial plantings show little herbivore/pathogen damage.

A third potential problem is mobilisation and transport of metal off-site. Phytoremediation of metal-contaminated soils requires that metal be mobilised from the soil and placed into above-ground plant tissues. The mobility of herbivores implies that some amount of metal may be transported in the bodies or waste materials of herbivores to other sites in the vicinity of the remediation site.

The extent of this transport depends on many factors, including the amount of plant biomass involved, the nature (size, mobility, consumption rate) of the herbivore, etc. The gravity of this phenomenon may also depend upon whether the herbivore may later be targeted for human consumption (e.g. deer, quail), in which case the metal in an herbivore's body may directly enter human diets.

A final problem deals with the potential toxicity of metal hyperaccumulators to beneficial non-target organisms. This might involve folivores that are attracted to the remediation site. Consumption of hyperaccumulator foliage could cause illness or mortality, depending on the susceptibility of the animal and the dose received. Metals may also work their way up food chains. A potential danger here is bio-magnification, the increasing concentration of metals in higher trophic levels in an ecosystem. It is unclear at this time to what extent metal biomagnification occurs (Straalen and Ernst, 1991; Laskowski, 1991).

Toxicity to non-target organisms may also occur if pollinators or seed dispersers were to ingest pollen, nectar, or fruits of hyperaccumulating plants. Of course, some hyperaccumulator species may rely on abiotic vectors of pollen and seeds and thus will not attract animals to their flowers or fruits. However, a number of hyperaccumulators have showy flowers that seem likely to be visited by animals (see Reeves *et al.*,1995). Very little is known of the metal contents of pollen, nectar, and fruits of hyperaccumulators. Whole flowers of some hyperaccumulators have been analysed. For example, flowers of the nickel hyperaccumulators *Psychotria douarrei* (Jaffré and Schmid, 1974), *Streptanthus polygaloides* (Reeves *et al.*,1981) and *Stackhousia tryonii* (Batianoff *et al.*,1990) had 24,000, up to 16,400 μg/g, and 8400 μg/g nickel, respectively. But the metal content of pollen and nectar of hyperaccumulating species has not been investigated.

Similarly, entire fruits and seeds of some hyperaccumulators have been analysed, and these show high levels of metals. For example, *Streptanthus polygaloides* fruits can contain 5230 μg/g nickel (Reeves *et al.*,1981), and fruits of *Psychotria douarrei* (Jaffré and Schmid, 1974) and *Sebertia acuminata* (Jaffré *et al.*,1976) from New Caledonia were reported to contain 19,000-28,000 μg/g nickel and 3000 μg/g nickel, respectively. The dose of metal that a seed-dispersing animal might acquire if it were to consume hyperaccumulator fruits is unknown, but a relatively large dose is clearly a possibility.

Future Directions

It is clear that the unusual elemental make-up of metal hyperaccumulators is likely to have many ecological ramifications, yet the pertinent questions are just being articulated and there is very little actual research upon which answers can be formulated. Areas deserving of additional research effort are discussed below.

Herbivores and pathogens of hyperaccumulators

The need for field-based studies regarding the defensive hypothesis of metal hyperaccumulation is great. It would be particularly interesting to study the selective value of metal hyperaccumulation by using hyperaccumulating and non-hyperaccumulating individuals of a species in its native habitat.

Surveys for resistant herbivores and pathogens are needed, and will likely turn up metal-accumulating specialists. Metal tolerant invertebrates are known from metal-polluted sites (Heliovaara and Vaisanen, 1993). Our own research on the insect fauna associated with *Streptanthus polygaloides*, a nickel hyperaccumulator from California, has shown at least one insect containing up to 600 μg/g nickel.

We are conducting a similar study in New Caledonia at a site with six co-occurring nickel-hyperaccumulator species. Once metal-accumulating animal species are found, many questions regarding the physiological bases for metal tolerance in herbivores can be addressed. As mentioned previously in this chapter, we have also discovered that a vascular plant parasite of *S.polygaloides*, *Cuscuta* (see Table 8.2), has elevated nickel in its tissues.

Sublethal effects of metal at non-hyperaccumulation levels

The defensive effects of metals may exist at levels below the threshold for defining hyperaccumulation. This might mean that the defensive aspects of metals pertain to plants other than hyperaccumulators. Further, the demonstration of sublethal defensive effects may indicate a stepwise path of evolution of metal hyperaccumulation. Plants that were able to accumulate metals had to evolve metal tolerance and may have reaped a defensive benefit as well. The further evolution of hyperaccumulation ability may have increased the defensive benefit while relying on previously-evolved tolerance mechanisms to prevent metal toxicity.

Interaction of elemental defences with plant secondary chemicals

Many plant taxa that contain metal hyperaccumulators are known to produce other defensive compounds. For example, glucosinolates are common defensive chemicals in the Brassicaceae (Howe and Westley, 1988), a family containing hyperaccumulators of Ni, Cd, Tl, and Zn (Chapter 1, Table 1.1). One of the benefits of metal hyperaccumulation may be that it replaces a defence based upon a metabolically-expensive secondary compound. On the other hand, an elemental defence may simply be an addition to the defensive arsenal of these plants. If so, the defensive combination of metal and secondary compounds may have synergistic effects. This might mean that relatively low levels of metals may have a defensive function. Also, there may be a trade-off between metal-based and secondary chemical defences. For example, some hyperaccumulators may produce lower levels of secondary compounds when metals are available for use as a defensive strategy.

Variability: levels and scales

Much effort has been expended to identify metal hyperaccumulating plant species. Once identified, the variability of metal hyperaccumulation needs to be addressed at important ecological scales. Differences in the degree of hyperaccumulation between populations, among individuals within a population and between plant parts and tissues, can be ecologically important in determining the relationship between herbivores/pathogens and hyperaccumulating plants. Time is yet another scale that is currently poorly explored, yet the change in metal contents of plants and plant parts through time may be ecologically important. Finally, the consequence of herbivore/pathogen attack on metal levels in hyperaccumulators is in need of study. Some plant defences are induced by attack (e.g. Smith, 1996). Could the degree of metal hyperaccumulation in plants be influenced by the timing or severity of attack? Increased hyperaccumulation in response to plant damage is suggested by the recent results of Varennes et al. (1996). Their study, using the nickel hyperaccumulator *Alyssum pintodasilvae*, compared nickel contents of two successive harvests of these perennial plants. The second harvest contained 3- to 5-fold more nickel (depending on the nickel level of the soil used). The possible importance of the magnitude of this response to metal-sensitive herbivore/pathogens is apparent.

Hyperaccumulation: genetics and heritability

Pollard and Baker (1996) point out that an understanding of the evolution of hyperaccumulation must depend upon knowledge of the heritability of this trait. Unfortunately, very little is known about the genetic basis for hyperaccumulation or the heritability of hyperaccumulation. Initial work done by Pollard and Baker (1996) demonstrated a degree of heritability for the ability to hyperaccumulate zinc in *Thlaspi caerulescens*. Certainly, if genetic variability in degree of hyperaccumulation can be demonstrated, subjecting mixed plantings of hyperaccumulating and non-hyperaccumulating plants to herbivore pressure may be able to mimic the forces that have selected for metal hyperaccumulation in nature.

Mode of action of metals

There is some information regarding the impact of metals on invertebrates (Hopkin, 1989; Heliovaara and Vaisanen, 1993). The physiological effects of metals on herbivores/pathogens need to be elucidated in the context of the types and quantities of metal found in hyperaccumulators. This will help both to understand the toxicity of some metals and to discover how circumvention of elemental defences is achieved.

Acknowledgement: This chapter is to appear in the Alabama Agricultural Experiment Station Journal No.6-975831.

References

Agrios,G.N.(1988) *Plant Pathology*. Academic Press, London, 803 pp.

Antonovics,J.,Bradshaw,A.D. and Turner,R.G.(1971) Heavy metal tolerance in plants. *Advances in Ecological Research* 7, 1-85.

Ausubel,F.M.,Katagiri,F.,Mindrinos,M. and Glazebrook,J.(1995) Use of *Arabidopsis thaliana* defense-related mutants to dissect the plant response to pathogens. *Proceedings of the National Academy of Sciences USA* 92, 4189-4196.

Baker,A.J.M.(1981) Accumulators and excluders - strategies in the response of plants to heavy metals. *Journal of Plant Nutrition* 3, 643-654.

Baker,A.J.M.(1987) Metal tolerance. In: Rorison,I.H.,Grime,J.P.,Hunt,R., Hendry,G.A.F. and Lewis,D.H. (eds), Frontiers of Comparative Plant

Ecology. *New Phytologist* 106 *(Supplement)*. pp.93-111.

Baker,A.J.M. and Brooks,R.R.(1989) Terrestrial higher plants which hyper-accumulate metallic elements - A review of their distribution, ecology and phytochemistry. *Biorecovery* 1, 81-126.

Baker,A.J.M. and Walker,P.L.(1989) Ecophysiology of metal uptake by tolerant plants. In: Shaw,A.J.(ed.), *Heavy Metal Tolerance in Plants: Evolutionary Aspects*. CRC Press, Boca Raton, pp. 155-177.

Batianoff,G.N.,Reeves,R.D. and Specht,R.L.(1990) *Stackhousia tryonii* Bailey: a nickel-accumulating serpentinite-endemic species of Central Queensland. *Australian Journal of Botany* 38, 121-130.

Beard,J.(1990) Did nickel poisoning finish off the dinosaurs? *New Scientist* 126, 31.

Belovsky,G.E. and Schmitz,O.J.(1994) Plant defenses and optimal foraging by mammalian herbivores. *Journal of Mammalogy* 75, 816-832.

Berenbaum,M.R.(1995) The chemistry of defense: theory and practice. In: Eisner,T. and Meinwald,J.(eds), *Chemical Ecology: the Chemistry of Biotic Interaction*. National Academy Press: Washington, D.C. pp.1-16.

Borovik,A.S.(1990) Characterization of metal ions in biological systems. In: Shaw,A.J.(ed.), *Heavy Metal Tolerance in Plants: Evolutionary Aspects*. CRC Press: Boca Raton. pp. 3-5

Boyd,R.S. and Martens,S.N.(1992) The raison d'être for metal hyper-accumulation by plants. In: Baker,A.J.M.,Proctor,J. and Reeves,R.D.(eds), *The Ecology of Ultramafic (Serpentine) Soils*. Intercept, Andover, pp.279-289.

Boyd,R.S. and Martens,S.N.(1994) Nickel hyperaccumulated by *Thlaspi montanum* var. *montanum* is acutely toxic to an insect herbivore. *Oikos* 70, 21-25.

Boyd,R.S. and Martens,S.N.(1998). Nickel hyper-accumulation by *Thlaspi montanum* var. *montanum*: a constitutive trait. *American Journal of Botany* (in press).

Boyd,R.S.,Shaw,J. and Martens,S.N.(1994) Nickel hyperaccumulation defends *Streptanthus polygaloides* (Brassicaceae) against pathogens. *American Journal of Botany* 81, 294-300.

Brooks,R.R.(1987) *Serpentine and its Vegetation: a Multi-disciplinary Approach*. Dioscorides Press, Portland, 454 pp.

Brooks,R.R.,Shaw,S. and Marfil,A.A.(1981) The chemical form and physio-logical function of nickel in some Iberian *Alyssum* species. *Physiologia Plantarum* 51, 167-170.

Brown,M.T. and Hall,I.R.(1990) Metal tolerance in fungi. In: Shaw,A.J.(ed.), *Heavy Metal Tolerance in Plants: Evolutionary Aspects*. CRC Press, Boca Raton, pp.95-104.

Brown,P.H.,Welch,R.M. and Cary,E.E.(1987) Nickel: a micro-nutrient essential for higher plants. *Plant Physiology* 85, 801-803.

Dallinger,R.(1993) Strategies of metal detoxification in terrestrial invertebrates. In: Dallinger,R. and Rainbow,P.S.(eds), *Ecotoxicology of Metals in*

Invertebrates. Lewis Publishers, London, pp.245-289.

Dalton,D.A.,Russell,S.A. and Evans,H.J.(1988) Nickel as a micronutrient element in plants. *BioFactors* 1, 11-16.

Ernst,W.H.O.(1972) Ecophysiological studies on heavy metal plants in South Central Africa. *Kirkia* 8, 125-145.

Ernst,W.H.O.(1987) Population differentiation in grassland vegetation. In: Van Andel,J.,Bakker,J.P. and Snaydon,R.W.(eds), *Disturbance in Grasslands. Causes, Effects and Processes*. W.Junk, Dordrecht, pp.213-228.

Ernst,W.H.O.,Schat,H. and Verkleij,J.A.C.(1990) Evolutionary biology of metal resistance in *Silene vulgaris. Evolutionary Trends in Plants* 4, 45-51.

Farago,M.E. and Cole,M.M.(1988) Nickel and plants. In: Sigel,H. and Sigel,A.(eds), *Metal Ions in Biological Systems, Vol. 23, Nickel and its Role in Biology*. New York, Marcel Dekker, pp.47-90.

Gabrielli,R.,Mattioni,C. and Vergnano,O.(1991) Accumulation mechanisms and heavy metal tolerance of a nickel hyper-accumulator. *Journal of Plant Nutrition* 14, 1067-1080.

Gibbs,P.E.,Bryan,G.W. and Ryan,K.P.(1981) Copper accumulation by the polychaete *Melinna palmata*: an antipredation mechanism? *Journal of the Marine Biological Association UK* 61, 707-722.

Grubb,P.J.(1992) A positive distrust in simplicity - lessons from plant defences and from competition among plants and among animals. *Journal of Ecology* 80, 585-610.

Heliovaara,K. and Vaisanen,R. (eds)(1993) *Insects and Pollution*. CRC Press, Boca Raton, 393 pp.

Hopkin,S.P.(1989) *Ecophysiology of Metals in Terrestrial Invertebrates*. Elsevier Applied Science, London.

Howe,H.F. and Westley,L.C.(1988) Ecological Relationships of Plants and Animals. Oxford University Press, New York.

Jaffré,T.(1977) Accumulation du manganèse par des espèces associées aux terrains ultrabasiques de Nouvelle-Calédonie. *Comptes Rendus de l'Academie des Sciences (Paris) Série D* 284, 1573-1575.

Jaffré,T.,Brooks,R.R.,Lee,J.,and Reeves,R.D.(1976) *Sebertia acuminata*: a hyperaccumulator of nickel from New Caledonia. *Science* 193, 579-580.

Jaffré,T. and Schmid,M.(1974) Accumulation du nickel par une Rubiacée de Nouvelle-Calédonie, *Psychotria douarrei* (G.Beauvisage) Däniker. *Comptes Rendus de l'Academie des Sciences (Paris) Série D* 278, 1727-1730.

Klerks,P.L.(1990) Adaptation to metals in animals. In: Shaw,A.J.(ed.), *Heavy Metal Tolerance in Plants: Evolutionary Aspects*. CRC Press, Boca Raton, pp.313-321.

Krämer,U., Cotter-Howells,J.D.,Charnock,J.M.,Baker,A.J.M. and Smith, J.A.C.(1996) Free histidine as a metal chelator in plants that accumulate nickel. *Nature* 379, 635-638.

Kruckeberg,A.R.,Peterson,P.J. and Y. Samiullah,Y.(1993) Hyper-accumulation of nickel by *Arenaria rubella* (Caryophyllaceae) from Washington State.

Madroño 40, 25-30.

Laskowski,R.(1991) Are the top carnivores endangered by heavy metal magnification? *Oikos* 60, 387-390.

Lee,J.,Reeves,R.D.,Brooks,R.R. and Jaffré,T.(1978) The relation between nickel and citric acid in some nickel-accumulating plants. *Phytochemistry* 17, 1033-1035.

Marschner,H.(1995) *Mineral Nutrition of Higher Plants 2nd edn*, Academic Press, London, 889 pp.

Martens,S.N. and Boyd,R.S.(1994) The ecological significance of nickel hyperaccumulation: a plant chemical defense. *Oecologia* 98, 379-384.

Meharg,A.A.,Cumbes,Q.J. and M.R. MacNair.(1993) Pre-adaptation of Yorkshire Fog, *Holcus lanatus* L. (Poaceae) to arsenate tolerance. *Evolution* 47, 313-316.

Mesjasz-Przybylowicz,J.,Balkwill,K.,Przybylowicz,W.J., Annegarn,H.J. and Rama,D.B.K.(1996) Similarity of nickel distribution in leaf tissue of two distantly related hyperaccumulating species. In: van der Maesen,L.J.G. et al.(eds), *The Biodiversity of African Plants*. Kluwer Academic Publishers, Amsterdam, pp.331-335.

Morrison,R.S.,Brooks,R.R. and Reeves,R.D.(1980) Nickel uptake by *Alyssum* species. *Plant Science Letters* 17, 451-457.

Nicks,L.J. and Chambers,M.F.(1995) Farming for metals? *Mining Environmental Management* 3, 15-18.

Pollard,A.J. and Baker,A.J.M.(1996) Quantitative genetics of zinc hyper-accumulation in *Thlaspi caerulescens*. *New Phytologist* 132, 113-118.

Pollard,A.J. and Baker,A.J.M.(1997) Deterrence of herbivory by zinc hyper-accumulation in *Thlaspi caerulescens* (Brassicaceae). *New Phytologist (in press)*.

Posthuma,L. and Straalen,N.M.van(1993) Heavy-metal adaptation in terrestrial invertebrates: a review of occurrence, genetics, physiology and ecological consequences. *Comparative Biochemistry and Physiology* 106C, 11-38.

Price,P.W.,Bouton,C.E,Gross,P.,McPheron,B.A.,Thompson,J.N. and Weis,A.E.(1980) Interactions among three trophic levels: Influence of plants on interactions between insect herbivores and natural enemies. *Annual Reviews of Ecology and Systematics* 11, 41-65.

Proctor,J.,Phillips,C.,Duff,G.K.,Heaney,A. and Robertson,F.M.(1989) Ecological studies on Gunang Silam,a small ultrabasic mountain in Sabah, Malaysia. II. Some forest processes. *Journal of Ecology* 77, 317-331.

Reeves,R.D.,Brooks,R.R. and MacFarlane,R.M.(1981) Nickel uptake by Californian *Streptanthus* and *Caulanthus* with particular reference to the hyperaccumulator *S.polygaloides* Gray (Brassicaceae). *American Journal of Botany* 68, 708-712.

Reeves,R.D.,Baker,A.J.M. and Brooks,R.R.(1995) Abnormal accumulation of trace metals by plants. *Mining Environmental Management* 3, 4-8.

Rice,E.L.(1974) *Allelopathy*. Academic Press, New York, 422 pp.

Robertson,A.I.(1992) The relation of nickel toxicity to certain physiological aspects of serpentine ecology: some facts and a new hypothesis. In: Baker,A.J.M.,Proctor,J. and Reeves,R.D.(eds), *The Vegetation of Ultramafic (Serpentine) Soils*. Intercept, Andover, pp.331-336.

Schlegel,H.G.,Cosson,J.-P. and Baker,A.J.M.(1991) Nickel-hyperaccumulating plants provide a niche for nickel-resistant bacteria. *Botanica Acta* 104, 18-25.

Schlegel,H.G.,Meyer,M.,Schmidt,T. Stoppel,R.D. and Pickhardt,M.(1992) A community of nickel-resistant bacteria under nickel-hyperaccumulating plants. In: Baker,A.J.M.,Proctor,J. and Reeves,R.D.(eds), The Vegetation of Ultramafic (Serpentine) Soils. Intercept, Andover, pp.305-317.

Severne,B.C.(1974) Nickel hyperaccumulation by *Hybanthus floribundus*. *Nature* 248, 807-808.

Severne,B.C. and Brooks,R.R.(1972) A nickel-accumulating plant from Western Australia. *Planta* 103, 91-94.

Smith,C.J.(1996) Accumulation of phytoalexins: defence mechanism and stimulus response system. *New Phytologist* 132, 1-45.

Straalen,N.M.van and Ernst,W.H.O.(1991) Metal biomagnification may endanger species in critical pathways. *Oikos* 62, 255-256.

Varennes,A.de,Torres,M.O.,Coutinho,J.F.,Rocha,M.M.G.S. and Neto, M.M.P.M.(1996) Effects of heavy metals on the growth and mineral composition of a nickel hyperaccumulator. *Journal of Plant Nutrition* 19, 669-676.

Vázquez,M.D.,Poschenrieder,Ch.,Barcelo,J.,Baker,A.J.M.,Hatton,P. and Cope,G.H.(1994) Compartmentation of zinc in roots and leaves of the zinc hyperaccumulator *Thlaspi caerulescens* J&C Presl. *Botanica Acta* 107, 243-250.

Wild,H.(1978) The vegetation of heavy metal and other toxic soils. In: Werger,M.J.A.(ed.) *Biogeography and Ecology of Southern Africa*. Junk, Hague, pp.1301-1332.

Chapter nine:

Aquatic Phytoremediation by Accumulator Plants

R.R. Brooks and B.H. Robinson
Department of Soil Science, Massey University, Palmerston North, New Zealand

Introduction

Phytoremediation of metal-contaminated waters is readily achieved by use of aquatic and terrestrial plants because there is no problem of bioavailability of the target element. In terrestrial systems involving hyperaccumulator plants growing over polluted soils, the plant must first solubilise the target element in the rhizosphere and then have the ability to transport it to the aerial tissues. No such problem exists when the plant grows naturally, or is made to grow in, or on, an aqueous medium. It is not surprising therefore that some quite spectacular hyperaccumulations of heavy metals from aqueous systems can be achieved, and indeed this has been known for at least 50 years.

In the treatment of the subject of aquatic phytoremediation systems, there are two natural divisions that involve: 1 - purely aquatic plants such as the floating water hyacinth (*Eichhornia crassipes*); 2 - submersion of the rhizosphere of terrestrial plants in order to remove metal pollutants (*rhizofiltration*). Each of these two subjects will be discussed independently below.

There is a vast literature on the subject of trace elements in marine plants (mainly seaweeds) but these are described separately in Chapter 5.

Hyperaccumulation by Aquatic Plants

Introduction

A useful review of hyperaccumulation of elements by aquatic plants (Outridge and Noller, 1991) is the basis of much of the material in this first section of this chapter. Freshwater vascular plants (FVPs) when combined with macroscopic algae are known collectively as *macrophytes*. Their ability to concentrate elements from the aquatic environment was first reviewed by Hutchinson (1975) who reported that levels of potentially toxic elements such as cadmium, lead and mercury were at least an order of magnitude higher than in the supporting aqueous medium.

Building on the review of Hutchinson (1975), Outridge and Noller (1991) sought to establish the following facts:

1 - the "normal" concentrations of selected elements in natural fresh water systems;

2 - "normal" levels of elements in macrophytes growing in uncontaminated ecosystems;

3 - the median and maximum concentrations of elements in macrophytes growing in contaminated waters;

4 - pathways and rates of elemental uptake and excretion;

5 - environmental factors that control uptake of elements by these plants;

6 - the importance of these plants in terms of biogeochemistry or entry of trace elements into food chains;

7 - elements that are the most toxic to FVPs and how this compares with their toxicity to algae;

8 - the significance of trace element uptake for the field of wastewater treatment and biomonitoring of pollution.

In practical terms, the last topic is perhaps of greatest interest to scientists engaged in the theory and practice of bioremediation of aqueous systems and will be given some prominence in this chapter. It may be appropriate at this stage to define two of the terms used.

Freshwater vascular plants comprise mainly angiosperms (flowering plants) with a few fern species. They have a vascular system (xylem and phloem) as well as well defined roots. They exclude macroscopic algae.

Heavy metals is a term often used to describe some trace elements. It is an imprecise term that can include quite "light" elements such as copper. It has been suggested by Nieboer and Richardson (1980) that heavy metals are often associated with toxicity to biota, although some "heavy metals" such as manganese and uranium are not known to be particularly toxic to life forms.

Life forms include:

1 - rooted emergents that are rooted in bottom sediments;

2 - free-floating emergents such as water hyacinth and duckweed that are not rooted in bottom sediments;

3 - rooted submergents with leaves and flowers under water;

4 - free-floating submergents;

5 - rooted floating-leaved plants such as water lilies;

6 - free-floating floating-leaved plants.

Heavy metals in natural uncontaminated fresh water systems

The elemental composition of uncontaminated fresh waters has been calculated by Turekian (1969) and is shown in the second column of Table 9.1. The values are of course rather imprecise, partly because of analytical error involved in estimations in the ng/g (ppb) range and partly because of the problem of extrapolating to the entire fresh hydrosphere from a relatively small number of rivers.

Table 9.1. Elemental concentrations (μg/mL [waters] and μg/g dry weight for freshwater vascular plants - FVPs).

Element	A	B	C	D	E	C/A
Ag	0.003	0.06	0.15	0.12	67	500
As	0.002	0.20	2.7	1.4	1200	1350
Cd	0.0002	0.64	1.0	1.4	90	5000
Co	0.0002	0.48	0.32	0.37	350	1600
Cr	0.001	0.23	4.0	2.8	65	4000
Cu	0.007	14	7.9	42	190	1128
Hg	0.0001	0.015	0.50	0.58	1000	5000
Mn	0.007	630	370	430	8370	52,857
Mo	0.001	0.90	12	-	-	12,000
Ni	0.0003	2.7	4.2	6.1	290	14,000
Pb	0.003	2.7	6.1	27	1200	2033
Se	0.0002	0.2	1.0	0.30	21	5000
U	0.00004	0.04	0.50	0.05	1.1	12,500
V	0.0009	1,6	3.6	-	-	4000
Zn	0.02	100	52	47	7030	2600

A - uncontaminated river waters (Turekian, 1969), B - terrestrial plants (Bowen, 1966), C - median values for uncontaminated FVPs, D - median values for contaminated FVPs, E - maximum value for contaminated FVPs, C/A - accumulation factor for FVPs growing in uncontaminated waters. After: Outridge and Noller (1991).

It will be seen from Table 9.1 that in natural uncontaminated fresh waters, elemental abundances are extremely low and in most cases are in the ng/mL (μg/L) range. Under these conditions, the FVPs appear to be invariably hyperaccumulators of the heavy metals even in the uncontaminated environment.

Heavy metals in freshwater vascular plants

Several patterns were established by Outridge and Noller (1991) using median values for the data shown in Table 9.1 and elsewhere in their paper.

1 - manganese was the most strongly absorbed element followed in descending order by zinc, molybdenum, copper and lead;

2 - FVPs contained higher concentrations of Ag, As, Cd, Cr, Hg, Ni, Pb, Se and U than did terrestrial plants, whereas Co, Cu, Mn and V were similar in both groups.

3 - concentrations of As, Cd, Co, Cr and U were generally higher in submerged plants than in other FVP life forms;

4 - roots usually contained higher concentrations of heavy metals than above-

ground plant parts;

5 - maximum elemental abundances in FVPs were usually one or two orders of magnitude higher than naturally occurring levels, though the median concentrations were not usually very different;

A further observation needs to be made on the findings of Outridge and Noller (1991). The generally higher levels of elemental concentrations in roots would seem to indicate that these heavy metals are absorbed from the sediments rather than the waters. However, the sediments are in most cases derived from the settling of particulates from the waters and there must be some degree of constant proportionality between the two phases.

Environmental and physiological factors affecting elemental toxicity and uptake by FVPs

Outridge and Noller (1991) have shown that heavy metal uptake and retention by FVPs and macrophytes in general, are controlled by four main factors:

1 - sediment geochemistry;

2 - water physicochemistry;

3 - plant physiology;

4 - genotypic differences.

The first two control metal speciation in sediments and waters, whereas the last two control the ability of plants to accumulate plant-available forms of the metals.

The question of speciation is very important in deciding to what extent a given elemental species will be taken up by macrophytes or be toxic to them. It is less understood that any given elemental species will have a whole range of different bioavailabilities because of physiological differences with respect to uptake sites and uptake mechanisms (Leppard, 1983).

The bioavailability of elemental species to FVPs are less well understood than in the case of other aquatic biota such as fish. However, it is known that FVPs usually accumulate heavy metals by absorption followed by passive or active transport across membranes (Forstner and Wittman, 1981; Smies, 1983).

In their summary of their benchmark paper, Outridge and Noller (1991) showed that physicochemical factors that increase metal solubility (such as lake acidification) also increase elemental concentrations in FVPs presumably because of greater solubility of metals in the sediments.

Aquo metal and soluble metal chelate complexes were the most plant-available forms, whereas metals in reducing sediments were almost completely unavailable to plants, perhaps because of the great insolubility of most metal sulphides.

Elemental levels in most macrophytes exhibited a typical seasonal pattern of a spring maximum followed by a steady decline during the summer. This was ascribed to environmentally driven physiological changes including translocation between above-ground and below-ground tissues. Iron root plaque displayed an important role in availability and uptake of elements from reduced sediments.

Macrophytes play an important role in sediment-water trace element cycling through root uptake, excretion into water and detrital absorption of metals that may be conveyed to the sediments. Elemental toxicity experiments have shown that Ag, As, Cr, Cu, Hg and Ni are about 10 times more toxic to macrophytes than are lead or zinc. Copper is one of the most toxic of these elements and its effects are visible at concentrations of 0.05-0.15 μg/mL. Compared with phytoplankton, macrophytes are 10-100 times less sensitive to most elements except copper to which they are equally sensitive.

Practical applications of FVPs for water purity monitoring and decontamination

Water purity monitoring

Freshwater vascular plants can be used for biomonitoring of contaminated waters and possess several innate advantages over algae:

1 - they have longer life cycles;

2 - they have a higher degree of tolerance to most elements including those heavy metals largely responsible for pollution;;

3 - their biomass is much larger than in the case of algae, so that a larger sample is available for chemical analysis.

Phillips (1977) has proposed the following essential requirements for a suitable FVP biomonitor:

1 - the species must be endemic to the area;

2 - it should be easily cultivated or abundant in the field;

3 - it should concentrate the elements to above the threshold of the limits of detection of the analytical method that is to be employed;

4 - there must be a statistically significant relationship between the abundance

Table 9.2. Freshwater vascular plants proposed for biomonitoring of trace elements in waters.

Species	As	Cd	Co	Cu	Hg	Mn	Ni	Pb	Zn
Callitriche platycarpa				+					
Ceratophyllum demersum		+						+	
Eichhornia crassipes	+	+			+			+	
Elodea canadensis	+		+	+			+		
E.nutallii		+	+					+	+
Equisetum arvense				+					+
E.fluviatale								+	+
Myriophyllum exalbescens									+
M.verticillatum	+		+	+		+			
Nuphar lutea				+					
Potamogeton perfoliatus						+			+
P.richardsonii			+	+				+	+

Sources: various, summarised by Outridge and Noller (1991).

of a given target element in the plant and its concentration level in the surrounding water.

Several FVPs have been suggested as suitable for biomonitoring of trace elements in waters. These are summarised in Table 9.2 above. Some will also be discussed later in this chapter under the individual species.

As pointed out by Outridge and Noller (1991), the species in Table 9.2 were sometimes selected on the basis of their very high accumulation of trace elements from the surrounding waters. This criterion is not necessarily valid as it is far more important that there be a statistically significant relationship between the plant-water system of elemental abundances.

There is not usually a good correlation between elemental abundances in sediments and those in rooted aquatic plants. Indeed Campbell *et al.* (1985) found fewer than 30 significant correlations in the 100 cases that they investigated.

For monitoring purposes, it seems logical that free-floating plants should be used in place of rooted species since such plants derive their nutrients solely from the water column. Sprenger and McIntosh (1989) found a significant relationship for the free-floating *Utricularia purpurea*.

Decontamination of polluted waters

The use of FVPs for removal of pollutants has been established for over 20 years following the pioneering work of Wolverton (1975), Wolverton and McDonald (1975a, 1975b) and Wolverton *et al.* (1975), several species, notably the water hyacinth (*Eichhornia crassipes*), have been proposed for this purpose and will be discussed further below.

There are two main ways in which FVPs may be used for remediation of polluted waters. The first of these involves monospecific pond cultures of free-floating plants such as water hyacinth. The plants accumulate the pollutants until a steady state of equilibrium is achieved. They are then harvested by removal from the pond.

There are various problems inherent in the above method. The first is associated with how to dispose of the waste matter. One solution was achieved at a sewage treatment plant where the toxic waste was used to generate methane gas. Further problems associated with the "free floating" procedures include the presence of unwanted pathogens that may destroy the whole of a monospecific culture and there is also the problem of continual harvesting that requires specialised equipment.

The second methodology employing FVPs for pollutant removal involves growing rooted emergents in trickling bed filters. Uptake of trace elements in these systems is usually caused by rhizosphere microbes with relatively little uptake caused by the plants themselves. The rhizosphere methods do not necessarily have to employ FVPs and are discussed below in a separate section.

There will now be a discussion of selected FVP species that have significant promise for phytoremediation of polluted waters.

The water hyacinth (*Eichhornia crassipes*)

The water hyacinth is perhaps one of the most commonly cited species for phytoremediation of polluted waters (Gupta, 1980; McDonald and Wolverton, 1980). The plant has a rapid growth rate and can hyperaccumulate nutrients (Cornwell *et al.*,1977) as well as heavy metals (Wolverton, 1975; Wolverton *et al.*,1975).

The water hyacinth has a number of problems that tend to hinder its commercial use. The first of these is that in many countries it is a noxious weed that has choked large areas of waterways. For example in Sudan it has completely covered a large area of the Nile River with a dense mat known as the *sudd*. A further disadvantage of this plant is that it will only grow in tropical or warm-temperate parts of the world that are frost-free in winter.

As pointed out by Kay *et al.* (1984), much of the research on the water hyacinth has been faulty insofar as estimates of biomass production have been carried out in the field in unpolluted waters, whereas uptake experiments are usually carried out in the laboratory where there is concomitant, but unmeasured, reduction in biomass.

There is little doubt that there is a rapid rate of removal of heavy metals following absorption by the water hyacinth. This is demonstrated in Table 9.3 that shows the rate of removal of six heavy metals from a system in which the biomass production was 600 kg/ha/day.

Table 9.3. Rate of removal of heavy metals from the aqueous phase by use of the water hyacinth (*Eichhornia crassipes*).

Element	mg/g of dry biomass/day	g/ha/day
Cadmium	0.67	400
Cobalt	0.57	340
Lead	0.18	90
Mercury	0.15	110
Nickel	0.50	300
Silver	0.44	260

After: Outridge and Noller (1991).

The effect of lead, copper and cadmium on metal uptake and growth of the water hyacinth was studied by Kay *et al.* (1984). Although lead had no effect on plant growth, they found that cadmium and copper were toxic to this plant and that the thresholds were respectively 0.5 and 1.0-2.0 μg/mL in the ambient water. Beyond the threshold concentrations the effects were chlorosis, suppressed development of new roots, and greatly reduced growth rates.

Muramoto and Oki (1983) studied the uptake of cadmium, lead and mercury by the water hyacinth. Their data are summarised in Table 9.4. The absorption factors appear to be quite low compared with the average of several thousand shown in Table 9.1. However, it must be remembered that the data were on a fresh-weight basis and that dry weights of aquatic plants are usually about 5 % of the fresh weight. This would give concentration factors of the order of 1000 for the tops of plants growing in cadmium and lead, and about 5000 for the roots.

It can also be seen that increasing the heavy metal content of the aqueous medium actually has the effect of significant reduction of concentration factors owing to the toxicity of the metals concerned.

When the laboratory data were extrapolated to field conditions in Japan, Muramoto and Oki (1983) calculated (Table 9.4) the maximum removal of heavy metals from heavily contaminated waters. The findings are perhaps somewhat overoptimistic because we are again confronted here with the problems of extrapolating from the laboratory to the field where the biomass reported, was based on cultivation under metal-free conditions.

Table 9.4. Fresh-weight concentration factors (plant/water) for water hyacinth grown in different concentrations of cadmium, lead and mercury.

Treatment		Tops	Roots	Max. removal*
Cadmium	1.0 μg/mL	44	101	6.6
	4.0 μg/mL	47	47	17.7
	8.0 μg/mL	35	29	5.6
Lead	1.0 μg/mL	42	296	47.5
	4.0 μg/mL	19	220	336
	8.0 μg/mL	20	127	627
Mercury	0.5 μg/mL	0.044	438	3.5
	1.0 μg/mL	0.032	488	105
	2.0 μg/mL	0.048	340	7.5

*These values are in g/m^2 of whole plants under field conditions assuming a biomass yield of 45 kg (fresh weight)/m^2. After: Muramoto and Oki (1983).

Before leaving the discussion of water hyacinth, some mention should be made of a proposal to use this plant to recover gold from tailings (Anon, 1976). The Gold Hill Mesa Corp. of Colorado Springs undertook a survey in which gold tailings were to be washed free of cyanide and stored in settling ponds to which the *Eichhornia* was to be added. The initial vat leaching procedure removes only 60 % of the gold and it was hoped that some of the remainder could be removed by the water hyacinth and later recovered by burning the plant material. No information is currently available about the success or otherwise of the operation.

Ceratophyllum demersum, Egeria densa *and* Lagarosiphon major

The above plants are extremely common in temperate waterways and are mentioned here because of their extraordinary accumulation of arsenic derived from geothermal activity in New Zealand. The unusual accumulation of arsenic by FVPs in this area was first reported by Reay (1972) who found 650 μg/g (dry weight) in *C.demersum*. Later, Liddle (1982) reported arsenic concentrations of 265-1121 µg/g (dry weight) for the same species collected from this area.

Fig.9.1. Map of the Waikato River system, North Island, New Zealand.

A more recent survey by Robinson *et al.* (1995) involved a study of *Egeria densa, Ceratophyllum demersum* and *Lagarosiphon major* taken from the Waikato River (Fig.9.1). Means and ranges (µg/g dry weight) were as follows: *Ceratophyllum demersum* 378 and 44-1160; and *Egeria densa* 488 and 94-1120. *Lagarosiphon major* from the Huka Falls upstream from the geothermal area contained 11 µg/g arsenic, whereas the same species from Lake Aratiatia below the geothermal activity, had 300 µg/g of this element. *Ceratophyllum demersum* occurred from Lake Aratiatia northwards. *Lagarosiphon major* occurred between Lake Taupo and Broadlands where it was replaced by *E. densa*. All species had arsenic concentrations of up to 1200 µg/g (0.12%) dry weight.

Egeria densa and *C. demersum* had arsenic concentrations that had a highly

significant inverse correlation ($r=-0.86[S**]$ and $r=-0.76[S**]$ respectively) with
the distance of the plant downstream (Fig.9.2). These results show that the above
aquatic macrophytes actively extract arsenic from the water in which they grow.

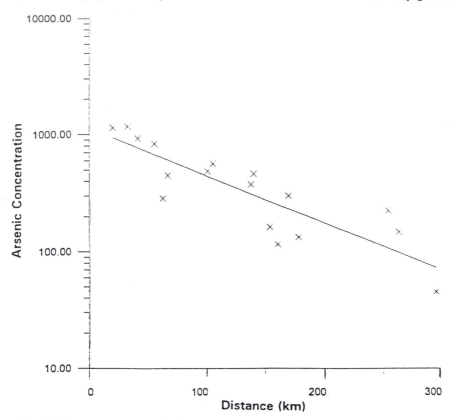

Fig.9.2. The arsenic content (μg/g dry weight) in *Ceratophyllum demersum* collected from the
Waikato River. Data are expressed as a function of distance (km) of the collection
site downstream from the source in Lake Taupo. Source: Robinson (1994).

The arsenic concentration of *C.demersum* showed a highly significant
($r=0.65[S*]$) positive correlation with the arsenic content of the water from
which the plant was taken. The arsenic concentration in this plant was about
10,000 times the arsenic concentration in the surrounding water. The arsenic
concentration in *E. densa* showed no significant correlation ($r=0.41[NS]$) with
that of the water from which it was taken. These results indicate that *C.demersum*
can also be used as a biomonitor of arsenic levels in waters.

The high concentrations of arsenic in the weeds may be the result of the
element being taken up by the same process as the uptake of an essential element.
Arsenic has some chemical similarities with phosphorus, which is an essential
plant macro nutrient. Accumulation of arsenic may be incidental to phosphorus
uptake. This possibility is supported by the observation that nearly all the plants

tested that grow in waters with elevated arsenic levels, accumulate arsenic to some degree.

The concentration of phosphorus in the water may affect the amount of arsenic accumulated by the plant. This may in turn affect the amount of arsenic in the river water. Benson (1953) showed that increasing levels of phosphorus decreased the toxicity of arsenic to barley plants by possibly competing for the binding sites that would otherwise have been occupied by this phytotoxic element.

It is normal farming practice to apply 400 kg/ha phosphate fertiliser to pumice soils (Hill, 1975) and some of the phosphorus leaches into the waterways of the area. The amount of phosphorus applied to farms around the Waikato River system may directly affect the amount of arsenic accumulated by plants, which will affect the arsenic concentration in the waters of this waterway.

Weed eradication programmes that involve the use of herbicides need to take into account the amount of arsenic released into the water as the weeds decay. The performance of carp and other herbivorous fish introduced in an attempt to control the weeds will be affected by the arsenic concentration of weeds in the Taupo Volcanic Zone. Arsenic in weeds may be toxic and/or unpalatable to the fish.

The weeds may also have a use as detoxification agents in waterways with high levels of arsenic. Arsenic may be removed from a body of water by growing, and periodically removing, the macrophytes in a particular area. Lakes such as Lake Rotoroa, which still contains large amounts of arsenic from a weed eradication programme 25 years ago, may be detoxified in this manner.

Water cress

Robinson et al. (1995) have found that New Zealand water cress Rorippa naturtium subsp. aquaticum collected from the Waikato River near the Ohaaki (Broadlands) power station (Fig.9.1) contained >400 μg/g (dry weight) of arsenic. Laboratory trials with water cress grown in tanks containing added arsenic confirmed the ability of this species to accumulate arsenic to a degree at least an order of magnitude higher than the ambient water.

Fig.9.3 shows the relationship between arsenic in the plants and in the surrounding waters. The correlation is extremely good and shows that water cress might also be used for biomonitoring of arsenic pollution in waters.

Arsenic levels in plants grown under laboratory conditions were neverthless about five-fold lower than in plants growing in the Waikato River. This was attributed to a number of factors including the likelihood that the laboratory plants were surrounded by reduced arsenic (As^{3+}) instead of the more strongly absorbed As^{5+} encountered under natural conditions. Another factor is that in the field, the water cress is rooted in sediment containing on average about 95 μg/g arsenic. It was recommended that water cress not be consumed in waters containing >0.05 μg/mL arsenic (essentially the Waikato River between Wairakei and Atiamuri). These studies carry an important message for health authorities.

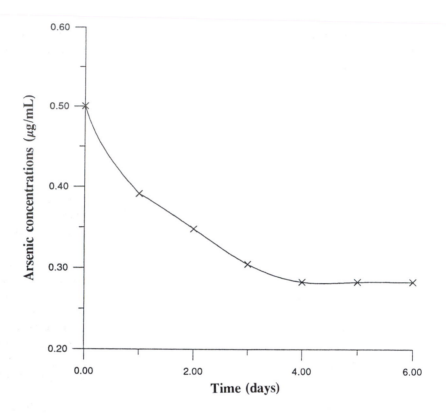

Fig.9.3. Laboratory study of arsenic concentrations in water cress compared with
the surrounding water. Source: Robinson *et al.* (1996).

Other floating FVPs

Before leaving the subject of metal uptake by specific plants, brief mention should
be made of use of duckweed (*Lemna minor*) and water velvet (*Azolla pinnata*).

Wahaab *et al.* (1995) studied copper and chromium (III) uptake by duckweed.
They found that the light incidence was a major factor in growth, with the plants
doubling in size every 3 days under the most favourable conditions. When grown
in water containing 0.25 and 1.0 µg/mL of copper and chromium (III)
respectively, uptake rates were 80-333 and 250-667 mg/day/m² of water surface.
Maximum levels of uptake were 1-2 g of metal/kg of dry matter.

A somewhat analogous series of experiments were carried out by Jain *et al.*
(1989) using duckweed and water velvet in which laboratory trials studied uptake
of iron and copper from solutions containing 1.0, 2.0, 4.0 and 8.0 µg/mL of
these two elements. Growth rates decreased slightly with increasing

concentrations of either element for both species. After 14 days, bioaccumulation coefficients (plant/water) approached about 1000 for both elements in both species for waters containing 8 μg/mL of the element concerned.

The authors suggested that both species would be useful for bioremediation of polluted waters and might be employed by passing effluent over a bed of these plants contained in ponds.

Rhizofiltration

Introduction

The discussion so far has been centred around freshwater vascular plants (FVPs) that are able to take up significant amounts of heavy metal pollutants from the aqueous medium. Although quite high concentration factors are achieved so that these species can be considered as hyperaccumulators, their biomass is usually very small so that the actual mass of pollutant removed is not large. In addition, these plants are often slow-growing.

Terrestrial plants usually have much larger root systems than FVPs. These roots are often fibrous and are covered with a multitude of root hairs that present a large surface area to the medium in which they grow. A new and burgeoning field has been described as *rhizofiltration* and is currently being studied at various institutions such as at Rutgers University, New Jersey, where I.Raskin is very active in this field. Much of this work at the university, and in their associated commercial enterprise (Phytotech Inc.), has been concerned with the possibility of extracting lead commercially from polluted waters using the root systems of terrestrial plants.

Rhizofiltration usually involves the hydroponic culture of plants in a stationary or moving aqueous environment wherein the plant roots absorb pollutants from the water. The principles of the technique have been described by Dushenkov *et al.* (1995) and by Salt *et al.* (1995).

Ideal plants for rhizofiltration should have extensive root systems and be able to remove metal pollutants over an extended period. Such plants should be capable of producing up to 1.5 kg (dry weight) of roots per month per m^2 of water surface.

Suitable candidates for rhizofiltration have included the Indian mustard (*Brassica juncea*) and sunflower (*Helianthus annuus*). Both of these species tend to concentrate heavy metals in the root systems and translocate only a small part of the metal burden to the shoots. If extensive translocation does occur, this has the effect of making the system less efficient.

Apart from physical absorption of heavy metals at root systems, plants are also able to precipitate metals such as lead at the root systems by exuding phosphate that can form the highly insoluble lead phosphate. Concentration factors (*bioaccumulation coefficients*) defined as elemental content of roots divided by that of the water, are as high as 60,000 for some elements (Salt *et al.*,1995).

Although the emphasis in this chapter is on bioremediation by metal uptake, a great deal of work has been carried out on removal of organic compounds and nutrient elements by rhizofiltration (Schnoor *et al.*,1995), often by the use of large trees such as willow and poplar. The same large trees can however also remove metals and will be discussed further below.

The field of rhizofiltration can be divided into three main divisions:

1 - the use of large annual or perennial herbs or grasses;

2 - the use of trees;

3 - the use of trees, large herbs, reeds or grasses to remove contaminated water whereby the water is recycled into the atmosphere and the metals retained within the plant.

Rhizofiltration with large annual or perennial herbs or grasses

Brassica juncea (Indian mustard)

Lead

Some of the earliest experiments on the use of *Brassica juncea* for rhizofiltration of lead were carried out by Dushenkov *et al.* (1995). They report the study of 24 different herbs, grasses and crop plants in order to discover which absorbed the greatest amount of lead in their root systems. Values ranged from 60 mg Pb/g of dry biomass to 136, 140 and 169 mg/g for *Brassica juncea, Helianthus annuus* and *Agrostis tenuis* respectively. The much smaller total biomass of the grass *Agrostis* precluded its potential use for rhizofiltration, though other grasses of greater biomass might be useful for this purpose.

Table 9.5. Shoot and root bioaccumulation coefficients for hydroponically grown specimens of *Thlaspi caerulesecens* and *Brassica juncea* exposed for 8 days to various metal solutions.

Metal	Initial conc.	BS	TS	BR	TR
Cd	5 μg/mL	175	59	20574	4258]
Cu	1 μg/mL	159	623	55809	60716
Cr	0.4 μg/mL	80	89	5486	8545
Ni	1 μg/mL	587	2739	11475	8425
Pb	5 μg/mL	3	29	1432	7011
Zn	3 μg/mL	49	770	1816	2990

BS-*Brassica* shoots, TS-*Thlaspi* shoots, BR-*Brassica* roots, TR-*Thlaspi* roots. After: Salt *et al.* (1995).

The uptake of lead and five other elements in roots and shoots of *Brassica juncea* was compared with corresponding values for *Thlaspi caerulescens*, a well known hyperaccumulator of cadmium and zinc. The results are shown above in Table 9.5 where it can be seen that differences in metal uptake are much greater for the shoots than the roots and that the latter have comparable uptakes for both species. The root biomass of *Brassica* is however, far greater than for *Thlaspi* and this is the greatest factor in its favour.

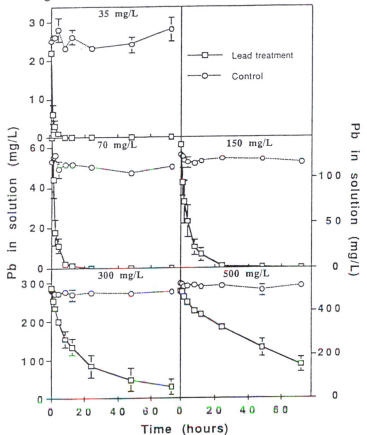

Fig.9.4. Removal of lead by roots of *Brassica juncea* grown hydroponically for up to 72 hours in test solutions containing 0-500 mg/L (µg/mL) lead. Source: Dushenkov *et al.* (1995).

Experiments described by Dushenkov *et al.* (1995) have demonstrated the rapid removal of lead from solutions containing 0-500 µg/mL (ppm) lead in which *Brassica juncea* was grown hydroponically (Fig.9.4). The roots were able to remove most of the lead in 24 hours for concentrations of lead up to 150 µg/mL. At higher concentrations, extraction was slower and less complete. There was a linear relationship between the initial lead concentration and the time needed for the *Brassica* roots to remove half of the initial lead content of the water. This is shown in Fig.9.5.

Other elements
Although most of the work described by Dushenkov *et al.* (1995) was devoted to lead uptake, they also described analogous experiments in which rhizofiltration

of cadmium, nickel, copper, zinc and chromium, as well as lead was described. These experiments are shown in Fig.9.6. The initial elemental concentrations (mg/L [μg/mL]) in the aqueous media were: Cd^{2+} (2), Ni^{2+} (10), Cu^{2+} (6), Zn^{2+} (100), Cr^{6+} (4) and Pb^{2+} (2).

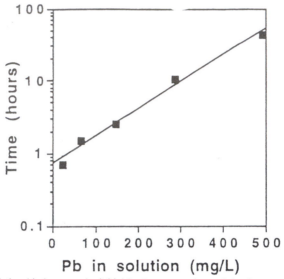

Fig.9.5. Relationship between the initial lead concentration and the time required for roots of *Brassica juncea* to reduce the lead level by 50%. Source: Dushenkov *et al.* (1995).

There were no visible signs of toxicity in the plants. After 8 hours there was a drastic reduction in concentration levels of all of the test metals. The bioaccumulation coefficients were as follows: Pb (563), Cu (490), Ni (208), Cr (179), Cd (134), Zn (131). When a mass balance was carried out, it was found that significantly less metal was found in the plant material than had disappeared from the solutions. This could be explained by precipitation of some of the ions by plant exudates such as phosphate (see also above).

It should be further mentioned that chromium extracted into plant roots was found to be present as the Cr^{3+} ion, thus indicating that the plant reduces chromium during the process of absorption.

Washed dried roots of *Brassica juncea* were much less effective at removing heavy metals from solution. For example, a solution initially containing 300 μg/mL lead decreased immediately to 250 μg/mL (due to precipitation) in all experiments and then decreased to about 20 μg/mL after 10 hours when live roots were used, and still had 200 μg/mL after dried roots had been employed.

Helianthus annuus *(sunflower)*

Extensive experiments have been carried out with the sunflower (*Helianthus annuus*). This large annual herb has about the same biomass as *Brassica juncea* and will readily absorb trace metals in its rhizosphere. Salt *et al.* (1995) have described experiments in which this plant was able to remove five elements from the solutions in which they were grown hydroponically (Fig.9.7).

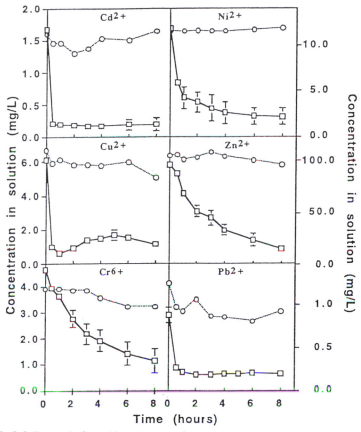

Fig.9.6. Removal of metal ions by roots of *Brassica juncea* (open squares) grown hydroponically in solutions of six elements for up to 8 hours. Controls (open circles) do not contain roots. Source: Dushenkov *et al.* (1995).

As can be seen from Fig.9.7, the sunflower was able to drastically reduce elemental concentrations of manganese, chromium, cadmium, copper and nickel to extremely low levels after only 24 hours. Similar results were obtained for uranium (VI), lead, zinc and lead.

Because rhizofiltration can be employed most economically in situations involving large volumes of water and relatively low levels of contaminants, it would seem to be a possible method of removing radioactive nuclides from polluted waters such as at Chernobyl. Cooney (1996) has reported on field tests carried out near Chernobyl by B.D.Ensley (Phytotech Inc.) and I.Raskin and others from Rutgers University. In these tests, a pond contaminated with radioactive strontium and caesium was covered with 1 m² styrofoam rafts in which seedlings of *Helianthus annuus* were inserted. After 4-8 weeks the plants were harvested, dried and analysed for both radioisotopes. The highest bioconcentration coefficient for caesium was found in sunflower roots and the highest for strontium was in the shoots.

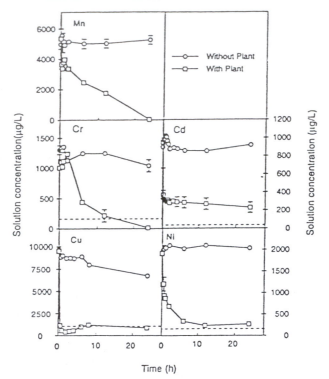

Time (h)

Fig.9.7. Removal of metal ions by roots of *Helianthus annuus* (open squares) grown hydroponically in solutions of five elements for up to 24 hours. Controls (open circles) do not contain roots. Source: Salt *et al.* (1995).

In tests carried out in Ashtabula, Ohio, sunflower plants had their roots submerged in waters contaminated with 100-400 ng/mL (ppb) uranium (Cooney, 1996). Within the first 24 hours, the uranium content decreased by 95% to below the EPA standard of 20 ng/mL.

Practical proposals for rhizofiltration using large herbs

An advertising leaflet issued by Phytotech Inc. proposes a commercial rhizo-filtration system in which polluted water runs through a series of cells in which plants are grown hydroponically and spray fertilised from above. The cleaned water then passes out of the system or can be recycled if not sufficiently pure. The benefits of the system according to the company's claims are that it allows *in situ* treatment minimising environmental disturbance. Further claimed benefits

are that the procedure is inexpensive yet more effective than comparable
technology, and that the system is suitable for removal of low-level radioactive
contamination.

It is not clear from the advertising brochure to what degree this system has
been proven by commercial applications, but the concept is impressive in its
design and obvious future potential.

Rhizofiltration by use of large trees

Rhizofiltration by use of large trees is also a viable, and better proven, emerging
technology. The technique is already well established for remediation of organic
contaminants such as TNT (Schnoor *et al.*,1995). There are certain essential
differences between the use of large trees and large herbs for rhizofiltration. The
trees are usually *in situ* so that the "mountain comes to Mohammed rather than
Mohammed to the mountain" insofar as circulating ground waters seek out the
rhizosphere of the plant rather than vice versa. Furthermore, the tree, unlike the
annual herbs described above, is not usually removed after rhizofiltration.

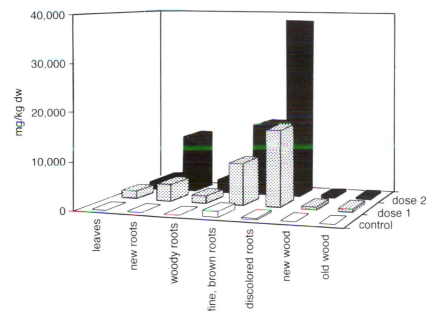

Fig.9.8. Zinc concentrations in poplars exposed to contaminated
waters. Source: Negri *et al.* (1996).

Nyer and Gatliff (1996) have reported work carried out by Negri *et al.* (1996)
in which zinc uptake by a hybrid poplar (*Populus* sp.) was investigated.
Laboratory experiments involved growing the plants in quartz sand to which

nutrients and varying amounts of zinc solutions circulated through the rhizospheres. A single pass of a solution containing 800 μg/mL zinc showed almost complete removal of the zinc in a period of 4 hours. For solutions containing > 1000 μg/mL of the metal, the leachates always had < 100 μg/mL zinc.

At the end of the experiment, the poplars were harvested and various plant organs analysed for zinc. The results are shown in Fig.9.8 from which it will be seen that by far the greatest concentration of zinc (38,055 μg/g) was to be found in the roots. In the aerial parts of the plants, the metal concentrations did not exceed 2250 μg/g in dry leaf tissue and 900 μg/g in the woody branches.

Rhizofiltration using plants to remove and purify contaminated waste waters

There are situations in which it is desired not only to purify waste water but also to reduce its volume. Phreatophytic trees such as poplar, willow and cottonwood are very efficient natural pumps that remove subterranean water and return it to the atmosphere by transpiration. Nyer and Gatliff (1996) have reported that a single willow tree uses and transpires as much as 22,500 litres (5000 gallons) of water in a single day.

Negri and Hinchman (1996a) have introduced the interesting concept of a plant-based system of water treatment that includes a bioreactor in which selected plants are grown hydroponically. The process was designed originally to deal with the large volume of salt-rich water emerging from drilling natural gas wells.

The water evaporates through plant transpiration and rapidly reduces the volume of water. At the same time the heavy metals and other salts are absorbed in the fine root system. The generated plant biomass is harvested periodically, dried, and composted or burnt to produce a small amount of ash that can be buried or even sold in some cases.

It is well known that fresh and saltwater marshes can act as efficient bioremediators of contaminated waters (Kirschner, 1995). One of the commonest plants in these marshes is the bulrush (*Scirpus validus*). This was one of the species selected by Negri and Hinchman (1996a) for their experiments.

Water passes through two cells in parallel containing hydroponically grown bulrush. The water enters with about 1.5% solutes (calculated as the sodium chloride equivalent) and on leaving has a salt content doubled to 3%. At this stage the water enters the final treatment cell containing the highly tolerant saltwater cordgrass (*Spartina alterniflora*). It leaves the unit with 6% salt content and a total volume reduction of 75% carried out in only 7.6 days.

A final assessment of rhizofiltration

Rhizofiltration, like the more general field of phytoremediation, is a new and burgeoning concept that is exciting worldwide interest because it is not only "green", but it appears to be far cheaper than many other alternatives, and in some cases there can even be a financial return if the final product is saleable. The initial euphoria of describing a new technique must be tempered by the realisation of the disadvantages as well as advantages of the new technology. These have been outlined by Black (1995) for phytoremediation in general rather than specifically for rhizofiltration. Nevertheless, the general picture is the same for both.

Advantages

1 - B.Ensley of Phytotech Inc. has estimated that the cost of remediating a cubic metre of contaminated soil would be about $US80 compared with $250 for conventional methods. No figure is given for contaminated waters but we could assume that rhizofiltration would cost about 25 % of that of other methods as was the case for soils;

2 - the basic simplicity and cheapness of phytoremediation might well speed up the current rate of hazardous waste remediation;

3 - the volume of biomass that has to be removed and stored or sold is several orders of magnitude less than the original volume of water;

4 - phytoremediation by rhizofiltration, particularly of water contaminated by toxic metals such as lead will do much to improve public health;

5 - rhizofiltration is a much more rapid method of bioremediation than techniques that involve growing and harvesting crops grown over contaminated soils where several years might be required.

Disadvantages

1 - Rhizofiltration is temperature-dependent. The plants will probably grow the whole year in tropical and warm-temperate regions, but in cool-temperate climates may well grow only half the year unless heating is installed;

2 - The right plants have to be discovered that will be suitable for the climate and tolerant to the heavy metals that are to removed;

3 - The method is still in its infancy and will need to be proven by extensive testing under field conditions rather than in the laboratory.

The future

It would seem that the future is bright for rhizofiltration though care must be taken not to overstate its potential at this early stage in case it suffers the fate of other new techniques that have initially been greeted with euphoria and then discarded after disillusionment. A new idea that has recently surfaced (Negri and Hinchman, 1996b) is the use of chelating agents to increase uptake of heavy metals by plants grown as crops over polluted soils. There is no reason why the same procedure might not be applied to rhizofiltration system with perhaps a drip-feed of chelators to the filtration unit.

We should look forward to other new ideas in the next few years that may serve to give rhizofiltration a due place in the armoury of those engaged in phytoremediation of the environment.

References

Anon.(1976) Hyacinths may help recover gold from tailings. *Engineering and Mining Journal* February, 32.

Benson,N.R.(1953) Effect of season, phosphate and acidity on plant growth in arsenic toxic soils. *Soil Science* 76, 215-224.

Black,H.(1995) Phytoremediation. *Innovations* 103, 1-6.

Bowen,H.J.M.(1966) *Trace Elements in Biochemistry*. Academic Press, London, 241 pp.

Campbell,P,G.C.,Tessier,A.,Bisson,M. and Bougie,R.(1985) Accumulation of copper and zinc in the yellow water lily *Nuphar variegatum*: relationships to metal partitioning in the adjacent lake sediments. *Canadian Journal of Fisheries and Aquatic Science* 42, 23-32.

Cooney,C.M.(1996) Sunflowers remove radionuclides from water on ongoing phytoremediation field tests. *Environmental Science and Technology News* 20, 194A.

Cornwell,D.A.,Zoltek,J.Jr.,Patrinely,C.D.,Furman,T.S. and Kim,J.I.(1977) Nutrient removal by water hyacinths. *Journal of the Water Control Federation* 70, 57-65.

Dushenkov,V.,Kumar,N.P.B.A.,Motto,H. and Raskin,I.(1995) Rhizofiltration: the use of plants to remove heavy metals from aqueous streams. *Environmental Science and Technology* 29, 1239-1245.

Forstner,U. and Wittman,G.T.W.(1981) *Metal Pollution in the Aquatic Environment 2nd Edition*. Springer, Berlin.

Gupta,G.C.(1980) Use of water hyacinths in wastewater treatment. *Journal of Environmental Health* 43, 80-82.

Hill,C.F.(1975) Impounded lakes of the Waikato River. In: Jolly,V.H. and Brown,J.M.A. (eds), *New Zealand Lakes*. Auckland University Press, Auckland, New Zealand.

Hutchinson,G.E.(1975) *A Treatise on Limnology Vol.3*. Wiley, London, pp.264-348.

Jain,S.K.,Vasudevan,P. and Jha,N.K.(1989) Removal of some heavy metals from polluted waters by aquatic plants: studies on duckweed and water velvet. *Biological Wastes* 28, 115-126.

Kay,S.H.,Hailer,W.T. and Garrard,L.A.(1984) Effects of heavy metals on water hyacinths (*Eichhornia crassipes* (Mart.) Solms.). *Aquatic Toxicology* 5, 117-128.

Kirschner,E.M.(1995) Botanical plants prove useful in cleaning up industrial sites. *Chemical and Engineering News* December 11, 22.

Leppard,G.G.(1983) Trace element speciation and the quality of surface waters: an introduction to the scope for research. In: Leppard,G.G.(ed.), *Trace Element Speciation in Surface Waters and its Ecological Implications*. Plenum Press, New York, pp.1-15.

Liddle,J.R.(1982) *Arsenic and other Elements of Geothermal Origin in the Taupo Volcanic Zone*. PhD Thesis, Massey University, New Zealand, 309 pp.

McDonald,R.C. and Wolverton,B.C.(1980) Comparative study of wastewater lagoon with and without water hyacinth. *Economic Botany* 34, 101-110.

Muramoto,S. and Oki,Y.(1983) Removal of some heavy metals from polluted water by water hyacinth (*Eichhornia crassipes*). *Bulletin of Environmental Contamination and Toxicology* 30, 170-177.

Negri,M.C. and Hinchman,R.R.(1996a) Plants that remove contaminants from the environment. *Laboratory Medicine* 27, 36-40.

Negri,M.C. and Hinchman,R.R.(1996b) Bioremediation of contaminated soils by enhanced plant accumulation chelation-assisted accumulation experiments. *Report of the Argonne National Laboratory*. 3 pp.

Negri,M.C.,Hinchman,R.R. and Gatliff,E.G.(1996) Phytoremediation: using green plants to clean up contaminated soil, groundwater and wastewater. *Proceedings of an International Topical Meeting on Nuclear and Hazardous Waste Management Spectrum 96 Seattle*. American Nuclear Society, pp. 1-5.

Nieboer,E. and Richardson,D.H.S.(1980) The replacement of the non-descript term "heavy metals" by a biologically and chemically significant classification of metal ions. *Environmental Pollution B* 1, 3-26.

Nyer,E.K. and Gatliff,E.G.(1996) Phytoremediation. *Ground Water Monitoring and Remediation*. 16, 58-62.

Outridge,P.M. and Noller,B.N.(1991) Accumulation of toxic trace elements by freshwater vascular plants. *Reviews of Environmental Contamination and Toxicology* 121, 1-63.

Phillips,D.J.H.(1977) The use of biological indicator organisms to monitor trace metal pollution in marine and estuarine environments-a review. *Environmental Pollution* 13, 281-317.

Reay,P.F.(1972) The accumulation of arsenic from arsenic-rich natural waters by aquatic plants. *Journal of Applied Ecology* 9, 557-565.

Robinson,B.H.(1994) Pollution of the Aquatic Biosphere by Arsenic and other Elements in the Taupo Volcanic Zone. MSc Thesis, Massey University, New Zealand, 127 pp.

Robinson,B.,Outred,H.,Brooks,R. and Kirkman,J.(1995) The distribution and fate of arsenic in the Waikato River system, North Island, New Zealand. *Chemical Speciation and Bioavailability* 7, 89-96.

Salt,D.E., Blaylock,M., Kumar,N.P.B.A., Dushenkov,V., Ensley,B.D., Chet,I. and Raskin,I.(1995) Phytoremediation: a novel strategy for the removal of toxic metals from the environment using plants. *Bio/Technology* 13, 468-474.

Schnoor,J.L.,Licht,L.L.,McCutcheon,S.C.,Wolfe,N.L.and Carreira,L.H.(1995) Phytoremediation of organic and nutrient contaminants. *Environmental Science and Technology* 29, 318A-323A.

Smies,J.(1983) Biological aspects of trace element speciation in the aquatic environment. In: Leppard,G.G.(ed.), *Trace Element Speciation in Surface Waters and its Ecological Implications*. Plenum Press, New York, pp.177-193.

Sprenger,M. and McIntosh,A.(1989) Relationship between concentrations of aluminum, cadmium, lead and zinc in water, sediments and aquatic macrophytes in six acidic lakes. *Archives of Environmental Contamination and Toxicology* 18, 225-231.

Turekian,K.K.(1969) Distribution of elements in the hydrosphere. In: *1969 Yearbook of Science and Technology*. McGraw Hill, New York.

Wahaab,R.A.,Lubberding,H.J. and Alaerts,G.J.(1995) Copper and chromium (III) uptake by duckweed. *Water Science and Technology* 32, 105-110.

Wolverton,B.C.(1975) Water hyacinths for removal of cadmium and nickel from polluted waters. *NASA Technical Memorandum TM-X-72721.*

Wolverton,B.C. and McDonald,R.C.(1975a) Water hyacinths and alligator weeds for removal of lead and mercury from polluted waters. *NASA Technical Memorandum TM-X-72723.*

Wolverton,B.C. and McDonald,R.C.(1975b) Water hyacinths and alligator weeds for removal of silver, cobalt and strontium from polluted waters. *NASA Technical Memorandum TM-X-72727.*

Wolverton,B.C.,McDonald,R.C. and Gordon,J.(1975) Water hyacinths and alligator weeds for final filtration of sewage. *NASA Technical Memorandum TX-X-72724.*

Chapter ten:

Revegetation and Stabilisation of Mine Dumps and other Degraded Terrain

R.R. Brooks[1], A. Chiarucci[2] and T. Jaffré[3]

[1]Department of Soil Science, Massey University, Palmerston North, New Zealand; [2]Dipartimento di Biologia Ambientale, Università degli Studi di Siena, Siena, Italy; [3]Centre ORSTOM de Nouméa, BP A5, Nouméa, New Caledonia

Introduction

It is only during the past two decades that real concern has been shown for the extensive areas of degraded and contaminated land that exist in most parts of the world. This degradation springs largely from mining activities, particularly opencast, and from other causes such as pastoral over-exploitation, or industrial activities in which enterprises such as smelters can blanket the surrounding countryside with toxic dust and fumes. Although a certain amount of window dressing could be seen in half-hearted attempts at remediation in the immediate post-war years, it was only the emergence of the "green movements" in the 1960s and 1970s that have compelled industries and governments to recognise the real concerns of their citizens to preserve the environment.

The situation has now changed radically and virtually all mining companies, smelters and other "dirty" industries operating in the western world now have environmental programmes that attempt to rehabilitate the damage that they now cause and at the same time even push back the clock to remediate previous devastation caused by their predecessors. This movement is especially strong in the USA where the "Superfund" regulations (see Chapter 1) now compel owners of contaminated land to remediate it, even if they had not originally caused the problem. The cost of such land restoration will be so great that this has spawned the development of many new remediation techniques purporting to provide ever less costly and ever "greener" alternatives to the classical remediation techniques.

The extent to which industrial concerns are prepared to undertake remediation is largely a question of geography. In many Third World countries, little or no effort is devoted to remediation due to the lack of public opinion that would enforce reform. An example of this can be seen in Fig.10.1 which is a view of the effects of opencast mining at Macedo near Niquelândia in Goiás State Brazil. The mine produces nickel from the huge Tocantins Ultramafic Complex and has no programme for revegetation of the area.

If one were to attempt to pinpoint a date at which the first steps were made towards an attempt at rehabilitation of polluted environments, we need to look no further than the work of Bradshaw (1952) who was one of the first to realise that abandoned mine sites support metal-tolerant populations of common grasses such as *Festuca rubra* and *Agrostis tenuis* that could be used to revegetate other mine dumps and colliery spoil heaps. Bradshaw's pioneering work led to the appearance of an important benchmark paper by Antonovics *et al.* (1971) that summarised much of this earlier work and set the scene for the significant advances that were made thereafter.

Fig.10.1. View of opencast mining for nickel at Macedo in Goiás State, Brazil. Photo by R.R.Brooks.

The emphasis of this book is of course on hyperaccumulators and it must freely be admitted that these specialised plants are no more adapted than other metal-tolerant species for the purpose of revegetation of degraded and contaminated terrain. They do however have one advantage over their non-accumulating cousins. If they are annuals that can be harvested and resown, or if they are perennials that can be cropped and recropped on an annual basis, there is thereby an opportunity not only to revegetate, but also to remove some of the metal burden of the soil. The removed biomass, if ashed, will have perhaps 15 times more of the metal than the dry material and might even provide a saleable sulphur-free *bio-ore* (see Chapters 14 and 16).

This chapter is divided into three sections. The first of these describes the inadvertent colonisation of anthropogenic environments by hyperaccumulating plants. The second examines the potential of fertilisation of naturally occurring metal-tolerant plants in order to accelerate the restoration of contaminated and degraded land. The final section examines the progress and technology of land restoration in New Caledonia where several hyperaccumulators of nickel and manganese have a potential role.

Inadvertent Colonisation of Anthropogenic Environments by Hyperaccumulators

Introduction

Colonisation of anthropogenic environments by plants has been described by Brooks and Malaisse (1985) in connection with the metal-tolerant flora of Shaba Province, Zaïre in Central Africa. Much of the material in this section is derived from this work. Colonisation of these man-made environments is either *advertent* or *inadvertent* depending on whether or not humans have been directly or indirectly responsible for the introduction.

Inadvertent processes involve colonisation of mining areas, roadsides dressed with gangue from mining activities, roads and railways contaminated by spillage from ore-carrying vehicles or wagons, and lastly, archaeological sites. The latter opens up the possibility of discovering such sites by the nature of their metal-tolerant vegetative cover (see Chapter 7). Advertent processes involve the intentional vegetating of unsightly mine dumps or industrial sites by selection of suitable metal tolerant plant species. This is a rapidly developing field in view of the current interest of conservationists in preserving and improving the environment. Both of the above types of colonisation of man-made environments will now be discussed.

Colonisation of mining sites

In examining the question of colonisation of mining sites, a clear distinction must be made between the natural vegetation existing at the site before exploitation of the mineral resources, and the subsequent adventive flora which colonised the site after human modification of the environment. In Shaba Province, Zaïre and in nearby Zambia, exploitation of copper deposits has in many places led to the destruction of most of the characteristic vegetation, so that all that remains is carpets of colonisers of sterile mine dumps such as those that occur in Shaba at Karavia, Kasombo, Kambove, Kakanda, Luiswishi and Lukuni (see Fig.7.12). These carpets are dominated by *Eragrostis boehmii*, *Rendlia cupricola* and *Haumaniastrum katangense*. The last two are hyperaccumulators of copper and/or cobalt. The *Haumaniastrum* is one of the most remarkable colonisers of copper deposits in Southcentral Africa and its behaviour has been documented by Malaisse and Brooks (1982).

Good examples of colonisation of modified mining environments are to be found over old mines in Shaba. For example, Malaisse and Grégoire (1978) studied vegetation over the Mine de l'Etoile near Lubumbashi which closed in 1926 and has since had a period of nearly 60 years to undergo the dynamics of changing vegetation patterns. The above authors reported distinct communities over mine tailings as opposed to the pre-existing natural vegetation. Common members of this adventive community include *H.katangense* together with *Cryptosepalum dasycladum*, *Pteris vittata*, *Rendlia cupricola*, *Bulbostylis mucronata* and *Crepidorhopalon dambloni*. Only the *Cryptosepalum* and *Pteris* are non-accumulators.

One of the best examples of disturbed mining ground is to be found at Mindigi about 100 km west of Likasi (see Fig.7.12) in Shaba Province, Zaïre (Duvigneaud, 1959). The area is riddled with pre-European workings and more recent excavations have brought to the surface small pebbles highly enriched in both copper and cobalt. The highly-toxic soils (3700 μg/g copper and 1900 μg/g cobalt) exposed by mining support a carpet of *Eragrostis boehmii* containing in addition, several highly metallicolous species such as members of the genera: *Ascolepis*, *Sopubia*, *Bulbostylis*, *Gladiolus*, *Icomum*, *Vernonia*, *Triumfetta* etc. many of whose constituent species are hyperaccumulators of copper and/or cobalt.

One of the most remarkable examples of the ability of metal-tolerant hyperaccumulators to colonise new metalliferous environments is demonstrated by the discovery of *Haumaniastrum katangense* at copper deposits near Kela (Malaisse and Brooks, 1982) some 25 km west of Kakanda (see Fig.7.12). The mining company Gécamines has opened up a network of roads crossing the open forest and leading to about 20 deposits between Kakanda and Kela. These deposits are usually situated on the flanks or at the summits of rocky hills. The presence of *H.katangense* at Kela can only be explained by transport of this species via supply lorries from a mine where this plant exists naturally. The nearest of such stations is some 60 km distant from Kela.

Colonisation of rail tracks and roadsides

In Zaïre, the extraction of copper from its minerals involves production of a large volume of sterile waste which is usually deposited near the sites of extraction and which forms large dumps. A secondary usage of this material is for the dressing of dirt roads during the wet season. During the dry season, the road traffic deposits dirt on both sides of the road creating a band of lightly-mineralised substrate. Malaisse and Brooks (1982) have reported the presence of a metal-tolerant community upon these road verges which includes the hyperaccumulators *Cryptosepalum maraviense*, *Bulbostylis mucronata*, *H.katangense* and *Crepidorhopalon dambloni* which are all annual cuprophytes.

The above authors have followed the west-northwest axis of the road from Lubumbashi for a distance of 8 km beyond the city limits and have observed these four taxa along a band of 50 km at 1 km intervals. A carpet of *Crepidorhopalon dambloni* along the roadside is shown in Fig.10.2.

Malaisse and Brooks (1982) have also observed the presence of *H.katangense* along railway tracks near Lubumbashi. At the beginning of the colonial epoch, a railway was built to connect the Etoile and Ruashi mines with the processing plant at Elizabethville (Lubumbashi). The minerals were transported in open wagons and today bands of *H.katangense* can be observed locally along the route of this railway for about 20 km as it crosses the city. The line was in operation from 1911 to 1926 when the mines were shut down. Along the Lubumbashi-Kipushi railway, traces of malachite are still to be found in the ballast.

Fig.10.2. Rose-coloured carpet of *Crepidorhopalon dambloni* colonising a metal-rich roadside near Lubumbashi, Shaba Province, Zaïre. Photo by R.R.Brooks.

An interesting example of inadvertent colonisation of a roadside is shown in Fig.10.3. which represents a carpet of the nickel hyperaccumulator *Alyssum euboeum* growing over an access road to a magnesite mine on the island of Euboea in Greece. The road was constructed from the nearby mine tailings and provides an ideal habitat for this plant which is not found **away** from the verges of the road.

Fig.10.3. The nickel hyperaccumulator *Alyssum euboeum* growing over a mine access road on the island of Euboea, Greece. The road was constructed from waste from the nearby magnesite mine and has provided an ideal habitat for this plant. Photo by W.Krause.

Colonisation of sites polluted by industrial effluent

It is well known that metal-tolerant taxa will readily colonise sites affected by industrial effluent. A good example of this is found in Central Europe where *Thlaspi rotundifolium* subsp. *cepaeifolium* is often found over river gravels heavily contaminated with zinc and lead (Reeves and Brooks, 1983).

A classical example of inadvertent colonisation of a former industrial site is to be found near the city of Lille in northern France. This is shown in Fig.10.4. The terrain had been contaminated by effluent from a nearby zinc smelter and the soil contains a very elevated concentration of this metal. The site has been colonised by a pink carpet of *Armeria maritima* subsp. *halleri* among which is interspersed stands of the zinc hyperaccumulator *Cardaminopsis halleri*.

Fig.10.4. Carpet of the metal-tolerant *Armeria maritima* subsp. *halleri* growing over soil contaminated by a zinc smelter near Lille in northern France. Interspersed within this carpet is the zinc hyperaccumulator *Cardaminopsis halleri*. Photo by B.H.Robinson.

In Shaba Province, Zaïre, the copper hyperaccumulator *Rendlia cupricola*, a hemicryptophyte, is now found outside of its original area and grows on alluvial sands enriched in heavy metals derived from water used at the Likasi ore treatment plant. The plant has also been found in grey colluvia near the River Mulunguishi, a site likewise polluted by metalliferous muds from the treatment plant. Another plant found in areas subject to industrial pollution is the non-accumulator *Bulbostylis mucronata* which is found on temporarily waterlogged soils rich in minerals, as on muds from concentration plants or even on soils subjected to extensive pollution from aerial fallout of zinc, lead, and copper from the Gécamines treatment plant.

A spectacular example of inadvertent colonisation of industrial effluent again comes from Zaïre. A carpet of *Haumaniastrum katangense was found to* colonise soil contaminated with copper derived from the fallout from the giant stack of the copper smelter at Lubumbashi, Zaïre.

Colonisation of archaeological sites

Inadvertent colonisation of archaeological sites in Zaïre and elsewhere has already been fully covered in Chapter 7. Only the very bare details will therefore be repeated here.

One of the most remarkable aspects of the distribution of *Haumaniastrum katangense* and other metallophytes in Zaïre is their presence over sites of furnaces traditionally used in precolonial days for the production of small copper crosses (Plaen *et al.*,1982).

The artisans used copper smelters which were constructed on ancient termite hills which thereby provided a gradient suitable for the flow of molten copper. They were surrounded by earth walls about a metre high which must have contaminated the moulds and smelting conduits. The furnaces were situated near dambos (periodically-inundated savanna) or near rivers which provided the necessary water. The sites were also near stands of *Pterocarpus tinctorius* which were also used to produce charcoal.

These 14th century artisans brought copper ore to the smelter sites and abandoned the activities after a few seasons. After a long period of time the old smelters weathered to ground level and were colonised by *H.katangense* and other plants that served as indicators for the sites and for artefacts buried beneath them. The artefacts include pottery, ancient pipes used for smelting and copper *"croisettes"* (crosses) that were used for currency at that time (see Chapter 7 for further details of these discoveries).

Revegetation of Degraded and Polluted Land by Fertilisation of Native Plants

Introduction

The problem of vegetating old mine dumps is one which is receiving a great deal of attention today. Mine dumps are not only unsightly, but provide unwelcome publicity for the mining industry as well as polluting the environment with wind-blown toxic waste which can affect agriculture as well as animal and human health.

Degradation of metalliferous environments does not necessarily always result from mining or industrial activity. Failed agricultural programmes in which the original vegetation cover was removed and the land abandoned after unsuccessful attempts at cropping, can also result in impoverished land that could take decades to be restored to its original condition. In some cases these attempts failed because the agronomists did not recognise the basic infertility of serpentine soils.

Most degraded sites are situated over ultramafic (serpentine) soils because these represent much greater areas (1% of the earth's surface) than the far smaller extent of other forms of mineralization such as sulphides. The most obvious way to restore degraded vegetation would seem to be fertilisation of the soil in order to increase plant productivity.

There are however, very few examples of employment of such a strategy except some fertilisation experiments on natural serpentine vegetation in Scotlaand (Carter *et al.*,1988; Looney and Proctor, 1989), California (Koide *et al.*,1988; Huenneke *et al.*,1990), and Tuscany (Robinson *et al.*,1997; Chiarucci, 1994; Chiarucci *et al.*,1994, 1997a, 1997b). There is clearly a need for such studies because increasing the biomass of the plant community by fertilisation might have the result of decreasing plant diversity by elimination of rarer species and encouraging the appearance of weeds that would not normally have tolerated the extreme edaphic conditions of unfertilised serpentine soils.

Fertilisation experiments over Tuscan serpentine soils

Introduction

Chiarucci *et al.* (1997a) have recently studied the effect of fertilisation on species richness and cover in two serpentine sites of southern Tuscany. In one of these sites, located in the Ombrone Valley near Murlo south of the city of Siena, the influence of fertilisation on biomass production was also investigated (Chiarucci *et al.*,1997b). The vegetation community belongs to the *Armerio-Alyssetum* phytosociological association which is the most typical vegetation of ultramafic soils of Tuscany (Chiarucci *et al.*,1995). This vegetation type, mainly formed by chamaephytes and hemicryptophytes is illustrated in Chapter 3 (Fig.3.2). This figure shows both of the type species: *Armeria denticulata* (bottom left) and the nickel hyperaccumulator *Alyssum bertolonii* (top right).

Thirty-five 1 m^2 quadrats were selected randomly in a gentle slope covered by the *Armerio-Alyssetum bertolonii* association and 5 replicates of each were fertilised with the following treatments:

1 - control;
2 - nitrogen at 10 g/m^2 as ammonium nitrate;
3 - phosphorus at 10 g/m^2 as sodium dihydrogen phosphate;
4 - combined nitrogen and phosphorus as above;
5 - nitrogen and phosphorus as above, and potassium (the latter at 10 g/m^2 as potassium chloride;
6 - calcium at 100 g/m^2 as calcium carbonate;
7 - calcium, nitrogen, phosphorus and potassium as above.

The plots were fertilised in October 1994 and 1995 and cover data were collected in Spring 1994, 1995 and 1996 by the points-quadrat method (Moore and Chapman, 1986) in which the intersections of the vegetation were recorded with a square grid of 441 points/m^2. After harvesting, a soil sample was taken over each plot and the soluble fractions of each soil determined by extraction into 1M ammonium acetate.

After the collection of cover data, the plants were harvested at ground level and sorted by species. The material was dried at 80°C for 48 h and then weighed. Total biomass and the biomass of each species and life form were transformed logarithmically and submitted to an analysis of variance (ANOVA).

Effect of fertilisers on the plant-availability of various elements

Soil pH and element availability data are summarised in Table 10.1 below. The pH showed a significant reduction where nitrogen was added and a significant increase after amendment with calcium. The availability of calcium and potassium increased in the quadrats where they had been added, whereas sodium appeared to be reduced in such plots. The availability of nickel was relatively constant except where the pH had been raised after addition of lime. Manganese availability increased in the NKP- and to a lesser extent in the N- and NP-treated plots. The copper increased in the NPKCa- and NP- and NPK-treated quadrats.

Table 10.1. Mean (n=5) availability at pH 7 of elements ($\mu g/g$ in original soil) of fertilised and control experimental quadrats (35) over serpentine soils in the Murlo district, Tuscany, Italy. Letters after each value indicate statistically significant differences of the means from other quadrats for the same element.

Element/pH	O	N	P	NP	NPK	NPKCa	Ca
pH	7.37a	7.05bc	7.41ab	6.93c	6.86c	7.75d	7.94d
Ca	837.5a	790.2a	817.0a	822.3a	959.0a	3718.4b	3800.7b
Co	0.095a	0.082a	0.084a	0.088a	0.072a	0.090a	0.072a
Cr	0.048a	0.026a	0.036a	0.012a	0.026a	0.012a	0.034a
Cu	0.078a	0.090a	0.076a	0.138ab	0.138ab	0.194b	0.092a
Fe	0.988a	0.480a	0.426a	0.190a	0.236a	0.470a	0.228a
K	27.9a	27.4a	24.4a	23.0a	81.8b	57.3ab	22.8a
Mg	1555a	1702a	1918a	1856a	1808a	1148a	1274a
Mn	2.08a	2.68ab	2.12a	2.98ab	3.59b	1.93a	1.86a
Na	1990a	1643a	1871a	1808a	2024a	1105b	1238b
Ni	4.43a	4.51a	3.53ab	3.64ab	4.37a	2.40b	2.23b
V	0.373a	0.380a	0.378a	0.378a	0.386a	0.382a	0.386a
Zn	0.103a	0.130a	0.176a	0.216a	0.140a	0.310a	0.344a

Source: Chiarucci *et al.* (1997a).

Effects of fertilisation on species richness and ground cover

Fertilisation induced marked increases in total ground cover and in the cover of some species, especially in the plots fertilised with at least nitrogen and phosphorus together, which showed a threefold increase in total ground cover.

The addition of potassium to the two other fertilisers induced another increase in ground cover, whereas calcium did not induce any significant effect. Species richness as slightly promoted in the fertilised plots. Although extremely strong changes were shown in the ground cover of the vegetation, species composition had not changed significantly after two years of fertilisation. The species present in the plots increased in abundance after fertilisation but there was no significant colonisation by other species. On average the hyperaccumulator *Alyssum bertolonii* increased its cover tenfold (from 0.9-9.6 %) in the NPK fertilised plots.

Effect of fertilisers on biomass production

The effect of different fertiliser treatments on the total biomass of each quadrat is reported in Table 10.2. The biomass data give a good reflection of the results of the cover measurements. Total ground cover and biomass were, in fact, very highly mutually correlated (P < 0.001). Significant differences in biomass production were seen for fertiliser treatments involving at least nitrogen and phosphorus.

Addition of fertilisers also reduced the coefficients of variation within the plots submitted to the same treatments, implying a more homogeneous biomass production. The life forms showing the highest increase in biomass were woody species (chamaephytes and shrubs) and annual graminoids and forbs. Although perennial graminoids and forbs showed an apparent increase in biomass after fertilisation, these increases were not statistically significant. There was enhanced biomass of geophytes only in NPKCa-treated plots.

The hyperaccumulator *Alyssum bertolonii* gave the second-highest increase in biomass, increasing sixfold with the NPK treatment. In other experiments described in Chapter 15, pure stands of cultivated plants of this species can increase their biomass by a factor of three. This is of supreme importance for phytomining with this plant.

Table 10.2. Geometric mean of the total biomass (g) and biomass of different life forms (g) in fertilised and control experimental quadrats (35) over serpentine soils in the Murlo district, Tuscany, Italy. Letters after mean values indicate statistically significant (P < 0.05) differences of the means from other quadrats as determined by the SNK test. The geometric mean of the biomass of the hyperaccumulator *Alyssum bertolonii* is also shown.

	O	N	P	NP	NPK	NPKCa	Ca
Total plant	40.01a	51.02a	83.75a	186.46b	282.52b	303.85b	72.60a
Woody plants	19.88a	35.22ab	40.41ab	89.59bc	196.15c	166.15c	40.87ab
Per. grasses	7.87	4.84	9.46	5.24	12.47	16.46	11.05
Per. forbs	2.67	0.68	2.48	6.88	17.02	2.54	5.99
Geophytes	0.02a	0.02a	0.11a	0.06a	0.18a	1.04b	0.06a
Ann. grasses	0.31a	2.61ab	2.79ab	17.84b	8.23b	9.51b	0.43a
Ann. forbs	0.86a	4.34ab	10.86bcd	19.2bd	15.47bcd	27.78d	3.06c
A. bertolonii	2.95ab	5.63abc	1.67a	4.71abd	19.79cd	14.76bc	3.79ab

Source: Chiarucci *et al.* (1997b).

The most significant increases in biomass production were by the following species in descending order: *Aira elegantissima, Alyssum bertolonii, Centaurea aplolepa* subsp. *caruleana, Helychrysum italicum, Herniaria glabra, Psilurus incurvus, Sedum rupestre* and *Thymus acicularis* subsp. *ophioliticus*. The hyperaccumulator *Alyssum bertolonii* gave the second greatest increase in biomass.

Conclusions

It is clear from the above experimental work, that fertilisation of natural stands of native metal-tolerant plants can greatly increase the biomass and provide a means of quickly restoring degraded terrain. It is however less clear what effect fertilisation will have on species composition. In fact, annual species showed the highest growth after fertilisation, probably because of their seed availability in the soil. This might adversely affect the performance of perennial species. The significant increase of biomass of woody species, among which *Alyssum bertolonii* was the most important, was performed by plants already growing in the plots and not by colonisation by new individuals. To promote the abundance of this species for phytomining or phytoremediation might therefore require that it be sown and thereby reduce the growth of annual species.

The experiments reported above have only extended over a three-year period and have not taken into account the possibility of invasion by weeds once the fertility of the ultramafic soils has been improved by the fertilisation regime. However, a rapid colonisation by weeds or other undesirable species after fertilisation is unlikely because of the lack of plant propagules in the surroundings of the plots. A longer term of fertilisation together with a shorter distance from potential propagules would be required to colonise these fertilised plots by non-serpentinic species.

Advertent Colonisation of Anthropogenic Environments by Hyperaccumulators in New Caledonia

Introduction

New Caledonia is one of the world's largest producers of nickel, derived from lateritic *garnierite* (hydrated nickel silicate) mineral. About one third of the 16,600 km^2 of the main island and its subsidiary the Ile des Pins is covered with ultramafic rocks (Fig.10.5) that are hosts for 56 hyperaccumulators of nickel (see Table 3.3).

This total comprises the world's second highest abundance of nickel hyperaccumulators in a single country or territory: i.e. after Cuba. It is not surprising therefore that revegetation programmes in that country include several hyperaccumulators.

There is a pressing need for revegetation of opencast mine sites in New Caledonia because of the widespread devastation caused by mineral exploitation on that island (Fig.10.6).

In an attempt to revegetate opencast mining sites in New Caledonia, the mining company *Société Métallurgique le Nickel* (SLN) in collaboration with the French research institutes *ORSTOM and CIRAD Forêts* is undertaking an ambitious programme of reclamation (Fig.10.7).

This programme has been described by Jaffré et al. (1993,1994), Jaffré and Pelletier (1992), Jaffré and Rigault (1991) and Pelletier and Esterle (1995). Most of the material in the remainder of this chapter is derived from these and other papers.

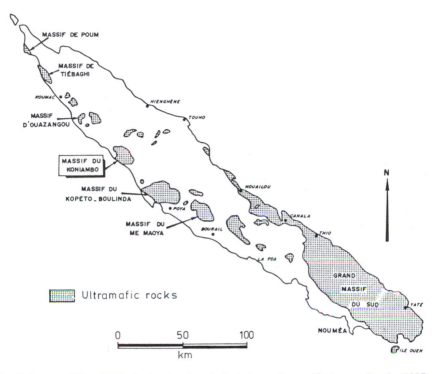

Fig.10.5. Map of New Caledonia showing the principal ultramafic massifs. Source: Brooks (1987).

Physical reconstitution and stabilisation of degraded terrain

The usual method of physical reconstitution of the environment consists of terracing the lateritic spoil from the opencast mining operations. The toxicity of the lateritic material is a result largely of its high magnesium content that produces a low Ca/Mg quotient that is highly unfavourable for plant nutrition. Although magnesium is the dominant element in the natural waters of the region, its actual concentration in these waters (typically 15 μg/mL) is not unduly high for an ultramafic environment and has not been increased by mining activities.

The terraces (see Fig.10.8) have a slope not exceeding 30° and have allowed the stabilisation of some 132 million tonnes of gangue formed in the period 1975-1990.

The base of the terraces is formed of a rocky drain over which the laterites are placed in compacted layers. The front walls of the terraces are composed of large rock boulders that retain the laterites in place. In front of the base of the terraces a rocky dam retains the percolating water in a diversion channel that leads the water away for safe disposal.

A channel just above the terraces diverts fresh water from above, in order to reduce the amount supplied for disposal through the diversion. The surface of the terraces is revegetated with appropriate plants as can be shown from Fig.10.9.

Fig.10.6. View of destruction caused by opencast mining in New Caledonia. Source: Jaffré and Rigault (1991).

Strategies of revegetation in New Caledonia

The two main problems of revegetation of opencast mine sites in New Caledonia are related to the poor nutrient status of the gangue coupled with the high content of magnesium (causing an unfavourable Ca/Mg quotient in waters and soil solutions - see above) and other phytotoxic metals such as nickel. Strategies that can be employed for revegetation and stabilisation are detailed below.

1 - Use of native plants that have a high resistance to unfavourable edaphic conditions and that can be put to good use. These include various pioneer species associated with nitrogen-fixing symbiotic bacteria, such as *gaïac* (*Acacia spiroorbis*) and *Casuarina collina* that will grow on the degraded material with very little extra attention.

2 - Amelioration of the substrate. In some cases, the sites can be remediated by adding fertile soils from other areas, or else lime can be added that will serve

to provide a favourable substrate for sowing with rapidly growing plants of economic use such as grasses and legumes. This strategy works well in flat areas but is useless for mountainous regions.

Techniques of revegetation

There are two main techniques of revegetation of mining sites in New Caledonia.

The first of these involves transplanting plants grown from seeds or as cuttings in some form of plant growth unit such as a greenhouse or shade house. This is by far the more costly of two procedures. Planting is performed manually in steep terrain or by machine in flat areas.

The young plants are surrounded by straw or a plastic film to preserve humidity in the early stages. About 6-8 months are required for the establishment of plants over the mine wastes.

Some of the plants used in revegetation are legumes such as *"gaiac"*, *Storckiella*, *Archidendropsis* and *Serianthes* and various species of Casuarinaceae such as *Casuarina collina* as well as several species of *Gymnostema*. Planting densities range from about 1 per m^2 for small trees or large shrubs and up to 5-10 per m^2 for the smaller cyperaceous species.

Fig.10.7. View of revegetation of degraded terrain in New Caledonia. Source: Jaffré *et al.* (1993).

Direct seeding is also feasible but less frequently used than direct planting. Use can be made of hydroseeding with a "seed cannon" in which a powerful pump projects a pulp composed of seeds and cellulose (finely chopped hay for example), as well as fertilisers and other materials. The cannon can project the seed as far as 500 m. A team of two persons can treat 1-5 ha/day.

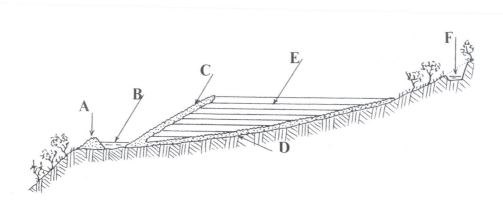

Fig.10.8. Sketch of a scheme for stabilisation of lateritic gangue derived from opencast mining in
New Caledonia. A - rocky dam to retain and divert water percolating through the laterite,
B - water-diversion channel, C - terrace wall composed of large rock boulders,
D - rocky drain below the laterites, E - layers of compacted laterite,
F - channel to divert fresh water away from the system.
Source: Jaffré and Pelletier (1992).

Table 10.3. Growth characteristics of seven New Caledonian nickel hyperaccumulators that could be
used for revegetating mine dumps and other degraded terrain.

Characteristic	GP	AL	BA	PA	PC	AD	HC
Ni content (% dry weight)	0.66	0.19	0.54	0.21	0.12	0.25	0.88
PREFERRED SOILS							
magnesian	3	1	2	1	2	3	3
ferralitic	1	-	-	1	1	-	-
alluvial volcanic/sedimentary	-	-	-	-	1	-	-
humid	2	0	0	2	1	1	1
dry	1	2	2	1	1	2	2
ALTITUDE (m)	0-700	nd	0-500	50-1000	0-1000	0-500	0-500
SEED							
maturation (month)	J,F	nd	J,F,M	nd	nd	J	J,F
latency (days)	6-12	nd	nd	nd	nd	nd	nd
germination	60%	nd	nd	nd	nd	nd	nd
CUTTINGS							
strike time (days)	nd	nd	nd	nd	nd	80	100
success rate	nd	nd	nd	nd	nd	45%	10%

GP - *Geissois pruinosa*, AL - *Argophyllum laxum*, BA - *Baloghia alternifolia*, PA - *Phyllanthus
aeneus*, PC - *P.chrysanthus*, AD - *Agathea deplanchei*, HC - *Hybanthus caledonicus*.
nd - no data, 0-3 ability to tolerate edaphic conditions where 0 = poor and 3 = very good. After
Jaffré and Pelletier (1992).

Fig.10.9. Aerial view of terracing of a lateritic mine discharge at Kongouhaou, New Caledonia. The surface of the laterite has been planted with *gaïac* and *bois de fer* (ironwood), both indigenous species adapted to mining environments. Source: Jaffré and Pelletier (1992).

A typical pulp would have the following composition: seed, 50-100 kg/ha (5000-8000 seeds); mulch, 2000 kg/ha; NPK fertiliser, 360 kg/ha; stabiliser (alginate), 200 kg/ha; lime (for pyritic gangue), 600 kg/ha; bacterial cultures. The purpose of the mulch is to increase the surface humidity, to even out the daily variation in temperature, to protect the microflora and microfauna, and to provide a reservoir of organic material from which the plants will derive nutrients.

Nickel hyperaccumulators that can be used for revegetation in New Caledonia

Jaffré and Pelletier (1992) have presented a large table of characteristics of 104 herbs, shrubs and trees that lend themselves to being grown for remediation of degraded land in New Caledonia. This list includes 7 nickel hyperaccumulators, details of which are shown in Table 10.3 above. Two of these plants (*Geissois pruinosa* and *Hybanthus caledonicus*) are illustrated in Figs 10.10 and 10.11.

It is perhaps fitting that this chapter should end with the above description of the rehabilitation work currently being carried out in New Caledonia. Where once this beautiful island was synonymous with human destruction of the environment, it has now become an example of what may be achieved when industry and governments work together to remediate a polluted and damaged land. The extent of the devastation is such that it would be difficult to believe that full remediation will ever be carried out, but at least a start has been made.

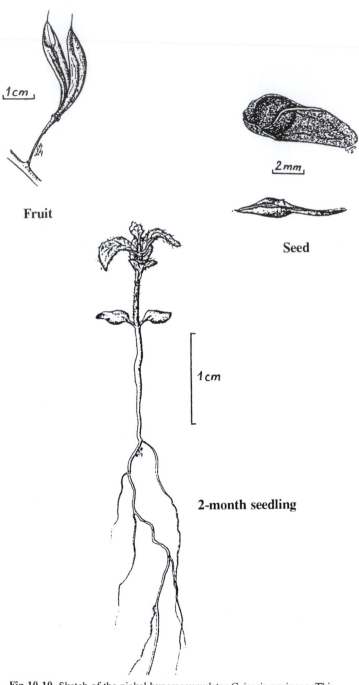

Fruit

Seed

2-month seedling

Fig.10.10. Sketch of the nickel hyperaccumulator *Geissois pruinosa*. This plant may be used for revegetating mine waste and contains about 0.60% nickel in its dry biomass. Source: Jaffré and Pelletier (1992).

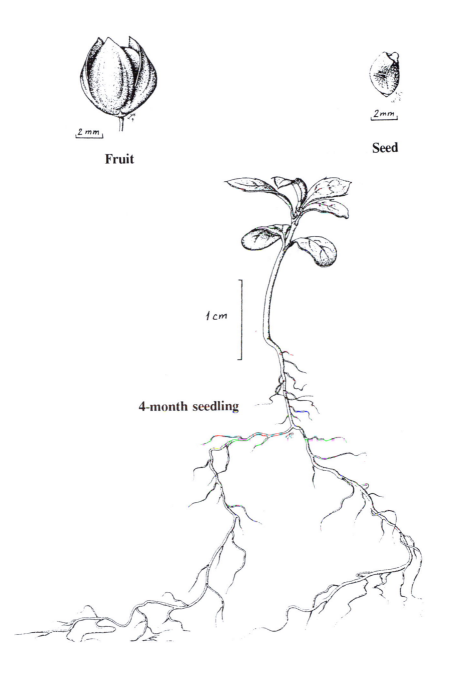

Fruit

Seed

2 mm

2 mm

1 cm

4-month seedling

Fig.10.11. Sketch of the nickel hyperaccumulator *Hybanthus caledonicus* that can contain nearly 1% of this metal on a dry-weight basis. This plant has potential for revegetation of mine sites. Source: Jaffré and Pelletier (1992).

References

Antonovics,J.,Bradshaw,A.D. and Turner,R.G.(1971) Heavy metal tolerance in plants. *Advances in Ecological Research* 7, 1-85.

Bradshaw (1952) Populations of *Agrostis tenuis* resistant to lead and zinc poisoning. *Nature* 169, 1098.

Brooks,R.R.(1987) *Serpentine and its Vegetation.* Dioscorides Press, Portland, 434 pp.

Brooks,R.R. and Malaisse,F.(1985) *The Heavy Metal Tolerant Flora of Southcentral Africa.* Balkema, Rotterdam, 199 pp.

Carter,S.P.,Proctor,J. and Slingsby,D.R.(1988)The effects of fertilisation on part of the Keen of Hamar serpentine, Shetland. *Transactions of the Botanical Society of Edinburgh* 45, 97-105.

Chiarucci,A.(1994) Successional pathway of Mediterranean ultramafic vegetation in central Italy. *Acta Botanica Croatica* 53, 83-94.

Chiarucci,A.,Bonini,I.,Maccherini,S. and De Dominicis,V.(1994) Remarks on the ultramafic garigue flora of two sites of the Siena province, Italy. *Atti Accademici Fisiocritici Siena* 13, 193-200.

Chiarucci,A.,Bonini,I.,Maccherini,S. and De Dominicis,V.(1997a) Effects of nutrient addition on species composition and cover on serpentine vegetation *Plant and Soil* (in press).

Chiarucci,A.,Foggi,B. and Selvi,F.(1995) Garigue plant communities of ultramafic outcrops of Tuscany (Italy). *Webbia* 49, 179-192.

Chiarucci,A.,Maccherini,S.,Bonini,I. and De Dominicis,V.(1997b) Effects of nutrient addition on community structure of "serpentine" vegetation (in press).

Duvigneaud,P.(1959) Plantes cobaltophytes dans le Haut Katanga. *Bulletin de la Société Royale Botanique de Belgique* 91, 111-134.

Huenneke,L.F., Hamburg,S.P., Koide,R.,Mooney,H.A. and Vitousek,P.M. (1990) Effects of soil resources on plant invasion and community structure in Californian serpentine grassland. *Ecology* 71, 478-491.

Jaffré,T. and Pelletier,B.(1992) *Plantes de Nouvelle Calédonie Permettant de Révégetaliser des Sites Miniers.* Société Métallurgique le Nickel, Nouméa, New Caledonia, 114 pp.

Jaffré,T. and Rigault,F.(1991) *Recherches sur les Possibilités d'Implantation Végétale sur Sites Miniers.* ORSTOM (New Caledonia) Report No.5, 78 pp.

Jaffré,T,Rigault,F. and Sarrailh,J.-M.(1993) *Essais de Revégétalisation par des Espèces Locales d'Anciens Sites Miniers de la Région de Thio.* ORSTOM (New Caledonia) Report No.7, 31 pp.

Jaffré,T,Rigault,F. and Sarrailh,J.-M.(1994) La végétalisation des anciens sites miniers. *Bois et Forêts des Tropiques* No.242, 46-57.

Koide,R.T.,Huenneke,L.F.,Hamburg,S.P. and Mooney,H.A.(1988) Effects of application of fungicide, phosphorus and nitrogen on the structure and productivity of an annual serpentine plant community. *Functional Ecology* 2, 335-344.

Looney,J.H.H. and Proctor,J.(1989) The vegetation of ultrabasic soils on the Isle of Rhum. Parts I and II. *Transactions of the Botanical Society of Edinburgh* 45, 335-364.

Malaisse,F. and Brooks,R.R.(1982) Colonisation of modified metalliferous environments in Zaïre by the copper flower *Haumaniastrum katangense*. *Plant and Soil* 64, 289-293.

Malaisse,F. and Grégoire,J.(1978) Contribution à la phytogéochimie de la Mine de l'Etoile (Shaba, Zaïre). *Bulletin de la Société Royale Botanique de Belgique* 111, 252-260.

Moore,P.D. and Chapman,S.B.(1986) *Methods in Plant Ecology 2nd Ed.* Blackwell Scientific Publications, Oxford, 589 pp.

Pelletier,B. and Esterle,M.(1995) Revegetation of nickel mines in New Caledonia. *Colloquium on French Research on the Environment in the South Pacific. Paris 28-31 March 1995*, 1-9.

Plaen,G.De,Malaisse,F. and Brooks,R.R.(1982) The copper flowers of Central Africa and their significance for archaeology and mineral prospecting. *Endeavour* 6, 72-77.

Reeves,R.D. and Brooks,R.R.(1983) Hyperaccumulation of lead and zinc by two metallophytes from a mining area of Central Europe. *Environmental Pollution Series A* 31, 277-287.

Robinson,B.H.,Chiarucci,A.,Brooks,R.R.,Petit,D.,Kirkman,J.H.,Gregg,P.E.H. and De Dominicis,V.(1997) The nickel hyperaccumulator plant *Alyssum bertolonii* as a potential agent for phytoremediation and phytomining of nickel. *Journal of Geochemical Exploration* 59, 75-86.

Chapter eleven:

Fertilisation of Hyperaccumulators to Enhance their Potential for Phytoremediation and Phytomining

F.A. Bennett, E.K. Tyler, R.R. Brooks, P.E.H. Gregg and R.B. Stewart
Department of Soil Science, Massey University, Palmerston North, New Zealand

Introduction

Although the concept of using plant hyperaccumulators for phytoremediation was proposed over a decade ago by Chaney (1983) and the first thorough scientific study appeared about four years ago (McGrath *et al.*,1993), the idea of using these plants to harvest a crop of a given heavy metal (phytomining) appeared only in 1995 (Nicks and Chambers, 1995 - see also Chapter 14). Both techniques are still in their infancy and will need to be proven over several years before there is complete scientific and commercial acceptance of their value.

Both phytoremediation and phytomining suffer from the same limitation; the degree of uptake of a given heavy metal by the chosen hyperaccumulator. If this uptake is too low, the effect on phytoremediation will be an increase in the number of croppings needed to effect a given degree of remediation and a concomitant increase in cost of the procedure. In the case of phytomining the situation is somewhat different. A low degree of uptake by the plant will mean that the crop of metal cannot be grown economically.

Few, if any, of the current studies in phytoremediation/ phytomining have been devoted to the possible uses of fertilisers in order to increase biomass and to establish whether this increase in biomass is at the expense of metal uptake. However, Robinson *et al.* (1997) studied the effect of fertilisation of natural stands of the nickel hyperaccumulator *Alyssum bertolonii, in situ* in Tuscany Italy.

Other studies (also in Italy) have been concerned with the use of fertilisers to improve the biomass and species composition of natural vegetation found over serpentine in Tuscany (Chiarucci *et al.*,1995 - see also Chapter 10) and have an obvious application to studies on phytoremediation and phytomining, there and elsewhere. Apart from these few reports, there is very little information on the nutrition of common hyperaccumulators and their response to fertilisers.

This chapter presents original unpublished data on pot trials carried out on three plant hyperaccumulators: *Alyssum bertolonii*, *Streptanthus polygaloides* and *Thlaspi caerulescens*. *A.bertolonii* was the first plant discovered to have the ability to hyperaccumulate nickel (Minguzzi and Vergnano, 1948).

Streptanthus polygaloides, was used by Nicks and Chambers (1995) to phytomine nickel (see Chapter 14).

Thlaspi caerulescens is a spectacular hyperaccumulator of both zinc and cadmium but has such a small biomass that the advantage of accumulation is offset to a large extent. However, the species is very variable and quite large specimens can be seen in its natural environment so that fertilisation might well have the ability to greatly increase its biomass.

Experiments with *Streptanthus polygaloides*

Materials and methods

The plants were grown in a mixture of 1:3 crushed serpentine rock and fine crushed bark. The original composition of the bark and serpentine are given below in Table 11.1. Sixteen different fertiliser treatments were used with 10 replicates of each treatment. Treatments for nitrogen were: N_0, N_1, N_2 and N_3 (10, 25, 50 and 100 $\mu g/g$ N as calcium ammonium nitrate), where N_0 was the control and N_2 (50 $\mu g/g$) was the recommended loading for nursery stock grown in peat or bark. N_1 and N_3 were respectively half and twice the levels of N_2. Each of the nitrogen treatments was combined with each of four phosphorus additions (P_0-P_3). This gave the following 16 mixtures: N_0P_0, N_0P_1, N_0P_2, N_0P_3, N_1P_0, N_1P_1, N_1P_2, N_1P_3, N_2P_0, N_2P_1, N_2P_2, N_2P_3, N_3P_0, N_3P_1, N_3P_2 and N_3P_3.

Table 11.1. Composition ($\mu g/g$ unless otherwise stated) of substrates used in pot trials with *Streptanthus polygaloides*.

Material	N	P	K	Ca	Mg (%)	Na	Ni
Bark	3	2	36	3	n.d.	14	n.d.
Serpentine	1	*130	83	3700	22.08	260	2198
3:1 mixture	2	34	49	927	5.75	270	550

*most of this was unavailable to plants and can be ignored in calculations.

It had originally been proposed that the phosphorus additions would have represented 0, 10, 20 and 40 $\mu g/g$ P (where 20 $\mu g/g$ is the recommended loading for nursery stock). However, phosphate was also present in the original serpentine (130 $\mu g/g$) and therefore in the mixture (34 $\mu g/g$) so that the final loadings should have been 34, 44, 54, and 74 $\mu g/g$ for P_0-P_3 respectively. However, the standard phosphate extraction test for plant-available phosphorus showed that only 1-2 $\mu g/g$ phosphorus was extractable from the serpentine and could be ignored in determining phosphorus loadings in the fertiliser amendments.

To give the required fertiliser loadings, a mixture of sand was prepared that contained 1.85% (w/w) calcium ammonium nitrate. A second mixture contained 2.2% (w/w) superphosphate. These mixtures were added to the starting mixture of bark and serpentine at a rate of 0, 0.5%, 1% and 2% (w/v) to give the required loadings within the range of 0-40 $\mu g/g$ P and 1-100 μg N.

Seeds were germinated and grown for about 2 weeks in standard commercial seed mixture and then transferred to cell trays each containing 30 units of 150 mL capacity. Over the next 3 weeks, plants that had died after the original transfer were replaced by others from the seed trays. The plants were then overhead watered three times a week until the end of a 20-week period. After 20 weeks the plants were harvested, washed, dried at 50°C, and weighed to determine biomass.

Dried samples were placed in 5 mL borosilicate test tubes and ashed overnight at 500°C in a muffle furnace. The resultant ash was redissolved in 5 mL of 2M hydrochloric acid and the target element determined by flame atomic absorption spectrometry.

Results and discussion

The experiments showed that *Streptanthus polygaloides* requires a moderate level of nitrogen and plants growing in the N_0 and N_1 series of pots (0 and 25 $\mu g/g$ N) did not survive the 20-week period. For those that remained, there was no apparent relationship between fertiliser treatment and biomass and nickel content. This is shown in Table 11.2 below. This table does, however, give the mean (geometric mean because the data were lognormally distributed) biomass and nickel contents of the individual groupings only within each fertiliser amendment.

The mean nickel content of the plants was well below the nickel content of 9750 (range 2350-3840) $\mu g/g$ dry weight reported for *Streptanthus* by Reeves *et al.* (1981). In the experiments reported in this chapter, the mean of individual plants was 4500 $\mu g/g$ nickel with a range of 1500-7100 $\mu g/g$. The experiments have certainly confirmed the hyperaccumulation ability of *Streptanthus polygaloides* even though the experiments were somewhat truncated because of insufficient nitrogen in some of the fertiliser treatments (i.e. N_0 and N_1). Since the curve in Fig.11.1 shows no sign of reaching a limiting value, it is clear that the level of fertilisation was still suboptimal even at the upper level of amendment used.

The product of biomass and nickel content is of course of great importance in assessing the suitability of a species for phytoremediation or phytomining. Taking the mean values in Table 11.2, the product was 92.7. Plants that are only moderate accumulators but have a high biomass may provide a higher overall

Table 11.2. Relationship between fertiliser treatment, mean biomass (g) and mean nickel content ($\mu g/g$ dry weight) for *Streptanthus polygaloides* plants.

Treatment (N = 10 for each)	Biomass	Ni
N_2P_0	0.022	4100
N_2P_1	0.015	6300
N_2P_3	0.015	3000
N_3P_0	0.024	3700
N_3P_1	0.026	4200
N_3P_2	0.038	4200
N_3P_3	0.016	4000
Overall mean	0.022	4214

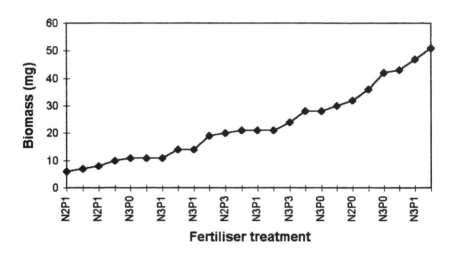

Fig.11.1. Effect of fertiliser treatment on the biomass of *Streptanthus polygaloides*.

metal yield than a hyperaccumulator with a low biomass.

Although there was no statistically significant relationship between the biomass and nickel content ($P > 0.5$) when the mean of individuals was grouped within each treatment, there was a gradual increase in the biomass for fertiliser treatments showing a progressive increase of nitrogen and phosphorus contents. This is demonstrated in Fig.11.1.

Experiments with *Alyssum bertolonii*

Materials and methods

The materials and methods were exactly the same as for the above experiments with *Streptanthus polygaloides*.

Results and discussion

Relationship between biomass and fertiliser treatment

There was a highly significant increase in biomass of *Alyssum bertolonii* with increasing addition of nitrogen during the 20 weeks of the experiments. This is shown in Fig.11.2.

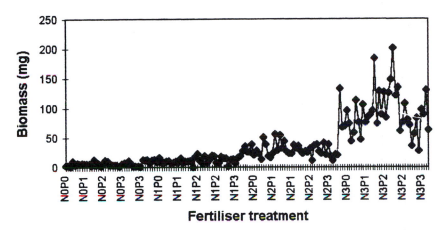

Fig.11.2. Effect of fertiliser treatment on the biomass of *Alyssum bertolonii*.

It is clear that the phosphorus level of treatment also had some effect in addition to that of nitrogen. As in the case of *Streptanthus polygaloides* above, there was a dramatic increase of biomass at the higher nitrogen level (N_3 = 100 μg/g N).

It is probable that there would have been a continuing increase of biomass if higher nitrogen concentrations had been used. From studies in the field by Robinson *et al.* (1997) it seems that this species is particularly susceptible to increasing its biomass with nitrogen fertilisers. These authors achieved an increase of 300% during a 12-month period.

Relationship between biomass and nickel content

The mean biomass and nickel concentrations for the four nitrogen treatments are summarised in Table 11.3 below. Means are geometric means because the data were lognormally distributed.

When the means of the biomass data were compared with each other using *t* tests, the following were found to be significantly different: N_0 vs. N_3 (0.05 > P > 0.02 = significant), N_2 vs. N_3 (P < 0.001 = very highly significant).

When all data points were ungrouped, a plot of nickel content vs. biomass was as shown in Fig.11.3.

Table 11.3. Mean biomass (g) and nickel concentrations (μg/g) for various nitrogen treatments (μg/g).

Treatment	Nitrogen	Biomass	Ni content
N_0	0	0.004	4182
N_1	25	0.010	4231
N_2	50	0.028	4728
N_3	100	0.086	3521

The plot below appears to show that there is a slight inverse relationship between the nickel content of the plants and the biomass. This is indeed confirmed by correlation analysis for which r = -0.342 (P < 0.001 = very highly significant for the inverse relationship). This statistical evaluation does not however show that the inverse relationship is in fact very slight and that the disadvantage incurred in increasing the biomass is more than offset by the overall much higher increase in the mass of nickel accumulated. This is highlighted in Table 11.4.

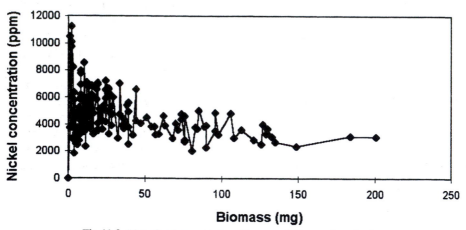

Fig.11.3. Plot of nickel content vs. biomass of *Alyssum bertolonii*.

Table 11.4. Product of mean biomass (g) and nickel content (μg/g) of *Alyssum bertolonii* with various treatments of nitrogen fertiliser.

Treatment	Biomass	Nickel	Product
N_0	4	4182	16.7
N_1	11	4231	46.5
N_2	28	4728	132.4
N_3	86	3521	302.8

The importance of the biomass/nickel product cannot be overemphasised because this quantity is the determining factor for the success or otherwise of phytoremediation and/or phytomining. To increase the biomass by fertilisation or the metal content by lowering the pH or adding complexing agents to the soil (Huang and Cunningham, 1996), are all strategies that should be followed in the continual striving to improve the methodology of this emerging technology. If both biomass and metal content can be increased without one being at the expense of the other, there can be little doubt that "green remediation" will be here to stay for the foreseeable future.

Experiments with *Thlaspi caerulescens*

Introduction

Thlaspi caerulescens is remarkable in that it can hyperaccumulate over 3.0% zinc (dry weight) and at the same time over 0.1% cadmium. Like many members of this genus, it has the ability to hyperaccumulate nickel when grown over ultramafic soils (Reeves and Brooks, 1983).

The practical use of this plant for phytoremediation/phytomining is restricted by its small biomass. For example McGrath *et al.* (1993) have shown that 15 annual croppings of this plant would be needed to reduce the zinc content of a polluted soil from 444 to 300 μg/g, the limit set by the Commission of the European Communities (CEC, 1986).

Although specimens commonly seen in the field have a low biomass, some individuals can have a biomass as much as ten times the norm and if selective breeding of this species could be carried out, it might well prove to be of practical use for phytoremediation, though not for phytomining given the low price of zinc. The experiments described below had the aim of testing the effects of fertilisers on the biomass and zinc content of plants grown over lead/zinc mine tailings as well as testing the reaction of this species to nickel-rich soils.

Materials and methods

Nickel-rich growth media

The growth media consisted of a mixture of bark and crushed serpentine rock in the volume ratios shown in Table 11.5. These mixtures were particularly high in magnesium as well as nickel and their phytotoxicity was probably linked to the former rather than to nickel.

Mine tailings growth media

The mine tailings originally contained about 400 μg/g zinc and 4000 μg/g (0.4%) lead (Morrell *et al.*,1995). This material was mixed with pumice to improve drainage (3 parts of tailings to one part of pumice) and then 2.5% (v/v) of lime as calcium carbonate was added to raise the pH from about 3 to 6.

Table 11.5. Growth media volumes and their nickel ($\mu g/g$) and magnesium (%) contents ($\mu g/g$) for experiments with *Thlaspi caerulescens* raised in a nickel-rich substrate. The nickel content ($\mu g/g$ dry weight) of the plants is also shown.

No.	Serpentine	Bark	Ni in medium	Mg in medium	Ni in plant
1	1	1	1499	11.04	nd
2	1	2	732	7.36	352
3	1	3	550	5.52	982
4	1	4	440	4.42	489
5	0	all	trace	trace	39

nd - no plants grew in this hostile environment. Probably because of the very high magnesium content rather than the effect of high nickel concentrations.

Table 11.6. Product of geometric means of biomass (g) and zinc contents ($\mu g/g$) of *Thlaspi caerulescens* with various treatments of nitrogen fertiliser.

Treatment	Biomass	Zinc	Product
N_0	0.095	2306	21.9
N_1	0.234	2075	485.6
N_2	0.205	2427	497.5
N_3	0.281	1311	368.4

The fertiliser additions were thereafter exactly as for *Streptanthus polygaloides* above with a total of 16 treatments and ten replicates of plants in pots for each treatment. The growing period was again 20 weeks.

Results and discussion

Experiments with mine tailings

The results of determination of biomass and zinc contents of *Thlaspi* plants are shown in Table 11.6.

There were no statistically significant differences in biomass for different additions of phosphorus at constant nitrogen content. However, the biomass increased dramatically from 0.095 to 0.234 g per plant after the first nitrogen addition and remained thereafter at the same level, there being no statistically significant differences ($P > 0.01$) of biomass between the treatments N_1, N_2 and N_3.

The zinc content remained fairly stable at over 2000 $\mu g/g$ and there appears to be a reduction after the N_3 treatment. The same pattern was evident for nickel in *Alyssum bertolonii* (Table 11.3).

When the zinc content of individual plants ($n = 102$) was plotted against biomass, there was a statistically significant decrease of the zinc concentrations with increasing biomass ($r = -0.326$ [$P < 0.001$]). The data are shown in Fig. 11.4.

As was the case with *Alyssum bertolonii*, the inverse correlation between biomass and metal content, though statistically significant, does not really present the true picture: i.e. increasing biomass is hardly at all at the expense of metal uptake. The defining factor for successful; phytoremediation or phytomining is the product of biomass and metal concentration in the plants.

Experiments with nickel-rich soils

As was shown Table 11.5, *Thlaspi caerulescens* was grown in artificial media containing varying concentrations of nickel and magnesium that simulated serpentine soils. The plants failed to grow in the 1:1 medium (1499 μg/g nickel and 11.04% magnesium).

The nickel content of this plant is shown in Table 11.5 above. There were insufficient data values for statistical treatment but nevertheless the pattern is quite clear. *Thlaspi caerulescens* may to all intents and purposes be considered as a hyperaccumulator of nickel.

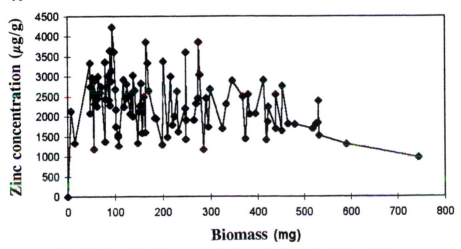

Fig.11.4. The relationship between the zinc content (ppm [μg/g] dry weight) and biomass (mg) of *Thlaspi caerulescens* grown in mine tailings.

Other specimens grown in the same medium grew to full size, flowered, and completed their growing cycle by setting seed. The plant is therefore adapted to growing in an ultramafic environment.

General Conclusions

A number of conclusions can be made about the above experiments:

1 - the addition of nitrogen fertiliser increases the biomass of all three species investigated. For *Streptanthus polygaloides* and *Alyssum bertolonii* the highest nitrogen level used (100 μg/g N), was not necessarily the optimum because in

both species there was a steady augmentation of biomass with increase of nitrogen, and with no sign of levelling off;

2 - for all three species there was no significant difference in biomass or metal uptake for varying concentrations of phosphorus at constant nitrogen levels of fertilisation;

3 - the *Alyssum* and *Thlaspi* showed a slight reduction in concentration of nickel and zinc respectively when the biomass of each plant was increased by the higher level (N_3) of nitrogen fertilisation. The trade-off of biomass against metal content was, however, slight;

4 - the product of biomass and metal content is the single most important quantity defining the suitability of these plants for phytoremediation or phytomining. Only the *Streptanthus* and *Alyssum* will be suitable for phytomining because the world price of nickel $7500/t is far higher than that of zinc ($1000/t);

5 - the concentrations of nickel and zinc obtained in these experiments are appreciably lower than those found in wild plants. The two main reasons for this may be the short growing period of the plants (20 weeks) in the growth units as well as the high growth rates achieved under the experimental conditions;

6 - *Thlaspi caerulescens* also has a phytoremedial potential as a hyperaccumulator of nickel, though this potential is limited by its small biomass.

7 - there is some evidence that an oversupply of nitrogen can affect the biomass and metal uptake by *Thlaspi caerulescens*. Nitrogen is particularly deficient in the natural environment of this species so that its requirement for optimum plant growth is below the maximum of 100 $\mu g/g$ administered in the N_3 series of experiments. The above experiments were restricted to a 20-week growing period and there is clearly a need for further experiments along the same lines in which the growing period could be extended as well as an increase in the amount of added nitrogen. The small biomass of *Thlaspi* may well mitigate against its use in phytoremediation, but the plant is unique in its extraordinarily high degree of uptake of zinc (and cadmium) and efforts should be made to identify the genes responsible for this high degree of accumulation and their transference to other plants with a much higher biomass such as *Helianthus annuus* (sunflower) or *Brassica juncea* (Indian mustard). If this can be achieved, phytoremediation will surely become of age.

References

CEC(1986) Commission of the European Community - Council Directive of 12 June 1986 on the protection of the environment, and in particular of the soil, when sewage sludge is used in Agriculture. *Journal of the European Communities* No.L181(86/278/EEC, pp.6-12.

Chaney,R.L.(1983) Plant uptake of inorganic waste constitutes. In: Parr,J.F.,Marsh,P.B. and Kla,J.M.(eds) *Land Treatment of Hazardous Wastes.* Noyes Data Corp., Park Ridge, pp.50-76.

Chiarucci,A.,Foggi,B. and Selvi,F.(1995) Garigue plant communities of ultramafic outcrops of Tuscany, Italy. *Webbia* 49, 179-192.

Huang,J.W. and Cunningham,S.D.(1996) Lead phytoextraction: species variation in lead uptake and translocation. *New Phytologist* 134, 75-84.

McGrath,S.P.,Sidoli,C.M.D.,Baker,A.J.M. and Reeves,R.D.(1993) The potential for the use of metal-accumulating plants for the *in situ* decontamination of metal-polluted soils. In: Eijsackers,H.J.P. and Hamers,T.(eds), *Integrated Soil and Sediment Research: a Basis for Proper Protection.* Kluwer Academic Publishers, Dordrecht, pp.673-676.

Minguzzi,C. and Vergnano,O.(1948) Il contenuto di nichel nelle ceneri di *Alyssum bertolonii. Atti della Società Toscana di Scienze Naturale* 55, 49-74.

Morrell,W.J.,Stewart,R.B.,Gregg,P.E.H.,Bolan,N. and Horne,D.(1995) Potential for revegetating base-metal tailings at the Tui mine site, Te Aroha, New Zealand. *Proceedings of the Pacific Rim Congress, 19-22 November 1995*, Auckland, New Zealand, pp.395-400.

Nicks,L. and Chambers,M.F.(1995) Farming for metals. *Mining Engineering Management* September, 15-18.

Reeves,R.D. and Brooks,R.R.(1983) European species of *Thlaspi* L. (Cruciferae) as indicators of nickel and zinc. *Journal of Geochemical Exploration* 18, 275-283.

Reeves,R.D.,Brooks,R.R. and McFarlane,R.M.(1981) Nickel uptake by Californian *Streptanthus* and *Caulanthus* with particular reference to the hyperaccumulator *S.polygaloides* Gray (Brassicaceae). *American Journal of Botany* 68, 708-712.

Robinson,B.H.,Chiarucci,A,Brooks,R.R,Petit,D.,Kirkman,J.H.,Gregg,P.E.H. and De Dominicis,V.(1997) The nickel hyperaccumulator *Alyssum bertolonii* as a potential agent for phytoremediation and phytomining of nickel. *Journal of Geochemical Exploration* 59, 75-86.

Chapter twelve:

Phytoextraction for Soil Remediation

S.P. McGrath

Department of Soil Science, IACR-Rothamsted, Harpenden, UK

Introduction

Phytoextraction is a term which is used to indicate the process by which plants can remove significant quantities of substances from their substrate. It could also be used for extraction of elements from water or possibly from air, but here it will be used to denote uptake from solid rooting media. Using plant roots to clean up contaminated water has been called *rhizofiltration*, which is covered elsewhere in this book (Chapter 9). This chapter focuses on uptake from soil or soil-like substrates.

It has been known for well over a century that plants extract significant quantities of the major nutrients phosphorus and potassium from soils. On the other hand, some time later, trace elements received that classification because they are normally taken up by most plants in very small quantities. The term is often used for those elements which are plant nutrients (B, Cl, Co, Cu, Fe, Mn, Mo, Ni, Zn - Welch, 1995). However, elements that are not essential to plants are also accumulated to a variable degree, and these are sometimes included in the literature along with trace elements: for example, Cd, Pb, Cr, Hg and As. Plants which hyperaccumulate trace elements were found early this century (Chapter 3), but their potential use for extracting potentially toxic elements from soils has taken a long time to come to the forefront of scientific attention. The stimuli for this have been the development of concern about environmental contamination, and the expense of cleaning up pollutants in soils.

Basically, phytoextraction depends on high concentrations of the target elements in plant biomass, especially in above ground portions (due to the practical problems of harvesting roots) and production of a relatively large biomass. These two features determine the efficiency of the process. High concentrations may be due to natural plant properties, especially hyperaccumulation, or to enhancement of the more usual concentration of an element in non-accumulator plants. Enhancement can be achieved by additions of substances which modify soil metal release and/or plant uptake characteristics.

It can be shown that yield is a less important trait than hyperaccumulation. Table 12.1 was constructed with zinc in mind, but the same kinds of consideration are relevant to all other elements. From this it can be seen that 25 t/ha, the maximum potential yield of an annual crop (or biomass which can be cut from a long-lived perennial) still only results in 12.5 kg/ha/yr removal, even if the non-tolerant non-hyperaccumulator has 500 $\mu g/g$ of zinc, which is the usual threshold for toxicity of this element in many species. In contrast, growing a hyperaccumulator with a biomass yield of 5 t/ha and a maximum concentration of 20,000 $\mu g/g$ (Brown et al.,1994, 1995b) results in a very effective 100 kg/ha removal.

Future developments in phytoextraction must strive for hyperaccumulation, rather than simply increased biomass without massive accumulation. Other advantages also accrue from metal hyperaccumulation: e.g. there is a reduction in the amount of material to be disposed of, and the concentration in the bio-ores are maximised, making their use more favourable for phytoremediation.

Table 12.1. Annual metal removal (kg/ha) by plant shoot harvests in relation to biomass yield (t/ha) and metal concentration in the plant ($\mu g/g$).

Type of plant	Concentration	Yield	Removal
Non-hyperaccumulator	50	5	0.25
	50	10	0.50
	50	15	0.75
	50	20	1.00
	50	25	1.25
	500	5	2.50
	500	10	5.00
	500	15	7.50
	500	20	10.00
	500	25	12.50
Hyperaccumulator	1000	5	5.00
	10,000	5	50.00
	20,000	5	100.00

So far, there have been few attempts to clean soils using phytoextraction. Growing interest and research activity is taking place in this area, however, and new technologies can be expected to be developed in the relatively near future. In this chapter, the phytoextraction of metals and other elements which have been investigated and the developments to date are reviewed, along with the more speculative prospects for removing metals and other contaminants from soils.

Removal of selenium from soils by plants

The first and perhaps the leading example of the use of plants to extract excess concentrations of an element in soils came about because of an environmental problem. That was the presence of unusually high concentrations of selenium of geochemical origin on the west side of the Californian San Joaquin Valley. Sediments containing selenium have been used for irrigated agriculture and the resulting drainage is rich in selenium and flows to reservoirs including the Kesterson Reservoir, and there were recorded cases of death or deformity in wildlife populations. Presence of selenium in drainage threatens the sustainability of agriculture in the area. Selenium is also potentially toxic to plants, and the two dominant forms are selenate and selenite, with the former being the more mobile and toxic. Selenium can occur in four oxidation states in soil: elemental selenium (Se^0), selenide (Se^{2-}), selenite (SeO_3^{2-}) and selenate (SeO_4^{2-}). Speciation of selenium in soil has an important influence on the concentrations of selenium accumulated by plants. It has been clearly shown that accumulating and non-accumulating plants always take up more selenium when presented with selenate than selenite (Bañuelos and Schrale, 1989; Bañuelos, 1996).

It was hypothesised by Bañuelos and colleagues that species that accumulate large quantities of sulphur would also accumulate selenium. Apparently, there is no discrimination of plant uptake mechanisms between the two elements when present in the oxidised sulphate and selenate forms. They quantified the uptake of selenium by *Astralagus, Atriplex* and *Brassica* species, *Festuca arundinacea* and *Beta vulgaris,* at first in pot experiments, and noted that the brassicas contained the highest concentrations of selenium (up to 1200 μg/g in broccoli florets). Importantly, in this first report (Bañuelos and Schrale, 1989), they also showed that not all the selenium added to the experiments was accounted for in the soil and plants. Most of the selenium was present in the tops rather than the roots of the plants, and wild mustard (*Brassica juncea*) contained the highest concentrations (2500 μg taken up in the plant when 1 kg of soil was treated with 3500 μg selenium).

As with metal hyperacumulation, the selenium accumulators need a second key character in addition to high uptake: selenium tolerance. They need mechanisms that prevent substitution of selenium in protein synthesis and the subsequent disruption this causes.

Unlike for metals (Chapter 3), no one seems to have defined a hyperaccumulation threshold for selenium. Even those who have worked for some years with selenium-extracting crops and wild plants use the term *accumulator* plants in the most recent literature (Bañuelos, 1996). However, earlier work described three groups of plants in terms of selenium accumulation (Rosenfield and Beath, 1964): group 1 accumulate 1000 - 10,000 μg/g selenium, and include species of *Astralagus, Brassica, Stanleya;* group 2 rarely accumulate more than a few hundred μg/g (*Atriplex, Grindelia, Gutierrezia*); group 3 include many agricultural crops and grasses (Poaceae), which do not accumulate more than 50 ng/g (ppb) selenium. Using the previously defined groups, group 1 may equate with selenium hyperaccumulation, and *B.juncea* would certainly qualify for this status.

The ability of plants to decrease soil selenium was examined in pot studies in which *Astralagus incanus*, *Atriplex semibaccata*, *Atriplex nummularia*, *Brassica juncea* and *Festuca arundinacea* were grown with addition of either 3.5 μg/g selenate or selenite to the soil. Again, significantly less selenium was taken up when selenite was added, and *B.juncea* contained the most selenium, followed by *Atriplex nummularia*, *A.semibaccata*, *F.arundinacea* and *A.incanus*. *B.juncea* reduced soil selenium by 76%, which was significantly more than the other species (Table 12.2).

Table 12.2. Approximate percentage distribution of 3.5 μg/g selenium added to soil as selenate.

Species	Soil Se	Plant Se	Unaccounted Se
Brassica juncea	24	58	18
Astragalus semibaccata	57	17	26
Festuca arundinacea	58	9	33
A.nummularia	63	21	16
A.incanus	78	4	18

After: Bañuelos and Meek (1990).

However, in soil which naturally contained 1.17 μg/g total selenium, *Festuca arundinacea*, *Brassica napus* and *B.juncea* all contained similar concentrations of selenium. All species appeared to increase the loss of selenium from the soil with time (harvest), and to decrease selectively the concentrations of extractable selenium (extracted from soil paste) in the soil by much more than the total selenium (Table 12.3). It is important to note that control (unplanted) soil also lost significant amounts of selenium during the study, presumably by microbial volatilisation (see Chapter 13).

Table 12.3. Percentage decreases in total and extractable selenium in Panoche fine loamy soil.

Species	Total Se		Extractable Se	
	Harvest I	Harvest II	Harvest I	Harvest II
Brassica juncea	24	41	73	90
B.napus	31	-	73	-
Festuca arundinacea	-	40	-	62
Unplanted control	14	23	48	56

After: Bañuelos *et al.* (1993b).

Decreased selenium in field soils was first shown in 1990-91 in Los Baños, California on a soil containing 0.1-1.2 μg/g Se in the soil. Two experiments were performed which included *Brassica juncea*, *Lotus corniculatus*, *Festuca arundinacea*, and *Hibiscus cannabinus* (Table 12.4). In the first year, there were no differences in selenium concentrations between the species, but in 1991,

B.juncea contained the greatest concentration of selenium (>1 μg/g dm). Averaged across the species, total soil selenium decreased by 55% in the first year and 16% in the second. Postharvest values were in all cases significantly different from the unplanted control, at least at the 5% (P=0.05) level.

Table 12.4. Mean concentrations of total selenium in soil (0-60 cm) at Los Baños.

Species	Presowing	Postharvest	% Change
1990			
Unplanted	0.49	0.43	12
Brassica juncea	0.50	0.10	80
Festuca arundinacea	0.46	0.24	52
Lotus corniculatus	0.39	0.12	69
1991			
Unplanted	0.88	0.86	2
B.juncea	0.86	0.63	27
F.arundinacea	0.65	0.55	15
L.corniculatus	0.82	0.71	13
Hibiscus cannabinus	0.75	0.61	19

After: Bañuelos *et al.* (1993a).

As is necessary for the development of any phytoextractive technology, Bañuelos and collaborators have also determined the impact of other factors on selenium uptake in a series of papers. First, sulphate is potentially an antagonist to selenate in plant uptake, and although broccoli florets contained the highest concentrations of both sulphur and selenium, selenium concentrations were negatively correlated with sulphate concentrations in leaf tissues (Bañuelos and Meek, 1989). Sulphate at high concentrations could, therefore, decrease the efficiency of selenium removal. Drainage water from irrigated land is often saline to variable degrees. Therefore, plants grown in these environments must be tolerant of salts when grown in either the soil environment or in salt-affected drainage waters. *B.juncea* is reasonably salt tolerant and accumulated selenium and boron when grown in water enriched with all three substances (Bañuelos *et al.*,1990). The influence of phosphate has also been examined, in the context of including phosphorus in drip irrigation water. Although phosphorus did not affect yields of *B.juncea,* it was shown to significantly increase the plant selenium concentrations and decrease soil selenium (Bañuelos *et al.*,1992a). Phosphate could therefore be a means of optimising selenium removal from soils, but the mechanism of the phosphate response was not investigated. The effect of cutting the plants is another important management factor which has been studied. It appears that clipping enhanced the loss of selenium from the soil in a greenhouse experiment (Bañuelos and Meek, 1990), but again the mechanism of this effect was not examined.

One important issue inevitably surrounding any type of bioremediation is the subject of verification under the variable conditions in the field. In the simplest sense, this takes the form of measurements of the elemental burden in the soils before and after the growth of the phytoextractive crops. As seen in the above examples for selenium, this can be difficult, especially in this case where the usual need to take spatial variation in the soils is compounded by the complex nature of the selenium removal processes.

Statistically, it is known that in an experimental block, the selenium content of the soil was 0.70 $\mu g/g$ with a coefficient of variation of 5% before planting 5 different species, and that after growth, this decreased to 0.4 $\mu g/g$ with a CV of 16%. However, the breakdown of this for each individual species was not given (Bañuelos *et al.*,1992b). Although *B.juncea* gave a 75% reduction in soil selenium, *L.corniculatus* 70% and *F.arundinacea* 48% respectively (Bañuelos *et al.*,1993b), there was no significant difference between the species, even though the latter two are not sulphur-rich brassicas. In the 1990 experiment, *B.juncea* had a much greater yield and selenium concentration than *F. arundinacea* (Bañuelos *et al.*,1993a). A combination of the natural variation in the soil/plant system in the field and the difficulties of assigning the decrease in total soil selenium to plant or microbial processes may have made it difficult to prove that *B.juncea* is superior in uptake and phytovolatilisation of selenium in the field.

The removal in the crop itself is readily quantified, but the volatilisation from the plant and soil compartments is not. The latter appears as *unaccounted* selenium on the balance sheet, for example in Table 12.2. However, to date the pathways of the loss have not been fully quantified in the field. Crops such as *B.juncea* are able to volatilise selenium, but so are soil microorganisms. Indeed, bioremediation by the stimulation of soil microbial activity has been investigated as a means of remediating selenium-laden soils (Frankenberger and Karlson, 1992). In this process the microbial activity is stimulated by adding carbon substrates such as agricultural wastes (Frankenberger and Karlson, 1992), water and nutrients. Growth of crops also provides these materials to the soil microbiota, due to standard farming practices and the process of carbon loss from roots called rhizodeposition. It is interesting that rhizodeposition increases after clipping plants (Terry and Zayed, 1994), and this could partly explain the enhanced selenium removal after cutting treatments, plus the enhanced biomass production.

Selenium emissions from crop plants decreased in the order (on a dry matter basis): broccoli, cabbage, rice (Terry *et al.*,1992). Less was volatilised by other crop and vegetable species. The volatilisation of 20 μM selenate was studied in chambers with root and shoot separation, and it was shown that the roots of broccoli volatilised selenium more than 20 times faster than shoots (Fig.12.1). Complete removal of the shoot increased the volatilisation by roots by 20-30 times that of intact plant roots. Effects of de-topping broccoli lasted for at least 72 h (Zayed and Terry, 1994). Similar results were obtained with rice, cabbage, cauliflower, Chinese mustard and Indian mustard.

Presence of increasing concentrations of sulphate in the medium decreased selenium volatilisation (Zayed and Terry, 1994), presumably because selenate is chemically similar to sulphate, and there is competition between the two for incorporation into amino acids and proteins which are precursors to volatile selenium compounds. It is also known that plants volatilise selenium faster when presented with selenite (reduced) and even more with selenomethionine (Zayed and Terry, 1994). These are the opposite to the effects of selenium speciation in soil on the rate of uptake by plants.

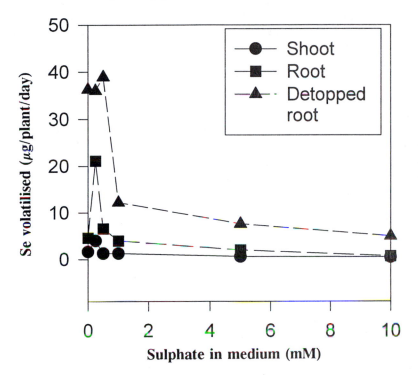

Fig.12.1. Volatilisation of selenium from shoot, root, and detopped roots of broccoli. After: Zayed and Terry (1994).

The compounds volatilised are dimethylselenide and, in selenium hyperaccumulators, dimethyldiselenide (Terry and Zayed, 1994). However, the exact mechanisms of selenium volatilisation remain unclear. It is possible that microbes either in the rhizosphere or inside the root free space enhance selenium volatilisation, and that these are stimulated by detopping. Terry and Zayed added prokaryotic and eukaryotic antibiotics to the root medium and showed that prokaryotics decreased volatilisation the most. In other words, bacteria seem to be more important than fungi. Volatilisation seems not to be solely bacterial, however, as the non-sterile nutrient solution alone, showed only little selenium volatilisation. Those authors speculated that bacteria may assist plants in chemical reduction or transport of selenium in some unknown way.

Despite the lack of information on mechanisms, selenium is still in many ways

a leader in the field of phytoextraction, because of the length and breadth of study
of the systems involved. It can be seen as a model for the areas which need
investigation in the phytoextraction of other elements, for example the following.

1. Bioavailable element stripping (BES)

This acknowledges that all of the element does not need to be removed
(McLaughlin and Hamon pers. comm.). Often it is the small bioavailable fraction
which needs to be removed from soil. In the case of selenium, this is selenate,
which is more mobile in crops and waters, and therefore into the food chain. To
make BES most efficient, the biomass has to be cut before the older leaves
senesce and fall on the soil, returning selenium (Bañuelos and Meek, 1990).

Extractable selenium in soil decreased by 34% more than the control at final
harvest when *B.juncea* was grown, but total selenium decreased only by 18%
(Table 12.2). However, the redistribution of selenium from non-soluble forms in
soil needs to be studied, and this takes time.

2. Effects on the food chain

It is known that concentrations > 3 $\mu g/g$ dm selenium are potentially harmful
to animals. Obviously, keeping animals out of fields in the case of selenium-
accumulating crops is a possible management option.

3. Effects on the ecosystem

At the Kesterson reservoir, selenium in the grassland growing on the dried out
evaporation ponds was studied, and the transfer from a primary consumer
(grasshopper) to a secondary consumer (praying mantis) quantified. The grassland
did show biomagnification of selenium in the food chain, but compared with the
similar wetland, the bioaccumulation was 90% less. Grassland was therefore a
safer type of environmental management, although areas with high concentrations
of selenium still lead to toxic concentrations of selenium in the food chain, and
may need remediation (Wu *et al.*,1995).

4. Toxicity of volatile species

Two factors mitigate the toxicity of volatile forms of selenium. One is that the
volatile species produced undergo large dilution with uncontaminated air prior to
any redeposition on the ground. The other is that the species dimethylselenide and
dimethyldiselenide exhibit markedly lower mammalian toxicity than the inorganic
forms (Frankenberger and Karlson, 1992).

5. Disposal or utilisation of plant materials

In grazing animals, selenium deficiency is more common than selenium
toxicity, and it has been shown that the selenium-enriched plant material can be
used as either a supplement to animal feed or can be incorporated in soil, and
release useful amounts of selenium to fodder crops (Bañuelos and Meek, 1990;
Bañuelos *et al.*,1992b). However, caution has to be exercised, as the margin

between selenium deficiency and toxicity in forages is narrow.

6. Presence of other deleterious elements in phytoextractive crops

Areas affected by selenium in California also contain other potentially toxic elements, and the latter have been quantified in the phytoextractor crops. Only cadmium, selenium and boron were found to be potentially limiting elements for the blending of *H. cannabinus* and *B. juncea* into animal feed.

Phytoextraction of boron

Many of the same processes and comments apply to boron in soils in arid and irrigated regions as those for selenium, except that gaseous species are absent. Boron accumulates in soils which are irrigated and experience high evapotranspiration rates. Boron can be toxic to plants over a narrow range of concentrations. For example, cotton (*Gossypium hirsutum*) and kenaf (*Hibiscus cannabinus*) grown in soil with 45 μg/g total boron and 7 μg/g water-soluble boron (in saturation paste) showed yield decreases of 27 and 15% respectively and contained 422 and 222 μg/g dm boron (Bañuelos *et al.*, 1996).

However, plants can accumulate 100-200 μg/g boron without yield decrease, and their growth has been shown to decrease soil total and extractable boron in the field (Table 12.5). A mass balance showed that *B. juncea*, *F. arundinacea* and *L. corniculatus* decreased soil total boron by about 7% compared with the control. The percentages of boron lost from plant tissues were 76, 92 and 57 respectively (Bañuelos *et al.*, 1993a). In contrast, extractable boron was decreased by a much greater extent; about 45% in 1990 and 25% in 1991. Use of any of the four species for BES would lead to decreases in extractable boron in soils, which is the most toxic and leachable species, thus alleviating two problems: toxicity to subsequent crops and the potential toxicity of drainage water.

Table 12.5. Percentage changes in total and extractable boron concentrations in bare or cropped Los Baños soil.

Species	Total B	Extractable B
1990		
Control (bare soil)	3.1	10.8
Brassica juncea	10.5	57.2
Festuca arundinacea	10.8	54.2
Lotus corniculatus	10.8	56.0
1991		
Control (bare soil)	3.0	3.9
B. juncea	7.3	31.5
F. arundinacea	5.7	26.0
L. corniculatus	9.8	28.3
Hibiscus cannabinus	10.8	26.0

After: Bañuelos *et al.*, 1993a.

One issue surrounding BES in this context is replenishment of soluble boron from the total pool (Peryea *et al.*,1985). In the studies of Bañuelos *et al.* (1993a), this did not occur on the uncropped soils, but even so, this may occur in the presence of a sink in the form of growing plants.

Based on pot experiments, Bañuelos *et al.* (1993b) contend that *B.juncea* is suited to short-term boron removal from soil, due to its high and constant boron uptake. However, *H.cannabinus* shows higher tissue boron concentrations (> 600 μg/g dm - Bañuelos *et al.*,1993c).

For animal nutrition, boron concentrations above 150 μg/g dm are not recommended and the accumulation of associated elements needs to be taken into account, as indicated in the last section.

Phytoextraction of zinc and cadmium

Zinc and cadmium are ubiquitous pollutants, which tend to occur together at many contaminated sites. Zinc is phytotoxic, and reduces crop yields before any potential harmful effects on the food chain. However, cadmium is a food chain toxin which rarely inhibits plant growth. Cadmium is included here with zinc, because even though it appears not to be an essential element, its chemical similarity to zinc means that both elements could be taken up by the same hyperaccumulator species.

Many species are known that hyperaccumulate the essential element zinc, but only one, *Thlaspi caerulescens,* is recorded as hyperaccumulating cadmium (> 100 μg/g dm). It may be that study of cadmium has been somewhat neglected and that more effort may reveal more hyperaccumulators of this element. Many of the zinc hyperaccumulators are in the family Brassicaceae, and are in the genus *Thlaspi*. In addition, *Cardaminopsis halleri* and *Viola calaminaria* are also known to hyperaccumulate zinc (> 10,000 μg/g dm zinc). In the case of cadmium, the threshold for cadmium hyperaccumulation has been set at 100 μg/g dm (Baker *et al.*,1994). Despite the similarity between these metals, there appears to be some discrimination of uptake, as shown by McGrath *et al.* (1993), who reported two populations of *T.caerulescens*, one of which took up more zinc than the other and *vice versa* for cadmium, even though they were grown on the same field plots.

First indications of the possibility of phytoextraction of zinc and other associated metals were given from pot experiments in background and sludge-treated soils, and estimates of the yield of *T. caerulescens* under field conditions (Baker *et al.*,1991). It was estimated that a crop of *T.caerulescens* could take up 34 kg/ha Zn, 0.16 Cd, 0.25 Ni, 0.22 Pb, 0.4 Cu and 0.27 Cr. Subsequent experiments were performed in the field on soil contaminated with metals due to twenty years of additions of sewage sludge applications aimed at quantifying metal removal in a number of known and putative hyperaccumulator species in comparison with two crop plants (McGrath *et al.*,1993). Additionally, this study attempted to determine whether uptake would decline over time. The soil was contaminated mainly with zinc (total metal concentrations (μg/g) were: 124-444 Zn, 26-138 Cu, 20-35 Ni, 2-13 Cd and 40-136 Pb), and the results for plant uptake are shown in Table 12.6. Several important points came out of this study, including:

1 - Zinc concentrations were maintained across a wide concentration range in soils. This result has good potential, because if a site is being cleaned up, each successive harvest lowers soil concentrations and if zinc accumulation in the biomass declined, the efficiency of clean up would fall with time.

2 - Zinc accumulation from relatively low soil concentrations occurred. In most plots this did not reach the 10,000 μg/g defined as hyperaccumulation of zinc, but 10,625 μg/g zinc was achieved in plant material sampled from a row in one of the plots containing 406 μg/g zinc in the soil.

3 - Cadmium accumulation, though not surpassing 100 μg/g, was high in *T.caerulescens* and *C.halleri*. Because the soil contained 2-13 μg/g total cadmium, removal at this rate would take some time, but considering instead soils with around 1 μg/g, due to either addition of sewage sludge or cadmium-contaminated fertilisers, which are of concern in some European and Australasian countries (McGrath *et al.*,1994; McLaughlin *et al.*,1996), cadmium remediation by phytoextraction seems possible.

Table 12.6. Maximum (A) and mean (B) concentrations of zinc (kg/t dm), maximum (C) and mean (D) yields (t/ha) and maximum (E) and mean (F) removals (kg/ha) in selected test species.

Year/Species	A	B	C	D	E	F
1991						
Thlaspi caerulescens (1)*	10.63	3.86	5.4	4.4	57.4	17.0
T.caerulescens (2)*	6.33	3.65	8.7	5.7	55.1	20.8
Cardaminopsis halleri	5.21	3.21	2.7	2.0	14.1	6.4
1992						
T.caerulescens (1)*	2.70	2.18	8.6	5.0	23.2	10.9
T.caerulescens (2)*	1.55	1.34	8.8	3.7	13.6	5.0
C.halleri	1.55	1.34	8.8	3.7	13.6	5.0
1993						
T.caerulescens (1)*	3.78	3.47	9.9	4.6	37.4	16.0
C.halleri	-	1.93	-	2.4	-	4.6

*Two different populations.

For example, *T.caerulescens* or *C.halleri*, with field removal rates of 150 and 34 g/ha/crop (McGrath and Dunham, 1997) could remove decades worth of cadmium accumulation from agricultural land where phosphorus fertilisers have added in the order of 3 g/ha/yr of cadmium. Also, if the process is thought of in terms of BES, even though the added cadmium can be very small (in the order of addition of 0.2-0.3 μg/g to a soil with total cadmium content of 0.2 μg/g), these small increases in the bioavailable fraction of cadmium still cause breaches of food regulations for crop and animal products (McLaughlin *et al.*,1996; Roberts *et al.*,1994). BES could in theory remove this important fraction and lower the environmental impact in about four harvests using these plants.

Brown *et al.* (1994) grew *T. caerulescens, Lycopersicon esculentum* and *Silene vulgaris* in a pot study to examine zinc and cadmium removal from soil at three sites impacted by a zinc smelter. *T. caerulescens* showed much greater tolerance than the other two species to zinc and cadmium. It contained 18,000 μg/g zinc and 1020 μg/g cadmium dm in the shoots with no effect on yields. In this study, BES was demonstrated by the change in 1:5 soil:water extractable metals. *T. caerulescens* significantly decreased water-extractable zinc in all three soils and cadmium in one of the soils (Table 12.7).

Table 12.7. Percentage decrease in water-extractable zinc and cadmium in three soils contaminated by a smelter after growth of *Thlaspi caerulescens*.

Site sampled	Zn	Cd
Farm	28*	10
Garden	17	22
Mountain	64*	70*

*Significantly different (P < 0.05) compared with the two other species grown in the same soils.
After: Brown *et al.* (1994).

McGrath *et al.* (1997) showed that the decreases in the fraction of zinc extracted by 1M NH_4NO_3 were 0.1-0.8 μg/g when *T. caerulescens* (Pryon) was grown in three different soils. These decreases did however account for 1-9% of the total uptake of zinc, considering only the shoot contents of the hyperaccumulator plant. In other words, more than 90% of the total zinc taken up, must have come from the non-extractable fractions in the soil. Although 1M NH_4NO_3 is a weak extractant that only removes the readily more mobile and exchangeable forms of zinc in soil, these results demonstrate that BES taps dynamic pools of metal in the soil, which probably re-equilibrate. The questions which remain to be answered concern whether this results in an overall decrease in the most toxic pool which is available to subsequent agricultural crops and the food chain, and whether this effect is lasting.

It was not possible to detect changes in the total concentrations of the two metals in the above pot experiments, because of the short-term nature of the trials (McGrath *et al.*,1997) and the large total concentrations of zinc and cadmium in these soils (Brown *et al.*,1994). It can be calculated that 1.44 μg/g zinc and 44 ng/g cadmium per week were removed from the soil. The authors offered a model of the phytoremediation of soil based on the results of this pot experiment which showed that it would take 28 and 15 years respectively to remove all the zinc and cadmium from the "farm" soil that had 2100 and 38 μg/g total zinc and cadmium respectively. However, there are two points missing from the original discussion:

1 - it is not desirable to remove all the metals (especially the essential nutrient zinc), and BES may be required to strip only a fraction of the total which may be little more than that which was water soluble and easily desorbable;

2 - the removal rate is not constant, and may vary with the age of the plant and cannot continue during winter, unless artificial inputs of heat and light are supplied.

Obviously, what is needed is information from the field, under more realistic conditions where there is likely to be variation in other nutrients, water, disease etc.

McGrath and Dunham (1997) showed, using data from field experiments in the UK, that *T. caerulescens* and *C. halleri* could remove a maximum of 41 and 0.15 kg/ha/yr of zinc and cadmium (i.e. a hypothetical crop with the largest mean yields observed in the field plots and average concentrations of metals). However, the average removals were 15 and 0.05 kg/ha/yr respectively. It is important for phytoextraction to establish whether these removals would decline with time. From the evidence of two moderately contaminated sites (Brown *et al.*,1995a; McGrath and Dunham, 1997), this does not seem to be the case, but yields did vary over time, presumably due to climatic and crop growth variations (Table 12.6).

McGrath *et al.* (1993) calculated that to remove zinc from their most contaminated plot (440 μg/g zinc in soil) to an acceptable 300 μg/g would take nine years with one crop grown each year. This was assuming maximum observed yields of two rows of plants in the field and their concentrations of zinc. It also assumed that the uptake would be constant with decreasing soil concentrations, which seems to be the case, based on these data and those of Brown *et al.* (1995a). If this species were to be grown as a crop, these conditions would have to be met. The results show what could be achieved in future with a new crop which has similar performance in terms of metal accumulation but engineered for higher yields. The calculation is in line with that of Brown *et al.* (1995a) except that they targeted background zinc concentrations (40 μg/g) and quoted 18 growing seasons, which reduces to only eight if the target of 300 μg/g is accepted.

The question of yields and plant metal concentrations has been most discussed in relation to zinc (Chaney *et al.*,1997; McGrath and Dunham, 1997). Both are key variables in the efficiency of phytoextraction processes. The relative importance of these two factors for zinc and cadmium is illustrated in Tables 12.8 and 12.9 using theoretical but realistic values for non-hyperaccumulator and hyperaccumulator plants and the extent of soil contamination they may be employed to clean up.

Although it was stated above that cadmium and zinc are similar, it is now clear that there are differences in terms of their accumulation by hyperaccumulator plants. One obvious difference is the extent of accumulation. For example, the field study of Brown *et al.* (1995a) contained only a maximum of 180 μg/g zinc in the soil, and lettuce contained similar concentrations of cadmium as *T. caerulescens*, and the yield of lettuce was greater. However, they used the low cadmium population of *T. caerulescens* in their study (McGrath *et al.*,1993) and it must be remembered that lettuce does not have the tolerance to zinc which *T. caerulescens* has, and so it would not grow well on more contaminated soils. But this does raise the question in practice as to which plants would be best to phytoextract metals from soils which are mainly cadmium-contaminated, or soils which are only modestly polluted. In this case, high biomass crops such as maize may be useful, especially with addition of metal chelating or acidifying compounds which would boost the metal concentrations in the crops by desorbing the metals from the soil and overcoming any diffusional limitations to their transport to the site of uptake in the roots and translocation to the shoots. Development of phytoextraction must therefore be a process of matching the plants to the particular situation.

Table 12.8. Removal of zinc in a hypothetical 10 t/ha dm crop growing in soil* contaminated with 1000 μg/g zinc with a target of 50 μg/g, showing the importance of hyperaccumulation (>10,000 μg/g zinc).

μg/g in crop	kg/ha removed	% of soil total in 1 crop	Years to target
100	1	0.04	2470.0
1000	10	0.38	247.0
10,000	100	3.85	24.7
20,000	200	7.69	12.4
30,000	300	11.54	8.2

*Assuming a 20 cm deep layer with a density of 1.3.

Table 12.9. Removal of cadmium in a hypothetical 10 t/ha dm crop growing in soil* contaminated with 1 μg/g cadmium, with a target of 0.2 μg/g, showing the importance of hyperaccumulation (>100 μg/g in shoot tissue).

μg/g in crop	kg/ha removed	% of soil total in 1 crop	Years to target
10	0.1	4	20.80
100	1	38	2.08
1,000	10	385	0.21
10,000	100	3846	0.02

*Assuming a 20 cm deep contaminated layer of soil with a density of 1.3.

Cadmium and zinc are also expected to differ not only because of differences between populations in their Zn/Cd specificity in accumulation (McGrath *et al.*,1993), but also in relation to their response to concentrations in the soil. McGrath *et al.* (1993) showed that *T. caerulescens* responded to increasing soil zinc concentrations by maintaining consistent hyperaccumulation of the metal, but although Brown *et al.* (1995b) found the same for zinc, the uptake of cadmium was more of a function of soil conditions, especially pH and extractable cadmium concentrations. This contrast between cadmium and zinc again shows that there are basic differences in the mechanisms of accumulation of the two metals in hyperaccumulators, a factor which is important for the future development of phytoextraction techniques.

Finally, it has now been demonstrated that zinc hyperaccumulators have a higher requirement for zinc than most other plants (Shen *et al.*,1997). Plants of the Pryon population of *T. caerulescens* were grown for 30 days then transferred to DTPA-buffered solution culture for 3 weeks, which exposed them to a range of zinc activities from $10^{-5.2}$ to $10^{-12.0}$M as calculated by GEOCHEM-PC. The yields decreased sharply at between $10^{-5.2}$ and $10^{-9.6}$M, and remained very low at lower concentrations (Fig.12.2). Also the plants with smaller yields became yellow and the leaf chlorophyll concentrations decreased as determined by a Minolta SPAD meter (Fig.12.2).

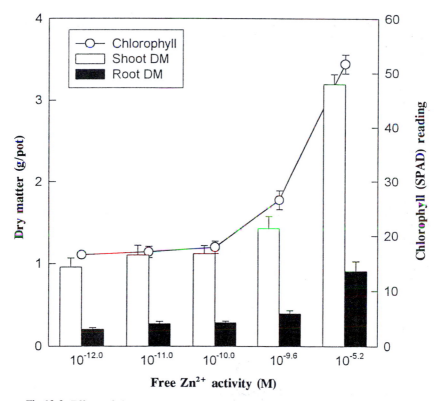

Fig.12.2. Effects of zinc treatments on foliar chlorophyll concentration and shoot and root (dry mass) of *Thlaspi caerulescens*. Vertical bars represent standard errors. Source: Zhao, Shen and McGrath - unpublished data.

The external concentration of zinc which gave optimal growth was found to be about five orders higher than that required by normal plant species. Even in the severely zinc-deficient plants grown on treatments with 10^{-10} to 10^{-12}M zinc, the shoots contained about 300-500 $\mu g/g$ dm zinc, which was far greater than the critical deficiency values of 15-20 $\mu g/g$ dm zinc widely reported for other plants. The minimum zinc concentration in shoots required for the growth of *T. caerulescens* is likely to be around 1000 $\mu g/g$ zinc (Zhao, Shen and McGrath, unpublished data). It is envisaged that the high internal and external requirement is due to the fact that the detoxification mechanism which ensures the extraordinary hypertolerance of *T. caerulescens* to zinc operates even when the external zinc supply is low. This may mean that when used on contaminated sites for phytoextraction, the plants would show a degree of containment on those sites because of their high zinc requirement, and would:

1 - indicate when available soil zinc was too low for their growth (by visible deficiency);

2 - leave sufficient available zinc in the soil for the subsequent growth of most other plant species.

This high requirement is likely to be a primary reason why hyperaccumulators of zinc are limited to metalliferous soils containing large amounts of the metal. Future phytoextractor crops may be engineered to possess similar hyper-accumulation and hypertolerance mechanisms for zinc. The possibility of the plants escaping and becoming a weed problem is therefore theoretically negligible.

Potential phytoextraction of lead

Lead is a difficult target metal for phytoextraction, because it is strongly held by soil organic matter and soil minerals. Relatively few plants have been recorded to hyperaccumulate lead, e.g. *Thlaspi rotundifolium* was reported to contain 130-8200 $\mu g/g$ (Reeves and Brooks, 1983), and there is some doubt about those which do, as surface contamination cannot be ruled out. It was indeed observed (R.R.Brooks -pers. comm.) that some specimens of this plant with very high lead contents were growing very near to a smelter chimney and might have been contaminated by metal-rich fumes from this source. Much of the lead accumulated in plants is held in the roots (Huang and Cunningham, 1996).

As established above for zinc, in order to make phytoextraction of lead contaminated soils feasible over relatively short periods, it is necessary to strive for plants with 10,000 $\mu g/g$ lead and > 20 t/ha/yr shoot dm. Kumar *et al.* (1994) screened the major crop brassica species for lead accumulation in sand/perlite, and found a range from 1416 to 18,812 $\mu g/g$ dm in the shoots, with the highest in *B.juncea*.

Huang and Cunningham (1996) studied uptake in a wider range of crops, and wild species growing on lead contaminated sites, and found the most lead in *Zea mays*, after either two weeks in solution culture or 4 weeks in contaminated soil (Table 12.10). A field trial with the same soil gave broadly similar results for *Zea mays* and *Helianthus annus* (Huang *et al.*,1997).

The most promising method of increasing plant lead uptake found to date seems to be mobilising lead from the soil particles using chelates. Huang and Cunningham (1996) applied 0.2% HEDTA (EDTA as the acid) to a sandy loam soil, pH (H_2O) of 5.1, containing 2500 μg/g total lead and found that lead concentrations in the shoots increased by a factor of 265 after one week, at which time the plants had died. Later, they tried a range of compounds, and EDTA proved the best desorbing agent (Huang et al.,1997).

Table 12.10. Concentrations of lead (μg/g) achieved in the shoots of various crop and wild species.

Species	Hydroponic culture	Soil culture
Zea mays	375	225
Brassica juncea (211000)	347	129
B.juncea (426308)	329	*
B.juncea (531268)	241	97
Thlaspi rotundifolium	226	79
B.juncea (175607)	176	*
Triticum aestivum	139	120
Ambrosia artemisiifolia	96	75
B.juncea Cern.	65	45
Thlaspi caerulescens	64	58
B.juncea (180269)	59	*
B.juncea (184290)	32	30

*Lead toxicity prevented growth. Source: Huang and Cunningham (1996).

One of the nutritional problems encountered by plants on lead-polluted soils is the lack of phosphate. This is due to the formation of insoluble lead phosphate (pyromorphite). The lack of phosphorus can be overcome by application of phosphate to the aerial parts of the plant, increasing the biomass and total amount of lead in the shoots by >400% and 115% respectively, although the lead concentration in the biomass decreased (Huang et al.,1997).

To date there is only one published field experiment on phytoextraction of lead (Huang et al.,1997), although trials are now being conducted at numerous sites (Cunningham and Berti, 1993). It seems likely that the commercialisation of lead phytoextraction will include the following components:

1 - application of a chelate such as EDTA to the soil/plant,

2 - a system for containing the leakage of water through the soil, so as to avoid groundwater pollution by metals mobilised by the chelates,

3 - phosphate supply to the plant by foliar spraying,

4 - modifying the growth of the plants in some way, for example increasing root density or depth of rooting,

5 - harvesting young, or killed older plants with higher biomass,

6 - reducing the volume of plant material to be disposed by physical, chemical or biological means.

Any compounds which are used for mobilising metals in soils to increase plant uptake need careful control. Their degradation and potential for toxicity to the ecosystem needs to be examined. For example, EDTA has a half life when applied to soils at 1-2 g/kg of about 10-20 days (Means *et al.*,1980). At present, the major problem appears to be that plant growth is severely curtailed after the lead availability is enhanced by adding chelating compounds to the soil.

The published papers do not give any yields of the plants which accumulated concentrations of lead in excess of 10,000 μg/g, making it impossible to calculate the efficiency of phytoextraction. Therefore, either the non-tolerant plants have to be harvested when senescent, or lead-tolerance needs to be engineered into the plants which accumulate large concentrations of lead. One calculation is that *Zea mays*, with a shoot yield of 25 t/ha/crop, if cropped twice each year, containing 10,600 μg/g lead would take 7 years to reduce soil lead from 2500 μg/g to 600 μg/g, the acceptable soil concentration for non-residential areas in the state of New Jersey (Huang and Cunningham, 1996).

No explicit information is available on the BES of lead from soils. However, Huang and Cunningham (1996) state that the increase in lead in the plants after application of chelates relates to the amount of lead in soil solution. This indicates that prior to application, much of the lead was in more insoluble pools which were not phytoavailable (Cunningham *et al.*,1995). Again, it is likely that the pool containing bioavailable lead is small in many soils and would be relatively easy to take out. However, research on the buffering of these pools is required.

One other way of enhancing lead uptake by plants is to combine technologies. Thus, combining the chemical process of electroosmosis with plant uptake is said to lead to enhanced lead uptake (Huang *et al.*,1997). Perhaps this is a concept which should be explored for the phytoextraction of other elements.

Phytoextraction of Radionuclides

No cases of phytoextraction of radionuclides have been reported so far, although apparently many investigations are in progress (Black, 1995; Cornish *et al.*,1995). Remediation of the large areas of land globally which are contaminated with low levels of radionuclides would be prohibitively expensive using physical or chemical techniques (Entry *et al.*,1996). Most attention is directed at the longer lived nuclides ^{137}Cs and ^{90}Sr which persist in soils, although plant accumulation of uranium ions also require study. Caesium and strontium have similar atomic configurations to potassium and calcium, which are taken up in large quantities by plants.

Review of the compilation by Chapman (1966) reveals no species that hyper-accumulate these elements, that is to say contain concentrations which are orders of magnitude greater than other plants. Thus, it is unlikely that the hyperaccumulation strategy is possible, but there are plant species which accumulate substantial amounts of radionuclides, and their uptake could be enhanced by changing the environment around their roots.

It has been suggested from a knowledge of the uptake of ^{137}Cs, ^{90}Sr and ^{238}U by plants and the concentrations of these nuclides in soils at contaminated sites in the USA, that hyperaccumulation may not be necessary (Cornish *et al.*,1995). In a pot experiment, *Eucalyptus tereticornis* seedlings extracted 31% of the ^{137}Cs and 11% of the ^{90}Sr present in a sphagnum peat medium in one month (Entry and Emmingham, 1995).

Development of phytoextraction will require screening for plant species which accumulate the nuclides, and which are adapted to grow in the environments affected by weapons testing and fallout from nuclear reactors. Species under test include *Brassica juncea, B.napus, B.rapa* and *Helianthus annuus* (Black, 1995). Their uptake could be enhanced by addition of mobilising or chelating agents to the soil. Some fertiliser applications are likely to decrease plant uptake, as potassium and calcium would compete with uptake by ^{137}Cs and ^{90}Sr (Entry *et al.*,1996). On the other hand, nitrogen in particular is likely to enhance plant growth and uptake of the nuclides.

Different soils have different degrees of fixation capacity for caesium and strontium, so the applicability of phytoextraction is likely to vary according to soil properties. In general, plants grown in organic soils usually accumulate larger concentrations of these elements than those grown in clay soils which have a larger capacity to fix the elements.

Another factor which may increase uptake of radionuclides is mycorrhizal infection. The fungal hyphae in this symbiosis increase the surface area for absorption and exploration of the soil. The effect of mycorrhizas need to be investigated as this may enhance the process considerably (Entry *et al.*,1996).

Phytoextraction of Other Elements

Introduction

With the advent of reliable multi-element instrumental techniques for analysing plant materials, one potentially profitable use of elemental "fingerprints" (Markert, 1992) of plant species would be to find those which accumulate significantly more trace elements, platinum group elements, lanthanides or actinides than others. "Significant" here implies orders of magnitude greater concentrations.

Often the concentrations depend on the location in which the plants were growing, but if they are consistently high, these could be true physiological accumulators. These species if found, would have value in future phytotechnologies for extracting elements which cannot be extracted at present.

Nickel

The first realistic suggestions concerning phytoextraction of metals featured nickel (Chaney, 1983). From a knowledge of the existence of nickel hyperaccumulating plants, he made hypotheses based on a plant assumed to contain 1% nickel in fresh leaves and 25% in the ash, which is a rich nickel ore. If a soil containing 1000 $\mu g/g$ nickel in the surface 15 cm, were to be cropped with a hyperaccumulator containing 10,000 $\mu g/g$ dm yielding 5 t/ha dm, this would remove 50 kg/ha/yr, and that the site could be cleaned in 20 years. Recent research has used the high biomass species *Berkheya coddii* from South Africa, which has four times this biomass (R.R.Brooks, pers. comm.). This shows promise for rapid cleaning of nickel-affected soils.

Because nickel is a relatively high value metal, this concept has been tested recently in the field of phytomining (Nicks and Chambers, 1995; Robinson *et al.*,1997) . The difference between phytomining and phytoextraction is very subtle, as both remove substantial quantities of metals from soil or substrate, and will leave the substrate with lower amounts of bioavailable metals which were potentially injurious to the environment. The distinction appears to be related to the aim of the processes. Phytoextraction is for remediation of the environment, whereas phytomining looks primarily at the economic value of the metal recovered. On the other hand, the value of the metals in phytoextraction can be significant, and help to pay for the costs of clean-up when they are recycled.

Nickel is a metal which occurs naturally at large concentrations, mainly in serpentine soils (McGrath, 1995). It is unusual for it to be present at high concentrations solely as a result of industrial uses or in waste materials. The application of phytoextraction for nickel is therefore mainly in specific areas where it is mined and processed. Often the by-products of these processes cover thousands of hectares. More information on the phytoextraction of nickel is given elsewhere in this book (Chapters 14 and 15 on phytomining).

Mercury

Many areas are contaminated with mercury due to activities such as industrial bleaching, dredging of waterways containing mercury, past agricultural use of mercury as a pesticide, or the mining of gold. Mercury is strongly bound by soils, and is not readily available to plants, so no accumulators are known. Relatively recently, molecular biologists have engineered the genes for mercuric ion reductase (which reduces toxic Hg^{2+} to the relatively inert Hg^0) into *Arabidopsis thaliana*. The sequence, *merApe9*, was constructed from mutagenised m*erA* and needed plant regulatory sequences to function in the transgenic *A. thaliana* plants (Rugh *et al.*,1996). These plants absorbed Hg^{2+} and were tolerant to this ion and evolved considerable amounts of Hg^0 presumably because of its inherent volatility. So, any phytoremediation process which is developed from this approach is essentially phytoextraction followed by non-biological volatilisation. The above authors are determining whether these genes can function in other plants which produce greater biomass than the diminutive *A. thaliana*.

Interestingly, the above constructs required some mercury in the medium for optimum growth (Rugh *et al.*,1996) and this requirement, like the increased requirement for zinc, may mean that phytoextractive plants would be tied to the contaminated sites, and should not escape to pose a weed problem.

Arsenic

It is strange that similar approaches to those with selenium and mercury have not yet been reported for arsenic. This element is a problem in large areas, due to its production by metal smelters, and use as a herbicide. Plants are able to take up and tolerate arsenate, which is a similar anion to phosphate. No hyperaccumulators of arsenic are known, but methyl arsenic compounds or arsine gas could be formed by a combination of plant and/or microbial metabolism. Plants through photosynthetic activity have a relatively good supply of reducing power, which is only indirectly available to microbes involved in bioremediation. Microbes in symbiosis with plants or in the rhizosphere can tap into this energy and this could be a future way of removing arsenic and other elements which have volatile reduced compounds from soils using plants.

Thallium

A thallium hyperaccumulator (>1000 μg/g dm) has recently been reported (R.R.Brooks pers. comm. 1997). Thallium is potentially toxic and occurs at elevated concentrations in the environment due to emissions around cement works using pyritic smelting residues which contain thallium, smelting and metallurgical industries. The species *Iberis intermedia* is able to accumulate 2132 μg/g thallium in its leaves on a soil containing only 40 μg/g thallium. Hyperaccumulation occurred on mine tailings where lead had occurred in the sulphide form. The unfertilised yield of *I.intermedia* was less than 2 t/ha dm, but this could be increased by addition of fertilisers. Brassica species accumulated up to approximately 450 μg/g thallium dm near a cement plant in Lengerich, Westphalia in Germany (Kemper and Bertram, 1991). Apparently, scientists at Hohenheim in Germany are screening plants including *Zea mays*, *Brassica napus* and *B.juncea* for their ability to phytoextract thallium from soils impacted by a cement factory (Kurz *et al*, 1997).

Copper and Cobalt

Copper and cobalt hyperaccumulators exist (Baker and Brooks, 1989), but as yet there has been no demonstration of their use in phytoextraction. Some 24 copper hyperaccumulating species have been recorded in families as diverse as the Cyperaceae, Lamiaceae, Poaceae and Scrophulariaceae. Cobalt and copper hyperaccumulators tend to be found in the south central part of Africa in Zaïre and Zambia on copper- and cobalt-rich soils. There is some controversy about the ability of plants to hyperaccumulate copper (J.A.C.Smith pers. comm.), and it is difficult at present for researchers to obtain plant material, especially from Zaïre, due to civil unrest.

Salt

Accumulation of salt in surface soils threatens agricultural production in many parts of the world, and it seems logical to ask whether accumulator plants can help remediate the problem. This was examined by Hussain (1963) for saline-sodic soils and found to take 4-6 years with rice and *Sesbania aculeata*. However, *Leptochloa* reportedly reduced the SAR of the plough layer from 186 to 40 in one year (NIAB, 1987). It has been concluded that more time and water are needed for phytoextraction of salt than reclamation by chemical methods (Muhammed *et al.*,1990), but in view of the large areas affected it may perhaps be worth examining further with different plant species.

Other elements

It is instructive that the major perceived environmental threats from heavy metals have received most attention so far (lead, zinc), but that not much attention has been focused on soil cadmium,which can be a problem for the food chain. Also the developments in selenium extraction were because of the occurrence of a specific environmental problem. Undoubtedly, more research and development will find other elements which are removed by plants. It is possible that plants exist which take up elements not mentioned above, in quantities which are useful for environmental clean-up. More exploration and plant analyses are needed to find these.

Conclusions

Plants are available which can remove Zn, Cd, Pb, Se and Ni from soils at rates which mean that clean-up is medium to long-term, but rapid enough to be useful. One factor which drives phytotechnology is the costs. Costs are highly variable, and full costs will not be available until mature technology is offered on the market. All plant-based techniques are likely to be less costly than those using conventional technology. For example, it has been estimated that soil washing costs $US250/m^3$, but phytoextraction only $80/m^3$ of soil remediated (Black, 1995).

To date, there have been few demonstrations of phytoextraction under field conditions. This lack is likely to be remedied in the near future, as many research and development groups are now working in this area. As the research matures it is likely that we will see at least some of the following features emerge as phytoextraction becomes an applied technology in the next few years:

1 - agronomic inputs to increase productivity of those accumulator plants which now exist;

2 - use of soil amendments which increase the availability of metals to these plants by desorbing them from the soil solids and increase root-shoot translocation;

3 - use of growth-modifying hormones or other compounds to alter the roots and shoots of accumulator plants with the aim of maximising removal;

4 - breeding of higher accumulation by conventional methods, or by tissue culture techniques which release somaclonal variation in properties such as growth and element uptake;

5 - use of plant symbioses with bacteria or fungi (e.g. mycorrhizas) to stimulate accumulation;

6 - use of transgenic plants with altered uptake and transport of elements.

One other process that will be essential is the control of the environmental impact of the technology. Examples are minimising dispersal of seed or pollen, ensuring that metals are not mobilised down the soil profile, and preventing herbivores accumulating dangerous amounts of elements. The problem of flow of propagules could be avoided by simply harvesting the plants before they flower, during a period of maximum biomass and element yield. This is likely to be useful also, because senescence of old leaves returns the elements to the soil in a bioavailable form. Preadaptation to contaminated soil, as mentioned above for zinc, or tailoring of the plants to suit contaminated sites may also help in their containment. Growth of non-crop or non-palatable species is likely to reduce the chances of food chain transfer during phytoextraction.

Recycling of the accumulated elements will be important and must be done in an environmentally sensitive manner. More research is needed in this area also.

References

Baker,A.J.M. and Brooks,R.R.(1989) Terrestrial higher plants which hyper-accumulate metal elements - A review of their distribution, ecology and phytochemistry. *Biorecovery* 1, 81-126.

Baker,A.J.M,,McGrath,S.P.,Sidoli,C.M.D. and Reeves,R.D.(1994) The possibility of in situ heavy metal decontamination of polluted soils using crops of metal-accumulating plants. *Resources, Conservation and Recycling* 11, 41-49.

Baker,A.J.M.,Reeves,R.D. and McGrath,S.P.(1991) In situ decontamination of heavy metal polluted soils using crops of metal-accumulating plants - a feasibility study. In: Hinchee,R.L. and Olfenbuttel,R.F.(eds) *In Situ Bioreclamation*. Butterworth-Heinemann, Boston, pp.600-605.

Bañuelos,G.S.(1996) Managing high levels of boron and selenium with trace element accumulator crops. *Journal of Environmental Science and Health, Part A, Environmental Science and Engineering and Toxic and Hazardous Substances Control* 31, 1179-1196.

Bañuelos,G.S.,Cardon,G.,Mackey,B.,Ben-Asher,J.,Wu,L.,Beuselinck,P., Akohoue,S and Zambrzuski,S.(1993a) Boron and selenium removal in boron-laden soils by four sprinkler-irrigated plant species. *Journal of Environmental Quality* 22, 786-792.

Bañuelos,G.S.,Cardon,G.,Phene,C.J.,Wu,L.,Akohoue,S. and Zambrzuski,S.

(1993b) Soil boron and selenium removal by three plant species. *Plant and Soil* 148, 253-263.

Bañuelos,G.S.,Mackey,B.,Cook,C.,Akohoue,S.,Zambrzuski,S. and Samra,P. (1996) Response of cotton and kenaf to boron-amended water and soil. *Crop Science* 36, 158-164.

Bañuelos,G.S.,Mead,R.,Phene,C.J. and Meek,D.W.(1992a) Relations between phosphorus in drip irrigation water and selenium uptake by wild mustard. *Journal of Environmental Science and Health. Part A Environmental Science and Engineering* 27, 283-298.

Bañuelos,G.S.,Mead,R.,Wu,L.,Beuselinck,P. and Akohoue,S.(1992b) Differential selenium accumulation among forage plant species grown in soils amended with selenium-enriched plant tissue. *Journal of Soil and Water Conservation* 47, 338-342.

Bañuelos,G.S. and Meek,D.W.(1989) Selenium accumulation in selected vegetables. *Journal of Plant Nutrition* 12, 1225-1272.

Bañuelos,G.S. and Meek,D.W.(1990) Accumulation of selenium in plants grown on selenium-treated soil. *Journal of Environmental Quality* 19, 772-777.

Bañuelos,G.S.,Meek,D.W. and Hoffman,G.J.(1990) The influence of selenium, salinity, and boron on selenium uptake in wild mustard. *Plant and Soil* 127, 201-206.

Bañuelos,G.S. and Schrale,G.(1989) Crop selection for removing selenium from the soil. *Californian Agriculture* 43, 19-20.

Bañuelos,G.S.,Wu,L.,Akohoue,S.,Zambrzuski,S. and Mead,R.(1993c) Trace element composition of different plant species used for remediation of boron-laden soils. In: Barrow,N.J.(ed.), *Plant Nutrition from Genetic Engineering to Field Practice.* Kluwer Academic Press, Dordrecht, pp.425-428.

Black,H.(1995) Absorbing possibilities: phytoremediation. *Environmental Health Perspectives* 103(12), 6 pp.

Brown,S.L.,Chaney,R.L.,Angle,J.S. and Baker,A.J.M.(1994) Zinc and cadmium uptake by hyperaccumulator *Thlaspi caerulescens* and bladder campion for zinc- and cadmium-contaminated soil. *Journal of Environmental Quality* 23, 1151-1157.

Brown,S.L.,Chaney,R.L.,Angle,J.S. and Baker,A.J.M.(1995a) Zinc and cadmium uptake by hypraccumulator *Thlaspi caerulescens* and metal-tolerant *Silene vulgaris* grown on sludge-amended soils. *Environmental Science and Technology* 29, 1581-1585.

Brown,S.L.,Chaney,R.L.,Angle,J.S. and Baker,A.J.M.(1995b) Zinc and cadmium uptake by hyperaccumulator *Thlaspi caerulescens* grown in nutrient solution. *Soil Science Society of America Journal* 59, 125-133.

Chaney,R.L.(1983) Plant uptake of inorganic waste constituents. In: Parr,J.F., Marsh,P.B. and Kla,J.M.(eds), *Land Treatment of Hazardous Wastes.* Noyes Data Corp., Park Ridge, pp.50-76.

Chaney,R.L.,Malik,M.,Li,Y.M.,Brown,S.L.,Angle,J.S. and Baker,A.J.M. (1997) Phytoremediation of soil metals. *Current Opinions in Biotechnology (in press).*

Chapman, H. D. (1966). *Diagnostic Criteria for Plants and Soils.* University of California, Division of Agricultural Sciences.

Cornish,J.E.,Goldberg,W.C.,Levine,R.S. and Benemann,J.R.(1995) Phytoremediation of contaminated soils. In: Hinchee,R.E..Means,J.L. and Burris,D.R.(eds), *Bioremediation of Inorganics.* Battelle Press, Columbus, pp.55-63.

Cunningham,S.D. and Berti,W.R.(1993) Remediation of contaminated soils with green plants: an overview. *In Vitro Cellular and Development Biology* 29, 207-212.

Cunningham,S.D.,Berti,W.R. and Huang,J.W.(1995) Phytoremediation of contaminated soils. *Trends in Biotechnology* 13, 393-397.

Entry,J.A. and Emmingham,W.H.(1995) Sequestration of Cs-137 and Sr-90 from soil by seedlings of *Eucalyptus tereticornis. Canadian Journal of Forest Research* 25, 1231-1236.

Entry,J.A.,Vance,N.C.,Hamilton,M.A.,Zabowski,D.,Watrud,L.S. and Adriano,D.C.(1996) Phytoremediation of soil contaminated with low concentrations of radionuclides. *Water Air and Soil Pollution* 88, 167-176.

Frankenberger, W.T. and Karlson, U. (1992). Dissipation of soil selenium by microbial volatilisation. In: D.C Adriano (ed.) *Biogeochemistry of Trace Metals.* Lewis, Boca Raton, pp.365-381.

Huang,J.W.,Chen,J. and Cunningham,S.D.(1997) Phytoextraction of lead from contaminated soils. In: Kruger,E.L. *et al.* (eds) *Phytoremediation of Soils and Water Contaminants.* American Chemical Society, Washington, 14 pp.

Huang,J.W. and Cunningham,S.D.(1996) Lead phytoextraction: species variation in lead uptake and translocation. *New Phytologist* 134, 75-84.

Hussain, M. (1963) Research on reclamation of waterlogged, saline and alkaline lands. Research Publications Volume 2, Directorate of Land Reclamation, Lahore, Pakistan, pp. 70-71.

Kemper,F.H. and Bertram,H.P.(1991) Thallium. In: Merian,E.(ed.), *Metals and their Compounds in the Environment.* VCH Publishers, Weinheim, pp.1227-241.

Kumar,P.B.A.N.,Dushenkov,S.,Salt,D.E. and Raskin,I.(1994) Crop brassicas and phytoremediation - a novel environmental technology. *Cruciferae Newsletter* 16, 18-19.

Kurz,H.,Schulz,R.,Römheld,V. and Marschner,H.(1997) Evaluation of a possible phytoremediation of soils enriched in heavy metals. *Institute of Plant Nutrition Hohenheim University Report in World-Wide Web,* 3 pp.

McGrath, S.P. (1995) Chromium and nickel. In: Alloway,B.J. (ed.), *Heavy Metals in Soils,* 2nd edn. Blackie, Glasgow, pp.152-178.

McGrath,S.P.,Chang,A.C.,Page,A.L. and Witter,E.(1994) Land application of sewage sludge: scientific perspectives of heavy metal loading limits in Europe and the United States. *Environmmental Reviews* 2, 108-118.

McGrath,S.P. and Dunham,S.J.(1997) Potential phytoextraction of zinc and cadmium from soils using hyperaccumulator plants. *Proceedings of 4th International Conference on the Biogeochemistry of Trace Elements, Berkeley,* Ann Arbor Press, Ann Arbor (in press).

McGrath,S.P.,Shen,Z.G. and Zhao,F.J.(1997) Heavy metal uptake and chemical changes in the rhizosphere of *Thlaspi caerulescens* and *Thlaspi orchroleucum* grown in contaminated soils. *Plant and Soil* 188, 153-159.

McGrath,S.P.,Sidoli,C.M.D.,Baker,A.J.M. and Reeves,R.D.(1993) The potential for the use of metal-accumulating plants for the *in situ* decontamination of metal-polluted soils. In: H.J.P.Eijsackers and T.Hamers (eds), *Integrated Soil and Sediment Research: a Basis for Proper Protection.* Kluwer Academic Publishers, Dordrecht, pp.673-676.

McLaughlin,M.J.,Tiller,K.G.,Naidu,R. and Stevens,D.P.(1996) Review: the behaviour and environmental impact of contaminants in fertilizers. *Australian Journal of Soil Research* 34, 1-54.

Markert,B.(1992) Multi-element analyses in plant materials - Analytical tools and biological questions. In: Adriano,D.C.(ed.) *Biogeochemistry of Trace Metals.* Lewis, Boca Raton, pp.401-428.

Means, J.L., Kucak, T. and Crerar, D.A. (1980) Relative degradation rates of NTA, EDTA and DTPA and environmental implications. *Environmental Pollution, Series B,* 1, 45-60.

Muhammed, S., Ghafoor, A., Hussain, T. and Rauf, A.(1990) Comparison of biological, physical and chemical methods of reclaiming salt-affected soils with brackish groundwater. In: *Soil for Agriculture Development.* Proceedings of the 2nd National Congress of Soil Science, Faisalabad, 1988. Soil Science Society of Pakistan.

NIAB (Nuclear Institute for Agriculture and Biology)(1987) *Fifteen Years of NIAB.* Faisalabad, Pakistan, pp.133-134.

Nicks,L. and Chambers,M.F.(1995) Farming for metals. *Mining Environmental Management* September, 15-18.

Peryea, J.J, Bingham, J.T. and Rhoades, J.D. (1985) Mechanisms for boron regeneration. *Soil Science Society of America Journal* 49, 840-843.

Reeves,R.D. and Brooks,R.R.(1983) Hyperaccumulation of lead and zinc by two metallophytes from a mining area of Central Europe. *Environmental Pollution Series A* 31, 277-287.

Roberts,A.H.C.,Longhurst,R.D. and Brown,M.W.(1994) Cadmium status of soils, plants and grazing animals in New Zealand. *New Zealand Journal of Agricultural Research* 37, 119-129.

Robinson,B.H.,Chiarucci,A.,Brooks,R.R.,Petit,D.,Kirkman,J.H. and Gregg,P.E.H.(1997) The nickel hyperaccumulator plant *Alyssum bertolonii* as a potential agent for phytoremediation and phytomining of nickel. *Journal of Geochemical Exploration* 59, 75-86.

Rosenfield,I. and Beath,O.A.(1964) *Selenium. Geobotany, Biochemistry, Toxicity and Nutrition.* Academic Press, New York, 411 pp.

Rugh,C.L.,Dayton-Wilde,H.,Stack,N.M.,Thompson,D.M.,Summers,A.O. and

Meagher,R.B.(1996) Mercuric ion reductase and resistance in transgenic *Arabidopsis thaliana* plants expressing a modified bacterial *merA* gene. *Proceedings of the National Academy of Sciences of the United States* 93, 3182-3187.

Shen,Z.G.,Zhao,F.J. and McGrath,S.P.(1997). Uptake and transport of zinc in the hyperaccumulator *Thlaspi caerulescens* and *Thlaspi ochroleucum* grown in contaminated soils. *Plant Cell and Environment* 20, 88-906.

Terry,N.,Carlson,C.,Raab,T.K. and Zayed,A.(1992) Rates of selenium volatilization among crop species. *Journal of Environmental Quality* 21, 341-344.

Terry,N.and Zayed,A.M.(1994) Selenium volatilization by plants. In: Frankenberger,W.T.Jr. and Benson,S.(eds), *Selenium in the Environment.* Marcel Dekker, New York, pp.343-367.

Welch, R.M. (1995) Micronutrient nutrition of plants. *Critical Reviews in Plant Sciences,* 14, 49-82.

Wu,L.,Chen,J.,Tanji,K.K. and Bañuelos,G.S.(1995) Distribution and biomagnification of selenium in a restored upland grassland contaminated by selenium from agricultural drainage water. *Environmental Toxicology and Chemistry* 14, 733-742.

Zayed,A.M. and Terry,N.(1994). Selenium volatilization in roots and shoots: effects of shoot removal and sulfate level. *Journal of Plant Physiology* 143, 8-14.

Chapter thirteen:

Phytoremediation by Volatilisation

R.R. Brooks

Department of Soil Science, Massey University, Palmerston North, New Zealand

Introduction

The emphasis in this book has hitherto been concerned primarily with plants that are able to *hyperaccumulate* the chemical elements. The metals are transferred from the soil system to the plants wherein they are *stored*. It is but a short step in logic to envisage a situation in which the stored metals are converted into a volatile form and released into the atmosphere. The end result is the same, the soil has been remediated and the pollutant transferred from the lithosphere into the atmosphere without the need for troublesome harvesting, possible combustion of the plant material, and physical removal of the resulting ash or dry biomass for storage or cost recouping by sale.

A number of the elements in subgroups II, V and VI of the *Periodic Table*, mercury, arsenic and selenium, form volatile hydrides or methyl derivatives that can be liberated to the atmosphere, probably as a result of the action of bacteria or soil fungi. Perhaps one of the earliest indications of the volatility of organic derivatives of these three elements was provided by the observation of O.A. Beath and his colleagues at Laramie, Wyoming (Beath *et al.*, 1939a, 1939b, 1940; Trelease and Beath, 1949 - see also Chapter 3) that *Astragalus* plants, hyperaccumulators of selenium, emit a characteristic garlic-like odour that can be sensed even from a moving vehicle.

Extensive studies have now been performed on the possibility of removing all of the above elements from polluted soils by some form of phytoremediation. The majority of these studies have been devoted to selenium because this element poses a serious problem in many parts of the world where there are significant areas of selenium-rich soils such as in the Joaquin Valley of Central California. The three elements, mercury, arsenic and selenium will be considered separately below. In this discussion I would like to acknowledge two important publications that have been of great assistance to me as sources of reference material; namely, Frankenberger and Losi (1995) and Terry and Zayed (1994).

Mercury

Mercury is often a contaminant of soils located in areas associated with geothermal activity. It is for example, common in the Central Volcanic Plateau of New Zealand where it concentrates in the aquatic biosphere and is found in trout of the local lakes and streams (Robinson, 1994) to such an extent that most of these fish contain well over the 0.5 $\mu g/g$ (wet weight) of mercury allowable in World Health Organisation guidelines for human health.

Mercury in the environment exists in at least five forms: Hg^{2+}, Hg^+, Hg^0, $HgCH_3^+$ and $Hg(CH3)_2$. This is also the order of increasing toxicity to living biota with the methyl forms being by far the most toxic. Methylation of mercury occurs readily in the presence of organic matter and is due to bacterial activity. Methylation can occur under both *aerobic* and *anaerobic* conditions. The bacteria responsible are called *methanogenic*. A non-methanogenic organism is *Clostridium cochlearium* that can methylate various forms of mercury (Yamada and Tonomura, 1972). Sulphate-reducing bacteria such as *Desulfovibrio desulfuricans* can methylate Hg^{2+} in anoxic marine sediments (Compeau and Bartha, 1985).There are several methods whereby mercury can be hyper-accumulated from solution by various bacteria but these involve extraction from the aqueous phase and a description of them is more appropriately given in Chapter 6.

Bioreduction of Hg^{2+} to gaseous elemental mercury can also be carried out by use of *Pseudomonas* spp. enteric bacteria, *Staphylococcus* spp. and *Cryptococcus* sp. (Ehrlich, 1990). Once again reference should be made to Chapter 6 for further details.

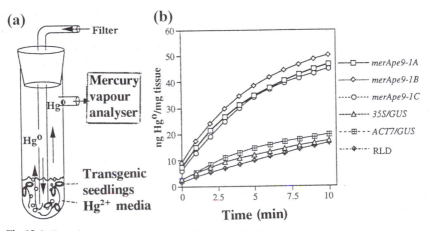

Fig.13.1. Experiments demonstrating volatilisation of Hg^0 from a solution of Hg^{2+} by use of transgenic seedlings of *Arabidopsis thaliana* (a). The graph shows a plot of mercury evolution by various plants including three with the *merApe9* gene, one control (RLD) and two with other transgenic plants expressing other genes.
Source: Rugh *et al.* (1996).

There are few references to the use of higher plants to accumulate mercury and to reduce it to the elemental form for discharge into the atmosphere. However, Rugh *et al.* (1996) have inserted a bacterial *merA* gene into the brassicaceous *Arabidopsis thaliana* plant as an approach to mercury sequestration and removal.

Mercuric ion reductase, *merA* converts Hg^{2+} toxic ions to less-toxic metallic mercury. The bacterial *merA* sequence was mutated to *merApe9* that modified the flanking region and 9% of the coding region. This sequence was placed under control of plant-regulatory elements.

Transgenic *Arabidopsis thaliana* seeds expressing *merApe9* germinated, flowered and set seed in a medium containing $HgCl_2$ concentrations of 25-100 μM (5-20 $\mu g/mL$). This concentration was toxic to several control species. The mutated seedlings evolved considerable amounts of volatile elemental mercury compared with the other control species. The experiments were performed in a mercury vapour analyser as is shown in Fig.13.1 above. The same figure shows the much improved performance of three transgenic seedlings expressing the *merApe9* gene, compared with the control (RLD) and plants expressing other genes.

Arsenic

Introduction

Arsenic occurs in Group VB of the Periodic Table and has the properties of both a metal and non-metal. It is often included under the term "heavy metal" and is probably more metallic than non-metallic in its behaviour in the environment. Like mercury, it is found in soils associated with geothermal activity and is one of the commonest components of sulphide mineralization. This element is extremely toxic to most biota, particularly in its trivalent arsenite form.

As discussed by Frankenberger and Losi (1995) there are two main strategies whereby arsenic may be detoxified or removed from the soil as a result of bacterial activity. The first of these involves bacterial oxidation to the less toxic pentavalent arsenate. The second strategy involves reduction of arsenate to arsenite followed by methylation to dimethylarsine. Fungi are also able to convert arsenic compounds (both organic and inorganic) into volatile methylarsines. The latter involves aerobic reduction to arsenite, followed by stepwise methylation ending with trimethylarsines. Table 13.1 shows the boiling points of elements, hydrides and dimethyl compounds of mercury, arsenic and selenium.

Uptake of arsenic by vegetation

Unlike other heavy metals such as nickel and zinc, there is little evidence that terrestrial plants are able to accumulate arsenic to any great degree. For example

Table 13.1. Boiling points (°C) of pure elements and hydrides, and methyl compounds of arsenic, mercury and selenium.

Compound	Arsenic	Mercury	Selenium
Element	615	356	688
Hydride	-55	-	-42
$MH(CH_3)_2$	36	95	58

Robinson (1994) found a maximum of only 54 $\mu g/g$ dry weight in leaves of *Leptospermum scoparium* in highly arseniferous soils in New Zealand. However, Robinson *et al.* (1995) found arsenic levels of around 1000 $\mu g/g$ in aquatic plants from the Waikato River, New Zealand. The corresponding arsenic content of the river water was of the order of 0.05 $\mu g/mL$. Further discussion of this unusual hyperaccumulation belongs to Chapter 9.

Bacterial oxidation of arsenic

Bacterial oxidation is one of the strategies that can be adopted for detoxifying contaminated soils. The theory is that oxidation of arsenite to arsenate results in

Table 13.2. Bioremediation of arsenic-contaminated soils with bacteria and fungi.

Strategy	Organism(s)	References
Bio-oxidation	Mixed bacterial culture	Mattison (1992)
Bio-oxidation	*Alcaligenes faecalis*	Phillips and Taylor (1976)
Bio-oxidation	*Alcaligenes* sp.	Osborne and Ehrlich (1976)
Methylation	*Methanobacterium* strain M.O.H.	McBride and Wolfe (1971)
Methylation	*Flavobacterium* sp.	Chau and Wong (1978)
Methylation	*Penicillium brevicaule* (= *Scopulariopsis brevicaulis*)	Challenger (1945)
Methylation	*Candida humicola, Gliocladium roseum* and *Penicillium* sp.	Cox and Alexander (1973)
Methylation	*Candida humicola*	Cullen *et al.* (1984)
Methylation	*Alcaligenes* sp. and *Pseudomonas* sp.	Cheng and Focht (1979)
Methylation	*Penicillium* sp.	Huysmans and Frankenberger (1991)

After: Frankenberger and Losi (1995).

a less toxic product, though the arsenic still remains *in situ*. Table 13.2 lists this and other strategies discussed below.

Frankenberger and Losi (1995) have described several microbially mediated oxidation reactions that can convert arsenite to arsenate. They include *Bacillus*, *Pseudomonas* and a strain of *Alcaligenes faecalis* extracted from raw sewage. Oxidation of arsenite by heterotrophic bacteria can result in a 78-96% conversion and plays an important part in detoxification of acidic arseniferous environments such as mine waters where arsenic concentrations can be found in the range 2-13 μg/mL (Wakao *et al.*,1988).

Bacterial methylation of arsenic

Bacterial methylation of arsenic (see Table 13.2 above) can be carried out by use of methanogenic bacteria that are found in a morphologically diverse group consisting of coccal, bacillary and spiral forms that are unified by production of methane as the end product. At least one species of *Methanobacterium* is capable of methylating inorganic arsenic to produce volatile dimethylarsine.

Under anaerobic conditions, biomethylation of arsenic proceeds only as far as dimethylarsine which is stable in the absence of oxygen but is rapidly oxidised under aerobic conditions. The system seems therefore unlikely to be suitable for widespread remediation of large areas of contaminated soils and is more suitable for a bioreactor system (McBride and Wolf, 1971).

A study by Cheng and Focht (1979) demonstrated that resting cell suspensions of *Pseudomonas* and *Alcaligenes* spp. incubated with arsenite or arsenate under anaerobic conditions produced arsine but no other intermediates.

Fungal methylation of arsenic

Resistance to heavy metals by bacteria and fungi has been known since well before World War II, and indeed even Agricola (Hoover and Hoover, 1950) reported the presence of "fungi" over mineralised veins. The earliest well-documented work is that of Challenger *et al.* (1933) who investigated the biotransformation of arsenic by *Penicillium*. This process also occurs with yeasts and other moulds, though to a lesser degree, and without the assistance of plasmids.

The mysterious death of Napoleon Bonaparte while in exile on St Helena in 1821 has resulted in a fascinating story that involves biotransformation of microorganisms. At a later date, a lock of his hair was analysed and was found to contain inordinately high amounts of arsenic. It was suggested that he had been poisoned deliberately by his captors.

Later in the century it was observed that people of Victorian England and elsewhere in Europe died of a mysterious disease similar to arsenic poisoning. The reason for this was the use in wallpaper of the green pigment copper arsenide, discovered by Scheele in 1775 (Scheele's green). When the wallpaper is dry, it is perfectly safe, but when it becomes damp, moulds such as *Scopulariopsis brevicaulis*, will grow on the paper using the cellulose or glue as the nutrient medium.

Fig.13.2. Longwood House, St Helena where Napoleon was incarcerated. Source: Richardson (1974).

Fig.13.3. A piece of the original arseniferous wallpaper from Napoleon's bedroom on St Helena. Source: Jones (1982).

Equipped to avoid the toxic affects of the arsenic, the moulds convert this element to the volatile, and highly-poisonous trimethyl arsine $[(CH_3)_3As]$ that permeates the atmosphere of the room and caused several deaths in 19th century England and elsewhere.

The link between Napoleon's death and arsenic has been studied by Jones (1982) who was fortunate enough to be able to obtain a piece of the original wallpaper present at Longwood House, St Helena (Fig.13.2) where the French emperor had been imprisoned. The wallpaper (Fig.13.3) was found in the scrapbook of a Ms. Shirley Bradley, and certainly contained large amounts of arsenic. Furthermore, the house itself was very damp and needed to have the wallpaper changed at frequent intervals due to the development of mould.

This story is a fascinating piece of historical detective work, and it seems quite probable that arsenic poisoning caused or contributed to Napoleon's death. The villain of the piece may not have been the British authorities, but rather the humble wallpaper mould *S.brevicaulis* (*Penicillium brevicaule*).

Several species of fungus are known to volatilise arsenic. For example, Cox and Alexander (1973) showed that *Candida humicola, Gliocladium roseum,* as well as *Penicillium*, were capable of converting arsenate and arsenite to trimethylarsine. Soil fungi play an important role in the transformation and volatilisation of arsenic chemicals used in agriculture. For example, *C.humicola* is able to reduce the level of arsenic in the wood preservative copper arsenate through volatilisation (Cullen *et al.*,1984). Soils treated with inorganic and methylated arsenic herbicides can readily lose this arsenic by the action of soil fungi that produce dimethyl- and trimethylarsine (Baker *et al.*,1983).

Selenium

Introduction

Selenium volatilisation is by far the most important process involving the three elements studied in this chapter. Important reviews have been published by Frankenberger and Losi (1995) and by Terry and Zayed (1994). The interest in selenium has been stimulated by the widespread threat to animal and human life posed by elevated concentrations of this element in soils throughout the world, particularly in Central California. The Kesterson Reservoir of the San Joaquin Valley of this state has such high levels of selenium that it has caused death and deformities in wildfowl and has been suggested (Valoppi and Tanji, 1988) as threatening food crops.

Selenium is readily volatilised by bacteria and fungi to produce compounds (see Table 13.1) considerably more volatile than the original selenite (SeO_3^{2-}) or selenate (SeO_4^{2-}). Selenium is emitted from soils even if no plants are present, but Zieve and Peterson (1984) have shown that there was much greater evolution of selenium from the soil after barley had been sown. Duckardt *et al.* (1992) found that the emission rate from soil was 225% greater when the selenium hyperaccumulator *Astragalus bisulcatus* was planted.

The release of volatile selenium compounds from higher plants was first reported by Lewis *et al.* (1966) in work with *Astragalus racemosus*. They also discovered subsequently that selenium can also be liberated by non-accumulator plants such as alfalfa. Evans *et al.* (1968) were the first to identify a volatile selenium compound released by plants and found that *A.racemosus* produced dimethyldiselenide (DMDSe). Lewis (1971) later showed that live cabbage leaves produced dimethylselenide (DMSe) and that volatilisation appeared to be metabolic rather than bacterial. She proposed that DMSe might have been produced by enzymatic cleavage of the Se-methylselenomethionine selononium salt.

Direct uptake of selenium by plants

There has already been some mention of hyperaccumulators of selenium in Chapter 3 and little more needs to be said except that the soils that give rise to these hyperaccumulators, a threat to livestock, are usually alkaline and contain free calcium carbonate (Lakin, 1973).

Frankenberger and Losi (1995) have divided selenium-accumulating plants into three categories:

1 - containing >100 μg/g dry weight such as many species of *Astragalus* as well as *Haplopappus, Machaeranthera* and *Stanleya;*

2 - secondary accumulators with 50-100 μg/g selenium in dry matter such as some *Astragalus* and *Aster, Atriplex, Castilleja, Grindelia, Gutierrezia* and *Mentzela;*

3 - plants that contain <50 μg/g selenium in dry matter.

This system of classification is much less limiting than the levels recommended in Chapter 3, but clearly shows the general distribution of selenium among some common species of the western United States.

Strategies for bioremediation of selenium

Vegetation uptake

Unlike in the case of arsenic, bioremediation of selenium can also be effected by using hyperaccumulator plants that, after harvesting, permit removal of some of the original selenium from the soil. This and other strategies in selenium bioremediation are summarised in Table 13.3.

One problem of direct harvesting is disposal of the highly toxic vegetation. Frankenberger and Losi (1995) have suggested that this problem might be solved by adding the tissue to animal feed deficient in selenium.

Agroforestry is another bioremediation technique that is being investigated. It involves the use of salt-tolerant trees such as *Eucalyptus* that reduce the volume of drainage water and permit cropping to remove selenium and other elements.

Bioreduction of selenium

In nature, selenium exists mainly in the reduced state. Anaerobic cell extracts of *Micrococcus lactylicus, Clostridium pasteurianum* and *Desulfovibrio desulf-uricans* can reduce selenite but not selenate, to Se^{2-} (Woolfolk and Whiteley, 1962). However, washed cells of *D.sulfuricans* subsp. *aestuarii* can reduce

SeO_4^{2-} to Se^{2-}. Selenite can be reduced directly to elemental selenium (Se^0) by aerobically grown *Salmonella heidelberg* (McCready *et al.*,1966) as well as by cell-free extracts of *Streptococcus faecalis* and *S.faecium* (Tilton *et al.*,1967).

Table 13.3. Strategies for bioremediation of selenium-contaminated soils.

Method	Organisms	References
Plant uptake	*Astragalus, Haplopappus, Machaeranthera* and *Stanleya*	Rosenfield and Beath (1964)
Plant uptake	*Atriplex patula, Bassia hyssopifolia, Melilotus indica* and *Salsola kali*	Wu (1994)
Agroforestry	*Eucalyptus*	Cervinka (1994)
Deselenification	*Acremonium falciforme, Penicillium citrinum, Ulocladium tuberculatum,* and *Alternaria alternata*	Karlson and Frankenberger (1988), Thompson-Eagle *et al.* (1989)
Bioreduction	*Clostridium, Desulfovibrio, Desulfotomaculum*	Kerr-McGee Patent US 4519912 (1985)
Bioreduction	*Desulfovibrio desulfuricans* subsp. *aestuarri*	Zehr and Oremland (1987)
Bioreduction	Algae and selenate-reducing bacteria	Gerhardt *et al.* (1991)
Bioreduction	Anaerobic granular sludge inoculum	Owens pers, comm, to Frankenberger and Losi (1995)
Bioreduction	*Thauera selenatis*	Macy *et al.* (1993)

After: Frankenberger and Losi (1995).

Conversion of selenate or selenite to the elemental form of selenium is an important mechanism for detoxification of selenium-contaminated soils. Several bioreactors have been devised and patented which can convert oxidised selenium to the elemental state. For example, the Kerr-McGee column precipitator uses an anaerobic soil column containing *Clostridium* sp. and is capable of reducing 1000 $\mu g/mL$ (0.1%) selenium in solution to under 50 $\mu g/mL$ at a rate of 100-175 $L/m^2/day$.

Deselenification by microbial methylation and volatilisation

There are many bacteria and fungi that can methylate selenium salts and convert them to various methyl forms of high volatility. The most common of the metabolites is dimethylselenide (DMSe). Other metabolites produced by this methylation are: dimethyldiselenide (DMDSe), methaneselenone ((CH_3)$_2SeO_2$), methaneselenol (CH_3SeH) and dimethylselenenyl sulphide (CH_3SeSCH_3). Under selected conditions, the original selenium content of a contaminated soil can be reduced significantly in a matter of months.

Frankenberger and Losi (1995) from their own experiments have shown that organisms naturally present in soils, saline and alkaline drainage water and in the water of settling ponds can methylate selenium compounds to DMSe and that the rate of methylation can be increased dramatically by the addition of amendments such as pectin and proteins.

In field trials at the highly seleniferous Kesterson Reservoir area (Merced County, California), Frankenberger and Karlson (1989) and Frankenberger et al. (1990) tested various nutrient additions and environmental conditions in soils containing 10-750 μg/g selenium.There was evidence of seasonal variation of selenium emission with the highest readings obtained in the late spring and early summer. The highest emission rates (40 times background) were 808 μg Se/m^2/h. After two years, 60% of the total selenium burden had been removed (Frankenberger and Karlson, 1989).

Frankenberger and Losi (1995) concluded that enhanced selenium volatilisation in the field was related to an available carbon source, aeration, moisture and high temperature. Tillage of the soil increased volatilisation as long as the soil was kept moist. The findings of the above workers are summarised in Fig.13.4 below.

The above technology seems on the face of it to be an excellent "green" solution to global pollution problems. There are however critics who will always be looking for some detrimental features of any system. The most obvious of these is the fate of DMSe in the environment once it has been liberated. Little is known about the complex reactions of DMSe with the atmosphere though it is known to react with hydroxyl and nitrate radicals as well as with ozone (Atkinson et al.,1990). The gas does not appear to be particularly toxic as Frankenberger and Karlson (1989) found that DMSe vapour was not toxic to animals at concentrations of up to 8037 μg/mL in air. Although the toxicity of DMSe to animals appears to be slight, the effect of this gas on the atmospheric ozone layer must be established very thoroughly before atmospheric scientists will be entirely happy.

Selenium uptake and toxicity in plants

Selenium can be accumulated by plants in various forms including selenate, selenite, elemental selenium, and even organic selenium (Terry and Zayed, 1994). Selenate readily mimics sulphate with which it has many properties in common. According to Leggett and Epstein (1956) selenate is absorbed by the same root carrier as sulphate. Sulphate and selenate are antagonistic to each other and this strengthens the evidence that they compete for the same root carriers. Sulphate levels may also affect the transport of selenium from the roots to shoots. **Singh et al. (1980)** showed that in *Brassica juncea*, selenium tends to accumulate in the roots but is translocated to the shoots when sulphate is added.

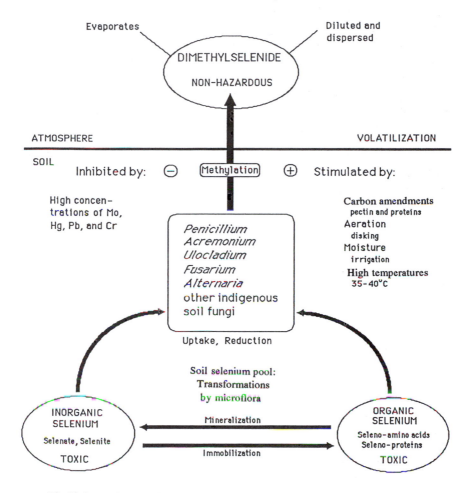

Fig.13.4. Accelerated microbial volatilisation used as a technology to bioremediate selenium contamination in soils. Source: Frankenberger and Losi (1995).

When selenate has entered the plant it is almost certainly metabolised enzymatically. This is because selenium mimics sulphur thereby forming selenium analogues of sulphur compounds that are substrates for the sulphur-assimilation enzymes.

The toxicity of selenium to plants is almost certainly related to competitive interactions between sulphur compounds and their selenium analogues. For example, inhibition of plant growth after addition of selenate can be counteracted by adding sulphate.

The main toxic effect of selenium is probably replacement of sulphur by this element in aminoacids of proteins thereby disrupting catalytic activity. Consistent with this hypothesis is the fact that selenium hyperaccumulators compared with non-accumulators limit the incorporation of endogenous selenoaminoacids into protein as a means of avoiding selenium toxicity (Brown and Shrift, 1981).

Factors affecting volatilisation of selenium by plants

Plant species

The volatilisation of selenium by crop species is heavily dependent on the species concerned. Terry and Zayed (1994) measured volatilisation rates for 15 crop species grown in growth chambers under controlled conditions. The results are shown in Table 13.4.

The plants were grown in quarter-strength Hoagland's solution with 20 μM selenium supplied as sodium selenate. The three best volatilisers were rice broccoli and cabbage, whereas the worst were sugar beet, bean, lettuce and onion. The authors emphasised that further tests under field conditions would be needed to verify the laboratory findings. Duckardt et al. (1992) measured the rates of selenium volatilisation from plants grown in soil containing 16.5 μg of selenium per kg. Astragalus bisulcatus and broccoli showed the highest rate of selenium volatilisation both on a leaf area and total biomass basis. These were followed in descending order by tomato, tall fescue and alfalfa. This agreement with the findings of Terry et al. (1992) for plants cultured hydroponically shows that the latter are a good indication of later findings in field trials.

Table 13.4. Rates of selenium volatilisation in terms of leaf area (μg Se/m^2/day) and of dry biomass (μg/kg/day).

Plant species	Leaf area	Plant biomass
Rice	340	1500
Broccoli	273	2393
Cabbage	221	2309
Carrot	102	548
Barley	102	486
Alfalfa	72	300
Tomato	70	742
Cucumber	57	752
Cotton	48	499
Eggplant	44	462
Maize	32	370
Sugar beet	22	246
Bean	14	217
Lettuce	11	179
Onion	n.d.	229

Source: Terry and Zayed (1994).

Chemical forms of selenium

The preceding sections have suggested that selenite is to be preferred over selenate as a precursor for bacterial reduction or methylation. That selenite is preferable has been demonstrated by experiments conducted by Asher *et al.* (1967) who measured the amount of volatile selenium compounds released from, oven-drying shoots or roots of *Medicago sativa* supplied with 15 μM selenite or selenate. Roots of plants supplied with selenite released 11 times more selenium than those treated with selenate.

In another experiment, Lewis *et al.* (1974) cultured cabbage plants in nutrient solutions and showed that those treated with selenite, liberated 10-16 times more selenium than those grown in selenate solutions. There is a certain logic in the volatilisation superiority of selenite over selenate. During bacterial reduction less energy will be needed for reducing the lower oxidation state of selenium to either its methylated or elemental form.

In unpublished work, N.Terry measure the rate of selenium emission from broccoli plants cultured in solutions containing selenium in the form of selenate, selenite, and selenomethionine. The latter treatment resulted in selenium volatilisation up to 4 times faster than in the case of selenite which in turn caused volatilisation up to 13 times faster than the selenate treatment.

Presence of other ions

Addition of Na_2SO_4, NaCl and $CaCl_2$ to soils containing selenium has been found to affect volatilisation of selenium by microbes (Karlson and Frankenberger, 1990). Zayed and Terry (1992) showed a decreasing evolution of selenium from broccoli plants when sulphate was added to a nutrient solution. The results are shown in Fig.13.5 where the rate of selenium volatilisation decreased from 97 to 14 μg Se/m^2/day. The above discussion has focused on the competition between selenium and sulphur and it is therefore easy to understand why increasing addition of sulphate should result in decrease of selenium emission.

Nitrate ion appears to have a similar effect to sulphate (N.Terry, unpublished data) but the effects are much less dramatic than in the case of sulphate and are difficult to explain in terms of competition.

Effect of temperature, pH and light

Believing that selenium volatilisation by plants is an enzymatically mediated process, Lewis *et al.* (1974) used cabbage leaf homogenates to study the effect of temperature, pH and light on selenium emission. The effect of temperature is shown in Fig.13.6. The optimum temperature was about 40°C. The optimum pH (Fig.13.7) was about 7.8.

In a greenhouse experiment with alfalfa, Lewis *et al.* (1966) demonstrated that after an initial hiatus, the rate of selenium emission increased with increasing

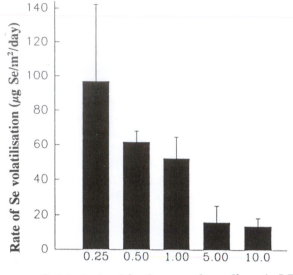

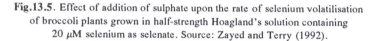

Sulphate level in the growth medium (mM)

Fig.13.5. Effect of addition of sulphate upon the rate of selenium volatilisation of broccoli plants grown in half-strength Hoagland's solution containing 20 μM selenium as selenate. Source: Zayed and Terry (1992).

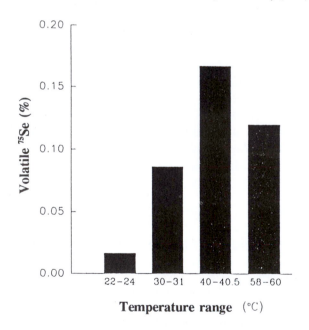

Temperature range (°C)

Fig.13.6. Effect of temperature on selenium volatilisation from cabbage leaf homogenates. Source: Lewis *et al.* (1974).

illumination. They attributed this to enhanced stomatal opening and/or heating of plant tissues so that transpiration and enzymatic reactions would be increased so that there was greater production of volatile selenium.

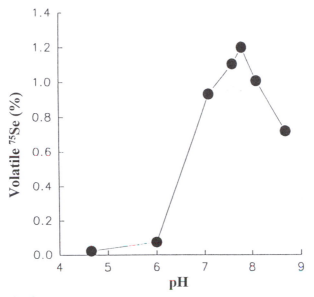

Fig.13.7. Influence of pH on selenium volatilisation from cabbage leaf homogenates. Source Lewis *et al.* (1974).

Effect of roots, shoots and shoot removal

Zayed and Terry (1994) compared root with shoot volatilisation of selenium again using broccoli plants. They found that roots volatilised 85.5% of the total selenium even though these comprised only 12.5% of the total plant biomass. On a unit weight base, the volatilisation rate was 7-20 times faster. This was in spite of the fact that the selenium content of the shoots was 2-5 times higher than that of the roots.

When shoots are excised, there is enhanced emission of selenium from the roots. This is illustrated in Table 13.5 which shows the effect for six different crops. In broccoli, selenium volatilisation increased progressively for 72 hours after shoot removal, attaining rates that were 20-30 times the rate of volatilisation from roots of non-excised plants. Thereafter the rate decreased to a limiting value after about 170 hours.

This is illustrated in Fig.13.8. Reasons for the enhancement effect are not fully established though Terry and Zayed (1994) have suggested two reasons:

1 - assuming a microbial role in selenium volatilisation, excising of shoots may cause a leakage of reduced carbon compounds into the rhizosphere thereby causing enhanced production of volatile selenium by rhizosphere microorganisms;

2 - studies have shown that plants with mechanically injured roots evolve

Table 13.5. Selenium volatilisation quotients (dry weight basis) of root/shoot (A) and detopped root/intact root (B).

Species	A	B
Rice	2.2	3.68
Broccoli	14.3	2.90
Cabbage	14.7	5.515
Cauliflower	18.6	1.58
Chinese mustard	10.1	2.18
Wild brown mustard	3.1	4.34

Source: Terry and Zayed (1994).

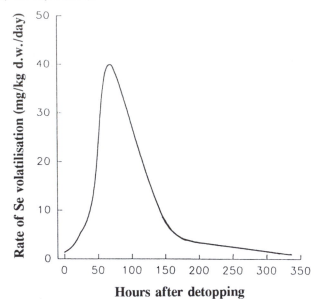

Hours after detopping

Fig.13.8. Time dependence of the enhancement of selenium volatilization rate of broccoli roots following excision of shoots. After: Zayed and Terry (1994).

greater amounts of volatile selenium than intact plants (Wilson *et al.*, 1978); since selenium appears to be metabolised via the sulphur-assimilation pathway, it is possible that mechanical injury resulting from shoot removal might also enhance selenium volatilisation.

Bañuelos and Meek (1990) noted the increase of selenium content of the whole plant of *Brassica juncea* (Indian mustard) and other plants (*Atriplex semibaccata, A. nummularia, Astragalus incanus* and tall fescue grass (*Festuca arundinacea*) after clipping of the original shoots. They attributed this to an indirect effect of carbohydrate and protein synthesis.

Experiments were also conducted under controlled conditions in which the total amount of selenium was measured in soil and plants. The data are shown in Fig.13.9 and refer to plants grown in selenate.

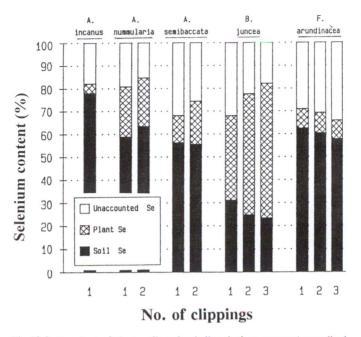

Fig.13.9. Bar chart of plant, soil, and volatile selenium content (normalised to 100%) for *Brassica juncea* and four other species grown under controlled conditions. Source: Bañuelos and Meek (1990).

In absolute terms, the *Brassica* contained the most selenium (1.20 μg/g dry weight) and *Astragalus incanus* the least (0.14 μg/g). Inspection of the figure shows that the whole plant selenium content increased from just under 40% of the total after one clipping to about 60% after three excisions. The percentage of selenium unaccounted for was presumably a volatile component and for the same species decreased from 32% to 18%. There is clearly a price to pay for clipping in that the amount in the plant increases to the detriment of the volatile fraction. Nevertheless the overall percentage removed from the soil increases. After one clipping the soil contained 31% of the selenium and this had decreased to only 23% after the third excision. The data of Bañuelos and Meek (1990) might appear to contradict the findings of Terry and Zayed (1994). However, it must be pointed out that other species in the experiments conducted by the former, such as *Festuca arundinacea*, showed an increase in the "volatile" selenium fraction after three excisions of shoots. This was coupled with a constant selenium concentration in the plant material together with a slight reduction of the selenium level in the soil from 62% to 58%.

Possible mechanisms of plant volatilisation of selenium

From the earlier part of this chapter it might be supposed that there is general agreement about the role of microorganisms in the liberation of volatile forms of selenium from plants. There is in fact no general consensus that this is the case, though most researchers in this field could believe that both microbial and metabolic processes can be responsible for the effect, either separately or in unison.

Microbes and fungi are always present in the soil around the rhizosphere where their abundance is usually greater than elsewhere in the soil profile. There is also evidence that soil microbes are able to penetrate plant tissue and reach the interior of the plant.

If we assume the truth of the hypothesis by Lewis (1971) that selenium may be volatilised by plant metabolic processes, a pathway for this conversion of selenium can be proposed. Fig.13.10 presents a possible mechanism proposed by Terry and Zayed (1994). The scheme depends on two alternatives based on whether or not the species is a hyperaccumulator of selenium.

There are five main processes proposed for the conversion of selenate to DMSe as follows:

1 - the first step is the formation of an activated form of selenate APSe (adenosine 5'-phosphoselenate) by combination of selenate with ATP (adenosine triphosphate). The reduction of APSe to selenite may proceed by a non-enzymatic reaction;

2 - according to Ng and Anderson (1979) selenite is reduced non-enzymatically to selenotrisulphide (GSSeSG) using reduced glutathione. GSSeSG is then reduced to selenide in two steps using NADPH;

3 - again according to Ng and Anderson (1979), inorganic selenide is converted to selenocysteine by cysteine synthase;

4 - selenocysteine is converted to selenomethionine (Burnell, 1981; Dawson and Anderson, 1988);

5 - according to Lewis (1971) selenocysteine is methylated to the methyl-selenomethionine selononium salt (MSeMS) leading to the final step of the process, the cleaving of MSeMS to DMSe and homoserine.

In selenium hyperaccumulators, the process is thought to be identical up to the formation of selenocysteine (Burnell, 1981). At this point selenocysteine is then methylated twice to form volatile DMDSe.

Selenium volatilisation in plants as a potential tool for practical bioremediation

According to Terry *et al.* (1992), plants can volatilise up to 40 g Se/ha/day. The selenium can be removed not only by the gas-phase process but also by mechanically excising the shoots of the plants. This increases the amount of selenium removed, but at the same time stimulates the roots to produce more volatile selenium. Duckardt *et al.* (1992) calculated the annual amount of selenium added to the soil by natural processes and time needed to remove half of this amount by soils alone, or soils in which plants were growing. The experiments were carried out in the Kesterson Reservoir of the San Joaquin Valley of California. They found that the soils alone required 461 days for removal of half of this annual amount whereas only 6-12 days were required when specific plants were present.

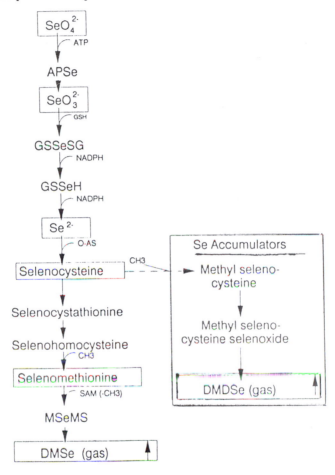

Fig.13.10. Possible mechanism of selenium volatilisation by plants. Source: Terry and Zayed (1994).

In summary, it would seem that bioremediation of selenium contamination in soils is probably a viable proposition if plants such as *Brassica juncea* can be grown and harvested on a commercial basis. Several croppings during the annual growing cycle could stimulate more removal of selenium either in biomass or as volatile DMSe.

The problem remains of what to do with the large seleniferous biomass (a frequent problem in bioremediation). Bañuelos and Meek (1990) have suggested a simple solution based on selling the biomass as an additive to selenium-deficient feedstock in other parts of the country.

Hypervolatilisation is the mirror image of hyperaccumulation in the plant kingdom and is a fitting counterpart to bioremediation strategies involving removal of soil contaminants by growing and harvesting hyperaccumulators in such substrates.

References

Asher,C.J.,Evans,C.S. and Johnson,C.M.(1967) Collection and partial characterisation of volatile selenium compounds from *Medicago sativa* L. *Australian Journal of Biological Science* 20, 737-748.

Atkinson,R.,Ashmann,S.M.,Hasagawa,D.,Thompson-Eagle,E.T. and Frankenberger,W.T.Jr.(1990) Kinetics of the atmospherically-important reactions of dimethylselenide. *Environmental Science and Technology* 24, 326-1332.

Baker,M.D.,Inniss,W.E.,Mayfield,C.I.,Wong,P.T.S. and Chau,Y.K.(1983) Effect of pH on the methylation of mercury and arsenic by sediment microorganisms. *Environmental Technology Letters* 4, 89-100.

Bañuelos,G.S. and Meek,D.W.(1990) Accumulation of selenium in plants grown on selenium-treated soil. *Journal of Environmental Quality* 19, 772-777.

Beath,O.A.,Gilbert,C.S. and Eppson,H.F.(1939a) The use of indicator plants in locating seleniferous areas in western United States. I. General. *American Journal of Botany* 26, 257-269.

Beath,O.A.,Gilbert,C.S. and Eppson,H.F.(1939b) The use of indicator plants in locating seleniferous areas in western United States. II. Correlation studies by states. *American Journal of Botany* 26, 296-315.

Beath,O.A.,Gilbert,C.S. and Eppson,H.F.(1940) The use of indicator plants in locating seleniferous areas in western United States. III. Further studies. *American Journal of Botany* 27, 564-573.

Brown,T,A. and Shrift,A.(1981) Exclusion of selenium from proteins of selenium-tolerant *Astragalus* species. *Plant Physiology* 67, 1051-1053.

Burnell,J.N.(1981) Selenium metabolism in *Neptunia amplexicaulis*. *Plant Physiology* 67, 316-324.

Cervinka,V.(1994) Selenium removal through agroforestry farming systems. In: Frankenberger,W.T.Jr. and Benson,S.(eds), *Selenium in the Environment*. Marcel Dekker, New York, pp.237-250.

Challenger,F.(1945) Biological methylation. *Chemical Reviews* 36, 315-361.

Challenger,F.,Higginbothom,C. and Ellis,L.(1933) The formation of organo-metalloidal compounds by microorganisms. Part I. Trimethylarsine and dimethylarsine. *Journal of the Chemical Society* 95, 1-31.

Chau,Y.K. and Wong,P.T.S.(1978) Occurrence of biological methylation of elements in the environment. *American Chemical Symposium Series 82, 39-53*.

Cheng,C.N. and Focht,D.D.(1979) Production of arsine and methylarsines in soil and in culture. *Applied Environmental Microbiology* 38, 494-498.

Compeau,G.C and Bartha,R.(1985) Sulfate-reducing bacteria: principal methylators of mercury in anoxic estuarine sediments. *Applied Environmental Microbiology* 50, 498-502.

Cox,D.P. and Alexander,M.(1973) Production of trimethylarsine gas from various arsenic compounds by three sewage fungi. *Bulletin of Environmental Contamination and Toxicology* 9, 84-88.

Cullen,W.R.,McBride,B.C.,Pickett,A.W. and Regalinski,J.(1984) The wood preservative chromated copper arsenate is a substance for trimethylarsine biosynthesis. *Applied Environmental Microbiology* 47, 443-444.

Dawson,J.C. and Anderson,J.W.(1988) Incorporation of cysteine and selenocysteine into cystathionine and selenocystathionine in crude extracts of spinach. *Phytochemistry* 27, 3453-3460.

Duckardt,E,C.,Waldron,L.J. and Donner,H.E.(1992) Selenium uptake and volatilization from plants growing in soil. *Soil Science* 53, 94-99.

Ehrlich,H.L.(1990) *Geomicrobiology* 2nd edn. Marcel Dekker, New York, pp.267-282.

Evans,C.S.,Asher,C.J. and Johnson,C.M.(1968) Isolation of dimethyldiselenide and other volatile selenium compounds from *Astragalus racemosus* (Pursh.). *Australian Journal of Biological Science* 21, 13-20.

Frankenberger,W.T.Jr. and Karlson,U.(1989) Dissipation of soil selenium by microbial volatilization at Kesterson Reservoir. United States Department of the Interior Bureau of Reclamation, December, Contract No.7-FC-20-054240.

Frankenberger,W.T.Jr,Karlson,U. and Longley,K.E.(1990) Microbial volatilization of selenium from sediments of agricultural evaporation ponds. California Water Resources Control Board Interagency Agreement No.7-125-250-1.

Frankenberger,W.T.Jr. and Losi,M.E.(1995) Applications of bioremediation in the cleanup of heavy metals and metalloids. In: Skipper,H.D and Turco,R.F.(eds), *Bioremediation: Science and Applications*. Soil Science Society of America *et al.*, Madison, pp.173-210.

Gerhardt,M.B.,Green,F.B.,Newman,R.D.,Lundquist,T.J.Tresan,R.B. and Oswald,W.J.(1991) Removal of selenium using a novel algal bacterial process. *Research Journal of the Water Pollution Control Federation* 63, 799-805.

Hoover,H.C. and Hoover,L.H.(1950) *De Re Metallica by Georgius Agricola*. Dover, New York.

Huysmans,K.D. and Frankenberger,W.T.Jr.(1991) Evolution of trimethylarsine by a *Penicillium* sp. isolated from agricultural evaporation pond water. *Science of the Total Environment* 105, 13-28.

Jones,D.(1982) The singular case of Napoleon's wallpaper. *New Scientist* 14th October, 101-104.

Karlson,U. and Frankenberger,W.T.Jr.(1988) Effects of carbon and trace element addition on alkylselenide production by soil. *Soil Science Society of America Journal* 52, 1640-1644.

Karlson,U. and Frankenberger,W.T.Jr.(1990) Alkyl selenide production in salinized soils. *Soil Science* 149, 56-61.

Kerr-McGee Patent US 4519912 (1985).

Lakin,H.W.(1973) Selenium in our environment. *Advances in Chemistry* 123, 96-111.

Leggett,J.E. and Epstein,E.(1956) Kinetics of sulphate absorption by barley roots. *Plant Physiology* 31, 222-226.

Lewis,B.G.(1971) Volatile Selenium in Higher Plants. The Production of Dimethylselenide in Cabbage Leaves by the Enzymatic Cleavage of Methylselenomethionine Selononium Salt. PhD Thesis, University of California, Berkeley.

Lewis,B.G.,Johnson,C.M. and Broyer,T.C.(1974) *Plant and Soil* 40, 107-115.

Lewis,B.G.,Johnson,C.M. and Delwiche,C.C.(1966) Release of volatile selenium compounds by plants: collection procedures and preliminary observations. *Journal of Agricultural and Food Chemistry* 14, 638-640.

McBride,B.C. and Wolfe,R.S.(1971) Biosynthesis of dimethylarsine by *Methanobacterium*. *Biochemistry* 10, 4312-4317.

McCready,R.G.,Campbell,J.N. and Payne,J.I.(1966) Selenite reduction by *Salmonella heidelberg*. *Canadian Journal of Microbiology* 1, 703-714.

Macy,J.M.,Lawson,S. and DeMoll-Decker,H.(1993) Bioremediation of selenium oxyanions in San Joaquin drainage water using *Thaurea selenatis* in a biological reactor system. *Applied Microbiological Technology* 40, 594-598.

Mattison,P.C.(1992) Bioremediation of metals - putting it to work. In: *Redox Treatment of Other Metals*. Cognis Inc., Santa Rosa, Chapter 6.

Ng,B.H. and Anderson,J.W.(1979) Light-dependent incorporation of selenite and sulphite into selenocysteine and cysteine by isolated pea chloroplasts. *Phytochemistry* 18, 573-580.

Osborne,F.H. and Ehrlich,H.L.(1976) Oxidation of arsenite by a soil isolate of *Alcaligenes*. *Journal of Applied Bacteriology* 41, 295-305.

Phillips,S.E. and Taylor,M.L.(1976) Oxidation of arsenite to arsenate by *Alcaligenes faecalis*. *Applied Environmental Microbiology* 32, 392-399.

Richardson,F.(1974) *Napoleon's Death: an Inquest*. William Kimber, London.

Robinson,B.H.(1994) Pollution of the Aquatic Biosphere by Arsenic and other Elements in the Taupo Volcanic Zone. MSc Thesis, Massey University, Palmerston North, New Zealand, 127 pp.

Robinson,B.H.,Outred,H.,Brooks,R.R.and Kirkman,J.D.(1995) The distribution and fate of arsenic in the Waikato River system, North Island, New Zealand. *Chemical Speciation and Bioavailability* 7, 89-96.

Rosenfield,I. and Beath,O.A.(1964) *Selenium: Geobotany, Biochemistry, Toxicity and Nutrition*. Academic Press, New York, 411 pp.

Rugh,C.L.,Dayton-Wilde,H.,Stack,N.M.,Thompson,D.M.,Summers,A.O. and Meagher,R.B.(1996) Mercuric ion reductase and resistance in transgenic *Arabidopsis thaliana* plants expressing a modified bacterial *merA* gene. *Proceedings of the National Academy of Sciences of the United States* 93,

3182-3187.

Singh,M.,Singh,N. and Bhandari,D.K.(1980) Interaction of selenium and sulfur on the growth and chemical composition of *Raya*. *Soil Science* 129, 238-242.

Terry,N.,Carlson,C.,Raab,T.K. and Zayed,A.(1992) *Journal of Environmental Quality* 21, 341-

Terry,N.and Zayed,A.M.(1994) Selenium volatilization by plants. In: Frankenberger,W.T.Jr. and Benson,S.(eds), *Selenium in the Environment*. Marcel Dekker, New York, pp.343-367.

Thompson-Eagle,E.T., Frankenberger,W.T.Jr. and Karlson,U.(1989) Volatilization of selenium by *Alternaria alternata*. *Applied Environmental Microbiology* 55, 1406-413.

Tilton,R.C.,Gunner,H.B. and Litsky,W.(1967) Physiology of selenite reduction by Enterococci. I. Influence of environmental variables. *Canadian Journal of Microbiology* 13, 1175-1185.

Trelease,S.F. and Beath,O.A.(1949) *Selenium: its Geological Occurrence and its Biological Effects in Relation to Botany, Chemistry, Agriculture, Nutrition and Medicine*. Trelease and Beath, New York, 292 pp.

Valoppi,L. and Tanji,K.(1988) Are the selenium levels in food crops and waters of concern? In: Tanji,K.K. *et al.*(eds), *Selenium Contents in Animal and Human Food Crops Grown in California*. Cooperative Extension of the University of California, Division of Agriculture and Natural Resources Publication No.330, University of California, Oakland.

Wakao,N.,Koyatsu,H.,Komai,Y.,Shimokawara,H.,Sakurai,Y. and Shiota,H.(1988) Microbial oxidation of arsenite and occurrence of arsenite-oxidizing bacteria in acid mine water from a sulfur-pyrite mine. *Geomicrobiological Journal* 6, 11-24.

Wilson,L.G.,Bressan,R.A. and Filner,P.(1978) Light-dependent emission of hydrogen sulfide from plants. *Plant Physiology* 61, 184-189.

Woolfolk,C.A. and Whiteley,H.R.(1962) Reduction of inorganic compounds with molecular hydrogen by *Micrococcus lactilyticus*. *Journal of Bacteriology* 84, 647-658.

Wu,L.(1994) Selenium accumulation and colonization of plants in soils with elevated concentrations of selenium and salinity. In: Frankenberger,W.T.Jr. and Benson,S.(eds), *Selenium in the Environment*. Marcel Dekker, New York, pp.279-325.

Yamada,M. and Tonomura,K.(1972) Formation of methylmercury compounds from inorganic mercury by *Clostridium cochlearium*. *Journal of Fermentation Technology* 50, 159-166.

Zayed,A.M. and Terry,N.(1992) Selenium volatilization in broccoli as influenced by sulfate supply. *Journal of Plant Physiology* 140, 646-652.

Zayed,A.M. and Terry,N.(1994). Selenium volatilization in roots and shoots: effects of shoot removal and sulfate level. *Journal of Plant Physiology* 143, 8-14.

Zehr,J.P. and Oremland,R.S.(1987) Reduction of selenate to selenide by sulfate-respiring bacteria: experiments with cell suspensions and estuarine sediments. *Applied Environmental Bacteriology* 53, 1365-1369.

Zieve,R. and Peterson,P.J.(1984) Volatilization of selenium from plants and soils. *The Science of the Total Environment* 32, 197-202.

Chapter fourteen:

A Pioneering Study of the Potential of Phytomining for Nickel

L.J. Nicks[1] and M.F. Chambers[2]

[1]PO Box 650, Fernley NV 89408, USA; [2]PO Box 13511, Reno NV 89507, USA

Introduction

Plants have long been associated with minerals and mining in the literature of biogeochemical and geobotanical methods of mineral exploration (Brooks, 1983, 1993; Ernst, 1974, 1993). It is somewhat surprising therefore that once the intriguing characteristics of hyperaccumulator plants became widely known, a flurry of activity on the use of hyperaccumulators as an alternative to mining did not occur. Essentially all reported research on utilisation of hyperaccumulators has been directed towards remediation.

This chapter describes a preliminary study which was the first reported (Nicks and Chambers, 1995) with the specific goal of determining the potential of using hyperaccumulators to recover a mineral economically from a natural deposit. A paper by Robinson *et al.* (1997) has also reported a study of *phytomining* in Italy with the nickel hyperaccumulator *Alyssum bertolonii*. A later paper by the same authors (in press) has been concerned with phytomining by the South African *Berkheya coddii*.

The American research was conducted from 1991 through 1994 out of the US Bureau of Mines (USBM) Reno Research Center in Reno, Nevada. Interest in the project was originally stimulated by the report of Reeves *et al.* (1981) that *Streptanthus polygaloides*, a member of the Brassicaceae, and endemic to serpentine soils in California was capable of accumulating nickel to concentrations greater than 15,000 μg/g (1.5%) in dry plant tissue. This species is unique in being the only species in this genus that has hyperaccumulation properties.

Exposures of ultramafic rocks in California and Oregon total more than 400,000 ha (Fig. 14.1). These dark-coloured, dense rocks are composed mainly of iron and magnesium silicates with high concentrations of associated nickel and chromium as well as manganese and to a lesser extent cobalt.

As summarised by Kruckeberg (1992), it is generally accepted that most, if not all, of the ultramafic rocks of the western US are of ophiolitic origin. Ophiolites

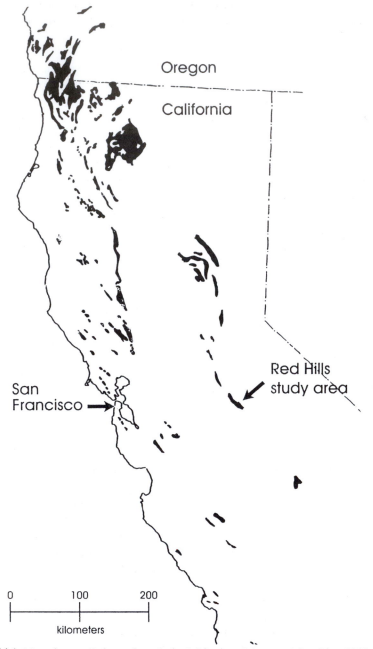

Fig.14.1. Map of exposed ultramafic rocks in California and Oregon. After: Rice (1957).

are fragments of oceanic crust and mantle that originally formed at mid-ocean ridges, or in association with island arcs and then became embedded in continental plates at areas of intense tectonic activity. The lower member of the ophiolite sequence is mantle material of ultramafic composition and is always at least partially hydrated to form minerals of the group known as serpentinite (Malpas, 1992).

Mineral exploration and mining have historically been very active in and near the ultramafic zones of California and Oregon. Whether mineralisation was associated with the original utramafic rocks or due to later hydrothermal activity along associated fault systems, these areas have produced significant quantities of magnesite, asbestos, talc, and jadeite. Mercury deposits at Almaden and New Idria were discovered when California was still a Spanish colony. Much of the gold produced in California was found in the fault systems along the western flank of the Sierra Nevada: faults which are probably remnants of ancient continental margins that received ophiolite emplacements.

Although nickel sulphide minerals, primarily pentlandite, are important sources, much of the world's supply of nickel comes from ultramafic rocks in which no identifiable nickel minerals exist. Ultramafics commonly contain minor amounts of nickel which occur by ionic substitution for magnesium in the primary magnesium minerals, especially olivine. Golightly (1979) states that the average nickel content of sulphide-free ultramafic rocks is about 3000 $\mu g/g$ (0.3%), which may be compared with the average crustal abundance of nickel of only 80 $\mu g/g$. Upon weathering, the nickel in ultramafic rocks remains in the disintegration products, usually at a concentration similar to that of the parent rock.

Under specific conditions of climate, topography, and geomorphology, a particular weathering pattern known as *laterisation* may produce an economic deposit with nickel concentrations as much as ten times greater than the parent rock. A small laterite deposit has been mined intermittently at Nickel Mountain, Oregon for the past 40 years. The nickel content of ultramafics in the western United States was discussed by Rice (1957) for California and by Ramp (1978) for Oregon.

The soils which develop over utramafic rocks are generically termed *serpentine soils*, or simply *serpentines*. These soils are commonly only sparsely covered with vegetation and unusual, often unique, systems of plant and animal life have been observed on isolated islands of serpentine soils worldwide. Kruckeberg (1984) and Brooks (1987) discuss at length what Jenny (1980) called the *serpentine syndrome*. Common chemical characteristics include high Mg/Ca quotients, pH usually above 6, relatively low concentrations of nitrogen, phosphorus and molybdenum, and often phytotoxic levels of nickel and chromium. A general view is that ultramafic rocks from the Earth's mantle are literally foreign to continental crust. Normal terrestrial plant communities evolved in soils derived from aluminosilicate and carbonate rocks typical of continental crust and they were faced with a challenge to utilise the light and moisture that were available in areas with inhospitable soils derived from mantle rocks. Most plants cannot survive in these soils. Some species did develop a tolerance to the foreign soils and became able to live on and off the serpentines. In other cases, new and distinct species evolved which are now found only on serpentines. *Streptanthus polygaloides* is an example of the latter. It is endemic to serpentine soils in the western foothills of the Sierra Nevada roughly between the Merced and Feather Rivers. Its ability to accumulate nickel and store it in a nontoxic form is apparently a part of its particular evolutionary adaptation that gave it a niche in the serpentine biota.

Experimental Observations and Discussion

Initial field work

The first phase of the work involved numerous field trips to California and Oregon to locate areas where *S. polygaloides* grew and to identify serpentines with sufficient nickel content to be attractive as potential phytomining sites. We found it useful to carry a field kit for semiquantitative determinations of nickel in soils and plant matter. Soil determinations were made by shaking 1 g of soil with 5 mL of 1M HNO_3 in a plastic vial for 5 min then adding 5 mL of 2M NH_4OH, shaking and allowing to settle. Five drops of the supernatant solution were placed on a spot plate and one drop of a saturated solution of dimethylglyoxime in ethanol was added. After some practice with nickel standards, we were able to estimate soil nickel concentrations between 100 and 5000 $\mu g/g$ by observing the intensity of red coloration of the nickel complex. To determine the presence of nickel in plants, we carried filter paper which had been soaked in the same indicator solution and dried. Crushing stems or leaves on the filter paper gave an intense red colour if the plant contained significant nickel. Selected soil and plant samples were returned to the laboratory for instrumental analysis by inductively coupled plasma emission spectrometry (ICP-ES) or by flame atomic absorption spectrometry (FAAS).

We found a wide variation in the nickel content of serpentine soils of different areas. This variability has also been noted by Kruckeberg in the references cited above and by Arianoutsou *et al.* (1993). Most of the serpentine sites visited in California were associated with extensive fault systems in the two parallel belts shown in Fig. 14.1. Within a particular block of ultramafic rock, soil characteristics, including nickel content, were found to be relatively consistent, but a nearby block often had a much different composition. The nickel values observed in the fault-associated serpentines ranged from 150-3820 $\mu g/g$. Soils varied in colour and texture from brick red with much clay development to free-flowing granular material with the grey-green colour of serpentinite. Most of the soils were thin and rocky.

The highest concentrations of nickel were found in the lateritic deposits of extreme northwestern California and southwestern Oregon, where several sites contained >1% nickel. The typical dark-red laterites also exhibited more deeply developed soil profiles. Depths of more than one metre, were common and depths of five metres or more were observed.

Fig.14.2. Serpentine flora at Red Hills, Tuolumne County, California.

As a result of the survey, we chose for further study a population of *Streptanthus polygaloides* growing in an area known as Red Hills in central California (arrow in Fig.14.1). Situated near the settlement of Chinese Camp in Tuolumne County, the Red Hills are the southeast terminus of a nearly continuous block of serpentinite extending approximately 22 km to the northwest with a maximum width of about 6 km. The overlying red, rocky soil is typically 5-10, and rarely 25 cm deep. It supports only a sparse flora dominated by digger pine *(Pinus sabiniana)* and buck brush *(Ceanothus cuneatus)* which stands in sharp contrast to surrounding grass and oak woodland (Figs 14.2 and 14.3). *Streptanthus polygaloides* is scattered over most of the serpentine area with the densest stands (Fig.14.4) frequently found on the rockiest and apparently poorest soil where it suffers little competition from other plant species. Initial analyses of plant specimens and soil samples collected in 1992 showed nickel concentrations of 1540-4810 μg/g in dried plant matter and 3250-3820 μg/g in soil. Plant seeds and bulk soil samples were returned to our laboratory for

Fig.14.3. Flora of non-serpentine soil 3 km from location in Fig.14.2.

greenhouse experiments carried out in cooperation with the University of Nevada, Reno.

Greenhouse studies

Pot trials were undertaken to determine plant response to soils with varying concentrations of nickel and to the addition of fertilisers. By mixing low-nickel (150 $\mu g/g$) serpentine soil collected near Grass Valley, California, with high-nickel soil from near Cave Junction, Oregon, blends with nickel concentrations of 3250, 6500, 9750, and 13,000 $\mu g/g$ were prepared. Plants were observed to grow well in all but the soil with the highest nickel content, where shoots were stunted. Addition of NPK (16-16-16) fertilisers to Red Hills soil with 3500 $\mu g/g$ nickel resulted in a >5-fold increase in biomass of 6-week-old plants with application of 50 kg/ha of each of nitrogen, phosphorus and potassium.

Plants were also grown in nutrient solutions with varying nickel concentrations. In an experiment with low nickel levels of 0, 1, 5, 10, 25, and 50 μM, weights of plants grown for four weeks in 5-50 μM solutions were 23 times greater than controls with zero nickel. This indicates that *S.polygaloides* actually requires nickel for optimum growth. A similar experiment with high nickel levels of 100, 500, and 1000 μM nickel gave evidence of the plant being overwhelmed at 1000 μM nickel where growth of both roots and shoots were inhibited by about 50%.

Final field study

We returned to Red Hills in 1994 to monitor wild plants throughout the growing season. The study site was a 2 ha flat 5 km southwest of Chinese Camp at an elevation of 375 m. The soil contained (in μg/g) 9320 Al, 1760 Ca, 174 Co, 1540 Cr, 97,100 Fe, 1510 K, 161,000 Mg, 5850 Na, 3340 Ni, and 382 P. The average pH of just-saturated soil was 7.95.

Fig.14.4. Wild stand of *Streptanthus polygaloides* at Red Hills. Distance between flags is 1 m.

After germination in late January, *S.polygaloides* grew slowly during the cool month of February, developing only a rosette of basal leaves. In the warmer weather of March, upright stems appeared and growth accelerated through April and into May. At the end of May, flowers began to open and the plants reached their maximum dry weight. During the flowering season, which continued as long as the plants survived through the hot and dry months of June and July, plant dry weight decreased dramatically although height slowly increased as new flowers appeared. The weight loss was partially due to shedding of lower leaves, but a more significant loss of stem mass was also observed. The plants apparently store resources in their stems during the period of rapid growth in spring then utilise those resources to ensure production of seeds during the dry summer. In addition to loss of plant mass, the concentration of nickel in the plants decreased as the plants began to flower.

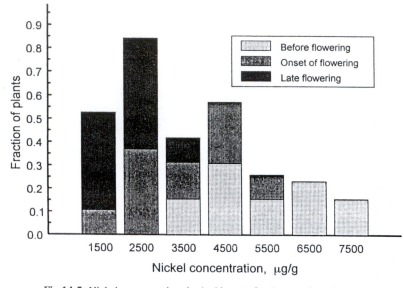

Fig.14.5. Nickel concentrations in the biomass for *Streptanthus* plants sampled at three different stages of the growth cycle.

Figure 14.5 is a histogram showing nickel concentrations in dry tissue for individual whole plant shoots sampled at three different times: about one week prior to flowering, at the onset of flowering, and from one to three weeks after the onset of flowering. Mean nickel concentrations for the three sampling times were 5306, 3372, and 2186 μg/g. To obtain maximum plant mass and nickel concentration, a crop of *S.polygaloides* should be harvested just before the onset of flowering.

Figure 14.5 also shows that individual plants sampled at the same time varied widely in nickel concentration. Plants collected one week before the onset of flowering contained 3280-7820 μg/g nickel. A similar variation was observed in individual plant sites; it was not uncommon to find two plants growing less than 20 cm apart to differ in weight by a factor of two to three. These variations make estimates of potential crop yields difficult but it is important to note that they also indicate a rich variety of genotypes in the wild population of *S.polygaloides*. Such genetic variety is favourable for successful application of selective breeding techniques to develop a high-yielding strain.

Estimates of overall yields of plant mass and contained nickel were made by stretching strings 1 m long through representative areas of the wild plants. Plants within 5 cm on either side of the string were counted and measured and specimens representing the average height of the 0.1 m² sample area were collected, dried, and weighed. Height and mass data for individual plants within one week of flowering onset fitted an exponential function of the form:

$$M = 0.37e^{0.031H}$$

where M is dry plant mass in grams and H is the plant height in cm. In Fig.14.6 each plotted point represents from two to ten individual plants and the dashed lines define the 95% confidence limits. The calculated yields of dry plant mass from 19 measured sample areas ranged from 215-1446 g/m².

To determine recoverability of nickel, bulk plant samples were ashed at temperatures from 600 to 900°C. The total ash obtained was consistently from 4 to 6% of biomass weight, with nickel concentrations in the ash proportional to their concentrations in the biomass. As an example, combustion of 100 g of dry plant matter containing 3600 μg/g nickel, yielded 4.5 g of ash containing 8.0% of this metal. After agglomeration with a binder such as clay, the ash should be suitable for direct feed to a nickel smelter. Alternatively, leaching of ash prepared at 700°C, using 5% sulphuric acid at 60°C, extracted 95% of the nickel in 30 min to give a solution suitable for hydrometallurgical treatment.

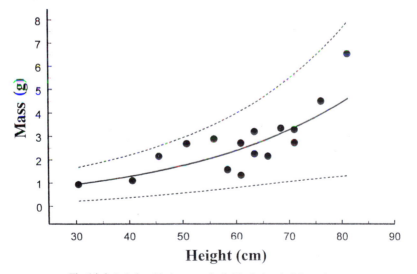

Fig.14.6. Relationship between individual plant height and mass.

Economic potential

Although the results of this limited investigation hardly constitute the basis for a thorough economic evaluation, we can speculate in general terms on the economic potential of "nickel farming". The biomass yield determined for small (0.1 m^2) areas of unfertilised wild plants ranged up to 1446 g/m^2. The number of plants per m^2 necessary to produce such a yield probably could not be sustained over a large area, but one third of that density should be achievable to give a yield of 482 g/m^2 or 4.82 t/ha with unfertilised plants.

Moderate application of fertilisers could certainly be expected to double the mass of individual plants, so our round number estimate of biomass yield for a crop of *S.polygaloides* is 10 t/ha. The concentration of nickel in wild plants harvested one week prior to flowering averaged 5306 μg/g. Using a nickel content of 5000 μg/g (0.50%) gives a conservative estimate of 50 kg of nickel per hectare from a 10 t/ha crop.

We know, however, that individual specimens of *S.polygaloides* can contain much higher levels of nickel than 5000 μg/g. We observed whole-shoot concentrations greater than 7500 μg/g and Reeves *et al.* (1981) found a mean value of 9750 μg/g for leaves from seven separate specimens and a maximum leaf concentration of 14,800 μg/g. It would be most unusual to begin cultivation of a new crop species without first conducting at least a minimal programme of selective breeding to develop a consistent strain with maximum desirable characteristics. It is not at all unreasonable to expect that selective breeding would produce a population with an average biomass nickel concentration of 1%. A 10 t/ha crop would then contain 100 kg of nickel. Incineration of the biomass would yield approximately 500 kg/ha of ash containing 20% nickel.

The average nickel price (USGS, 1996) on the London Metal Exchange for the second quarter of 1996 was US$7.65 per kg. If half of the nickel value were returned to the grower after deduction of metallurgical processing costs, the cash value of a crop containing 100 kg nickel per hectare would be US$382/ha. Furthermore, since incineration of the crop is a necessary step, processing large-scale farming of hyperaccumulators would justify a centrally located incinerator to recover part of the 17.5 GJ/t of heat typical for combustion of cellulose materials. Assuming a 25% recovery of energy with a value of US$3 per GJ, an additional US$131/ha could be realised. The total cash value of the crop would then be US$513/ha.

The annual cultivation cycle of planting, fertilisation, and harvesting by mechanical cutting for a hyperaccumulator crop would be similar to the operations of heat farming and should have similar costs. Therefore a comparison with the cash value of a wheat crop will serve as a useful indicator of the economics of phytomining. In 1995, the US national average yield for wheat (USDA, 1996) was 78.8 bushels/ha with a price of US$4.08 per bushel for a crop value of US$322/ha. Additional value for wheat straw increased the total crop value to US$333/ha. It therefore appears that if *S.polygaloides* can be cultivated to produce 100 kg of nickel per hectare, an economically viable operation should be possible. The later study by Robinson *et al.* (1977) on *Alyssum bertolonii* in Italy has produced comparable economic data.

Conclusions

The results of this small study can only serve as an indicator of possibilities. The indications are that phytomining can become, at least within a limited scope, a viable alternative to mining. More research will show how wide the scope may be. The question of whether phytomining *should* be practiced is certain to arise and is more difficult than the question of whether phytomining *can* be practiced.

To appease an insatiable appetite for more materials to support an ever-higher technology, humankind has dug minerals from holes in the earth. As the grade of available resources decreased, the holes have become deeper and the waste piles higher.

Perhaps we are reaching the maturity required to curb our appetite and to take more care to reuse the minerals we have already taken. Phytoremediation offers a tool to help clean up the mess we have already made and it seems to be a method acceptable by many people with strong ecological concerns. Phytomining is potentially also a tool that may be responsibly used to help provide for real future needs by selectively extracting minerals from low-grade resources without digging bigger holes.

Kruckeberg (1984) expressed concern that fragile serpentine ecosystems may be permanently destroyed even by seemingly low impact activities. We agree with him that representative areas should be withdrawn from all human activity. As for the danger of irreparable damage by phytomining, most of the serpentine areas we visited did not contain sufficient nickel to be of interest. Areas with higher nickel content that are of potential interest for phytomining have already been the object of moderate to intense mineral exploration and extraction activities as well as attempted agricultural development. In properly selected areas, the cultivation of serpentine-endemic plants would be a step nearer the original ecosystem.

At the conclusion of this work, the authors recommended to the USBM that the research be extended to:

1 - begin a programme of selective breeding to maximise plant mass and nickel concentration for *Streptanthus polygaloides;*

2 - study other, possibly higher yielding, hyperaccumulators such as the South African plant *Berkheya coddii;*

3 - investigate the possibilities of genetic manipulation to impart hyperaccumulator ability to robust, high mass plants that do not normally accumulate metals;

4 - begin small-scale cultivation studies with hyperaccumulator plants in carefully selected areas of the deep, high-nickel lateritic soils of northwestern California and southwestern Oregon. Unfortunately, in 1995 the USBM was desperately attempting to adapt to hostile political and economic conditions and our recommendations were ignored. The USBM has since been disbanded, but our hopes that research will continue along the lines we suggested have not diminished.

References

Arianoutsou,M.,Rundel,P.W. and Berry,W.L.(1993) Serpentine endemics as biological indicators of soil element concentrations. In: Markert,B.(ed.) *Plants as Biomonitors: Indicators for Heavy Metals in the Terrestrial Environment.* VCH Publishers, Weinheim, pp.179-189.

Brooks,R.R.(1983) *Biological Methods of Prospecting for Minerals.* Wiley, New York, 322 pp.

Brooks,R.R.(1987) *Serpentine and its Vegetation: a Multidisciplinary Approach.* Dioscorides Press, Portland, 454 pp.

Brooks,R.R.(1993) Geobotanical and biogeochemical methods for detecting mineralization and pollution from heavy metals in Oceania, Asia, and the Americas. In: Markert,B.(ed.) *Plants as Biomonitors: Indicators for Heavy Metals in the Terrestrial Environment.* VCH Publishers, Weinheim, pp.127-153.

Ernst,W. (1974) *Schwermetallvegetation der Erde.* Fischer Verlag, Stuttgart, 194 pp.

Ernst,W.(1993) Geobotanical and biogeochemical prospecting for heavy metal deposits in Europe and Africa. In: Markert,B.(ed.) *Plants as Biomonitors: Indicators for Heavy Metals in the Terrestrial Environment.* VCH Publishers, Weinheim, pp.107-126.

Golightly,J.P.(1979) Nickeliferous laterites - a general description. In: Evans,D.J.I., Shoemaker,R.S. and Veltman H.(eds) *International Laterite Symposium, New Orleans, Louisiana, Feb 19-21, 1979.* Society of Mining Engineers. American Institute of Mining, Metallurgical, and Petroleum Engineers, New York, pp.3-23.

Jenny,H.(1980) *The Soil Resource: Origin and Behavior.* Springer Verlag, Berlin.

Kruckeberg,A.R.(1984) *California Serpentines: Flora, Geology, Soils and Management Problems.* University of California Publications in Botany No.78, December 1984, 168 pp.

Kruckeberg,A.R.(1992) Plant life of western North American ultramafics. In: Roberts,B.A. and Proctor,J.(eds) *The Ecology of Areas with Serpentinized Rocks: a World View.* Kluwer, Dordrecht, pp.31-73.

Malpas,J.(1992) Serpentine and the geology of serpentinized rocks. In: Roberts,B.A. and Proctor,J.(eds), *The Ecology of Areas with Serpentinized Rocks: a World View.* Kluwer, Dordrecht, pp.7-30.

Nicks,L.J. and Chambers, M.F.(1995) Farming for metals. *Mining Environmental Management* 3, 15-18.

Ramp,L.(1978) Investigations of nickel in Oregon. *Oregon Department of Geology and Mineral Industries Miscellaneous Paper* No.20, 68 pp.

Reeves,R.D.,Brooks, R.R. and MacFarlane,R.M.(1981) Nickel uptake by Californian *Streptanthus* and *Caulanthus* with particular reference to the hyperaccumulator *S.polygaloides* Gray (Brassicaceae). *American Journal of Botany* 68, 708-712.

Rice,S.J.(1957) Nickel. *Mineral Commodities of California. California Department of Natural Resources, Division of Mines. Bulletin* 176, 391-399.

Robinson,B.H.,Chiarucci,A.,Brooks,R.R.,Petit,D.,Kirkman,J.H.,Gregg,P.E.H., and De Dominicis,V.(1997) The nickel hyperaccumulator plant *Alyssum bertolonii* as a potential agent for phytoremediation and phytomining of nickel. *Journal of Geochemical Exploration 59, 75-86.*

USDA(1996) Wheat production cash costs and returns, excluding direct Government payments 1994-95. *United States Department of Agriculture Publication.*

USGS(1996) Nickel in June 1996. In: *Mineral Industry Surveys Sept 25, 1996.* United States Department of the Interior - United States Geological Survey.

Chapter fifteen:

The Potential Use of Hyperaccumulators and Other Plants for Phytomining

R.R. Brooks and B.H. Robinson
Department of Soil Science, Massey University, Palmerston North, New Zealand

Introduction

The first suggestion that hyperaccumulator plants could be used for "metal mining" was made by Baker and Brooks (1989), but another eight years were to pass until Nicks and Chambers (1995 - see also Chapter 14) demonstrated the economic feasibility of growing a crop of nickel by use of the nickel hyperaccumulator *Streptanthus polygaloides* in California. They showed that the potential value of a crop of nickel was about the same as that of a crop of wheat provided that use could be made of some of the energy used in combustion of the dry material to produce the *bio-ore* containing about 15% nickel metal.

The benchmark paper by Nicks and Chambers (1995) was followed two years later by a study by Robinson *et al.* (1997) in Italy using the well known hyperaccumulator *Alyssum bertolonii*, the first plant of this type to be identified (Minguzzi and Vergnano, 1948). Salient details of this latter research are given below.

The term *phytomining* has been used to describe this emerging technology which at present is very far from being established commercially. Nevertheless, some new developments described below, may well serve to make the technique a viable proposition in the next decade. These include the following:

1 - discovery of hyperaccumulators with a large biomass;

2 - discovery of fertiliser amendments that can increase the biomass of hyperaccumulators without affecting metal uptake to any significant degree;

3 - utilisation of chelates that would permit higher metal uptake by hyperaccumulators or by plants of high biomass that do not usually hyperaccumulate the target metals;

4 - selection for plant breeding, of individual wild hyperaccumulators with

greater than average metal contents or biomass;

5 - use of biotechnology to introduce *hyperaccumulation genes* into non-accumulators of high biomass.

It should be emphasized that all of the above strategies are equally applicable to phytoextraction for remediation of polluted sites and that current research is mainly directed to this latter purpose. The essential difference between phytoremediation and phytomining is that economic considerations are paramount for the latter procedure, whereas they are of lesser importance in phytoremediation.

Potential Strategies for Enhanced Phytomining

Use of hyperaccumulators of high biomass

It is obvious that it is preferable to use hyperaccumulators with high biomass rather than those of lower biomass, provided that there is not the trade-off of a lower metal content. Unfortunately, most of the well known hyper-accumulators of heavy metals have a low biomass. This is especially true of plants such as *Thlaspi caerulescens*, that can contain over 3% zinc. Field trials with this plant over a soil containing 381 μg/g zinc showed that it contained about 6000 μg/g dry weight of this element (McGrath *et al.*, 1993). In these trials, the fertilised plant biomass was around 5.1 t/ha. Several of our field trials have shown that fertilising crops of many hyperaccumulators with NPK amendments usually increases the biomass by 300%.

Table 15.1 is a rough estimate of expected biomasses and potential metal yield of fertilised and unfertilised crops of selected hyperaccumulators of various metals. Though not an accumulator, *Zea mays* has been included for comparison since it affords one of the highest biomasses of any annual plant species.

Table 15.1. Unfertilised (A) and fertilised (B) biomasses (t/ha) and highest % metal content (C) of selected plants species. The final column shows the potential metal yield in kg/ha.

Plant species	Target	A	B	C	kg/ha
Zea mays	None	20.0	30.0	-	-
Berkheya coddii	Ni	12.0	24.0	0.60	144
Homalium kanaliense	Ni	12.0	36.0	*0.056	20
Alyssum tenium	Ni	7.7	23.0	0.34	78
A.lesbiacum	Ni	5.0	15.0	1.00	150
A.murale	Ni	4.6	13.8	0.71	98
A.bertolonii	Ni	3.0	9.0	1.34	121
Haumaniastrum katangense	Co/Cu	2.5	7.5	0.22	17
Thlaspi caerulescens	Zn/Cd	1.7	5.1	3.00	153
Cardaminopsis halleri	Zn	0.9	2.6	0.40	10

*Though leaves contain typically 0.6% nickel, the whole plant contains far less. Data based on own studies, McGrath *et al.* (1993), and Brooks *et al.* (1979).

The biomass of the plant must be related to its metal content in assessing its suitability for phytoremediation or phytomining. The final column of Table 15.1 shows (in kg/ha) the product of the biomass (t/ha) and the metal content (%) of the plant. This highest expected metal yield is more important than the actual biomass. It would appear from the table that the plants with the highest biomass are in descending order: *Homalium kanaliense, Zea mays, Berkheya coddii, Alyssum tenium, A.lesbiacum, A.murale, A.bertolonii, Haumaniastrum katangense, Thlaspi caerulescens* and *Cardaminopsis halleri*. In terms of expected metal yield (kg/ha) the order is somewhat different and, again in descending order, follows the sequence: *Thlaspi caerulescens, Alyssum lesbiacum, Berkheya coddii, A.bertolonii, A.murale, A.tenium, Homalium katangense, Haumaniastrum katangense* and *Cardaminopsis halleri*. It must be remembered that the essential difference between phytoremediation and phytomining is that in the latter technology, the value of the metal crop is of paramount importance. Whereas *Thlaspi caerulescens* would appear to extract the greatest mass of metal among the plants listed in Table 1.2 (Chapter 1), the commercial value of this metal (zinc) is so low that the plant could never be used for biomining. On the other hand, the Zaïrean copper/cobalt accumulator, *Haumaniastrum katangense* with a yield of only 17 kg/ha of each of copper and cobalt might well be able to provide an economic crop because of the current high price of cobalt ($48,000/t). This question will be addressed in a later subsection of this chapter.

Fertiliser amendments as a means of increasing biomass and metal yields

There has already been some discussion in Chapters 10 and 11 of the use of fertilisers to increase the biomass of selected hyperaccumulator plants. There will therefore be only limited discussion of the subject in this chapter. Table 15.2 summarises the increases of biomass achieved in various experiments with hyperaccumulator plants.

Table 15.2. Increases of hyperaccumulator plant biomass achieved by various fertiliser amendments.

Plant species	Fertiliser amendment	% increase	Reference
Alyssum bertolonii	N + P	760	Chapter 11
	Calcium	51	Robinson *et al.* (1997)
	Nitrogen	130	Robinson *et al.* (1997)
	Phosphorus	101	Robinson *et al.* (1997)
	N + P	189	Robinson *et al.* (1997)
	N + P + K	308	Robinson *et al.* (1997)
	As above + calcium	294	Robinson *et al.* (1997)
Streptanthus polygaloides	Nitrogen & phosphorus	153	Chapter 11
Thlaspi caerulescens	Nitrogen	195	Chapter 11

It is clear from the above table that quite dramatic increases of biomass can be achieved by fertilisation of the substrate. The experiments carried out by Robinson *et al.* (1997) were performed on naturally occurring populations of *Alyssum bertolonii* in Italy, whereas the other experiments were carried out as pot trials.

In the Italian work there was a highly significant positive correlation $(0.01 > P > 0.001)$ between nickel uptake in the reproductive structures of *A. bertolonii* and the extractable nickel content of the associated soils (Fig.15.1). This indicates that addition of fertilisers that increase this available nickel content should increase the nickel content of the plants.

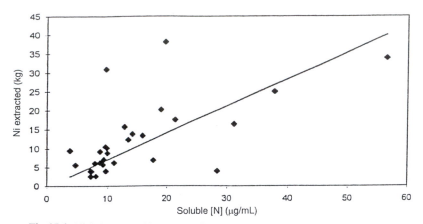

Fig.15.1. Nickel extracted by a crop of *Alyssum bertolonii* (kg/ha) as a function of the available nickel content of the soil as determined by ammonium acetate extraction at pH 7. Source: Robinson *et al.* (1997).

Table 15.2 shows that extractability of nickel was virtually unaffected by the nature of the treatment except in the case of $CaCO_3$ where the extractability of the former was halved. A similar result was obtained with the $N + K + P + Ca$ fertiliser. Robinson *et al.* (1996) have shown that the availability of trace elements in serpentine soils decreases exponentially as the pH is raised. This highlights the importance of avoiding $CaCO_3$ if a nickel "crop" is desired. There was a similar reduction in magnesium availability when calcium was used in fertilisers. Paradoxically, addition of $CaCO_3$, though reducing nickel and magnesium availability, has the effect of rendering the soil more fertile for non-serpentinic plants by increasing the Ca/Mg quotient from ca. 0.5 to 3.0. This increase would clearly improve a crop such as wheat or barley but reduce a "crop" of nickel.

With the addition of fertilisers, the maximum annual biomass increase (ABI) in this Italian study, was about 300% (Table 15.2). The highest individual increase (130%) was with nitrogen alone, and the highest combined increase (308%) was with $N + P + K$.

The nickel content of the Italian field specimens of *Alyssum bertolonii* remained fairly constant in the range 0.54-0.77% for all fertiliser amendments

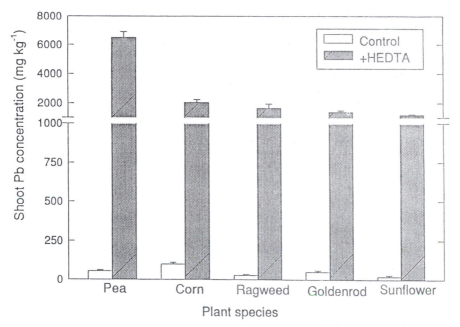

Fig.15.2. Lead accumulation in shoots of five plant species grown in lead-contaminated soil (2500 µg/g) after addition of 0.5 g/kg of HEDTA. Source: Huang *et al.* (1997).

except in the treatment with four fertiliser elements including calcium. It therefore appears that there is no appreciable decrease of nickel concentration with increase in biomass except when the pH is increased by addition of calcium. In other words there is no trade-off in reduced nickel concentration to offset the gain in biomass. Similar findings were made in work described in Chapter 11 (Figs 11.5 and 11.6).

The use of chelates in soils to increase elemental uptake by plants

The use of chelates in soils to increase metal uptake by plants represents perhaps the greatest step forward in the technology of phytoremediation. Early experiments by Norvell (1972), Wallace *et al.* (1977), Checkai *et al.* (1987) and Sadiq and Hussain (1993) showed that synthetic chelating agents could be used to increase uptake of various cations by plants. More recently, Huang and Cunningham (1996) and Huang *et al.* (1997) have shown that complexing of lead with EDTA renders this element much more available to large-biomass non-accumulators such as *Zea mays* so that a metal content of 1% (dry weight) can be obtained in the plant material. Harvesting of these plants permits a reduction of the lead content of the contaminated soil so that a satisfactory level can be reached in a small number of sequential crops.

One of the most serious environmental problems involving heavy metals concerns the accumulation of anthropogenic lead in soils throughout the world. Lead is normally very immobile to plants as it can precipitate at roots systems as the highly insoluble phosphate or sulphate and does not readily penetrate from roots to the stems.

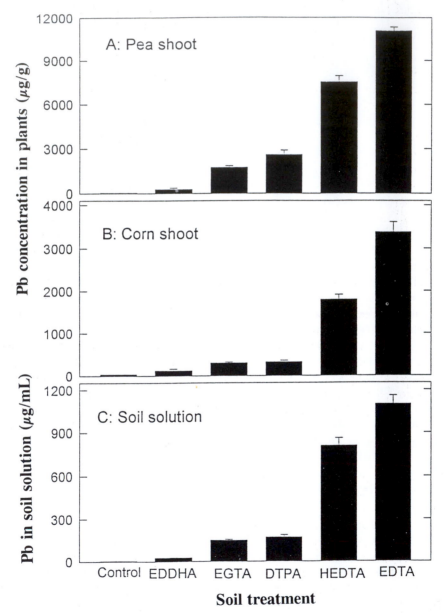

Fig.15.3. Lead uptake (mg/L = μg/g) by pea and corn plants treated for one week with 0.5 g of various complexing agents to 1 kg of soil. Values for the soil solution are also shown. Source: Huang *et al.* (1997).

However, Marten and Hammond (1966) found that ethylenediaminetetraacetic acid (EDTA) applied to lead-contaminated soils, increased the lead content of bromegrass from 5 to 35 μg/g.

Huang *et al.* (1997) have reported experiments in which EDTA (as the trisodium salt Na$_3$HEDTA) was added to lead-contaminated soils (2500 μg/g) at the rate of 0, 0.5, 1.0 and 2.0 g/kg. Within 24 hours the soluble lead content of the soil solution increased from almost zero to about 4000 μg/mL. Lead accumulation in shoots of five plant species grown in the same soil and with the addition of 0.5 g/kg of EDTA, is shown in Fig.15.2. It will be observed that the lead content of pea (*Pisum sativum*) reached over 6000 μg/g lead whereas that of corn (*Zea mays*) was only 2000 μg/g. However, the biomass of corn being at least ten times that of pea, shows that the former would be much more effective in removing lead from the soil. The same is true for the high-biomass sunflower (*Helianthus annuus*) with well over 1000 μg/g of the same element.

The effect of different complexing agents on the uptake of lead by pea and corn shoots is shown in Fig.15.3. It is clear that EDTA (trisodium salt) is by far the most efficient of the five chelates used, followed by HEDTA (N-(2-hydroxy-ethyl)ethylenediaminetetraacetic acid, DTPA (diethylenetriaminepentaacetic acid), EGTA (ethylenebis(oxyethylenenitrilo)tetraacetic acid), and EDDHA. EDTA has the further advantage of being the least expensive of the chelating agents (about $40/kg).

The above discussion has centred around complexing of lead in contaminated soils since this element is one of the most ubiquitous in sites polluted with heavy metals. The beneficial application of the technique to phytoremediation cannot be doubted, whereas for phytomining, the thrust of this chapter, the benefits are negligible because the low world price of lead (ca. $800/t) would militate against its economic recovery by growing plants. Nevertheless, the principles of chelate-assisted metal uptake by plants are equally applicable to other more valuable elements such as nickel, cobalt, or even gold. As far as we know, no other studies have yet been published in which chelates have been used to increase metal uptake by hyperaccumulators of these more expensive metals. Details of our experiments with chelating agents are given below.

Selection by plant breeding of specific cultivars and wild strains of hyperaccumulators

Among small-biomass hyperaccumulating plants such as *Streptanthus polygaloides* (Ni) and *Thlaspi caerulescens* (Zn/Cd) there is often a large difference of biomass and metal uptake among individual plants. In Chapter 14, field observations of *Streptanthus polygaloides* showed that it was not uncommon to find two plants growing less than 20 cm apart to differ in biomass by a factor of two to three. These variations made estimates of potential crop yields difficult but it is important to note that they also indicated a rich variety of genotypes in the wild population of this species. Such genetic variety is favourable for successful application of selective breeding techniques to develop a high-yielding strain.

In our studies on *Thlaspi caerulescens* growing over zinc/lead mine wastes at Les Avinières near Montpellier in southern France, we have observed a very wide range of biomass ranging from 1 to 64 g. This extremely variable species is also a good candidate for selection of high-biomass strains by plant breeding.

Kumar *et al.* (1995) have reported studies in which they tested 106 different cultivars of Indian mustard (*Brassica juncea*) for their ability to accumulate lead in roots and transport this element to the shoots. Some of their data are shown in

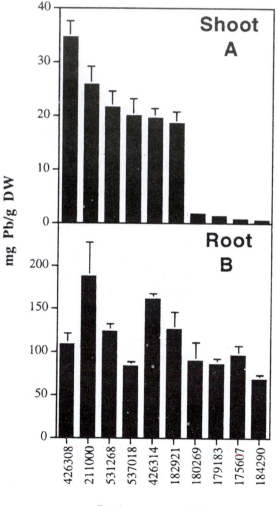

B. juncea cultivars

Fig.15.4. Lead content of shoots (a) and roots (B) of *Brassica juncea* cultivars grown for 14 days in a sand/Perlite mixture containing 625 µg/g lead. Source: Kumar *et al.* (1995).

Fig. 15.4. The most efficient cultivar (No.426308) contained 35 μg/g lead in the dry shoots compared with only 0.4 μg/g for the least efficient (No.184290).

The above observations clearly show the wide range of biomass and metal uptake that can be expected in wild strains and cultivars of many plant species, both hyperaccumulators and non-accumulators.

Use of biotechnology to introduce "hyperaccumulation" genes into plants of high biomass

The main problem in phytomining and phytoremediation technologies is that there are few plants that combine high biomass with a high degree of accumulation of the target heavy metals. In some cases, notably in Cuba and New Caledonia, there are indeed plants of high biomass such as the New Caledonian *Hybanthus austrocaledonicus* and *Sebertia acuminata*. These trees are however difficult to grow and have a slow rate of growth.

There is clearly a need for species that combine the extraordinarily high uptake of metals by low-biomass plants such as *Thlaspi caerulescens* with high-biomass fast-growing plants such as *Brassica juncea*. This problem has been discussed by Kumar *et al.* (1995) and by Cunningham and Ow (1996) who have suggested that the answer to the problem lies in genetic engineering to transfer the "hyperaccumulation genes" from plants of low biomass to fast-growing high-biomass plants such as *Brassica juncea* or *Helianthus annuus*. Such experiments are still in their infancy but encouraging results have been obtained by experiments in which scientists have cloned the gene for a vacuolar membrane transport pump that facilitates sequestering of the peptide-Cd complex (Ortiz *et al.*, 1995). Hyperproduction of this protein in the fission yeast *Schizosaccharomyces pombe* enhances tolerance to, and accumulation of, cadmium. Ow (1993) has suggested that hyperexpression of this yeast protein may yield similar results in a higher plan.

Cropping Sustainability of Phytomining Operations

Robinson *et al.* (1997) have examined the question of whether growing a crop of nickel will entail quick removal of the soluble fraction of this element from the soil in the way that a conventional crop will quickly remove plant nutrients. It would be scarcely feasible to grow a crop on an annual basis for a decade in order to answer this question. An alternative approach is the method of sequential extraction that we have developed in our laboratories. Using cumulative extraction of nickel by KH phthalate at pH 2, 4 and 6 shows that:

$$t_e = t - tc/(x+c)$$

Where: t_e = cumulative extracted concentration of nickel, t = concentration of potentially available nickel, x = number of extractions, and c = a constant dependent on the type of soil and the amount of nickel removed in a single

extraction. The cumulative extractions approach a limiting value where fewer
extractions are needed at lower pH values.

In experiments carried out on ultramafic rocks of the Murlo area of Tuscany
Italy (Fig.15.5), the limiting value was found to be 768 μg/g nickel in the soil
(approximately the amount of nickel removed in a single extraction with 0.1M
HCl). If we assume that removal of up to 30% of this limiting value would be
acceptable economically, a simple calculation shows how many crops could be
sustained by the site. Assuming that the soil is being phytomined to a depth of
0.15 m, the volume of a hectare of soil to this depth would be 1500 m³. For a
density of 1.3, the mass would be 1950 t. For a crop producing 72 kg nickel/ha,
the concentration of removable nickel would be 72,000/1950 g/t = 40 μg/g for
a single crop. The number of potential croppings before the soil was exhausted
would therefore be 30% of 768 divided by 40: i.e. 230/40 = 5.8 croppings.

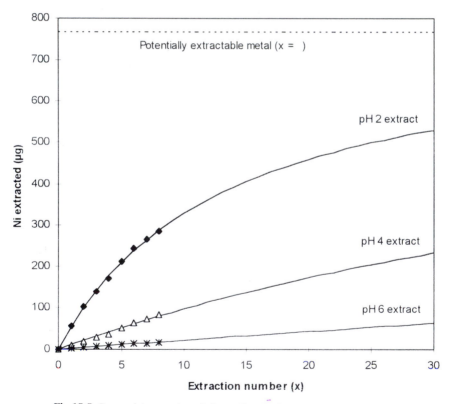

Fig.15.5. Sequential extraction of ultramafic soils from Murlo, Italy using
potassium hydrogen phthalate buffers at pH values of 2,4 and 6.

After each of the sequential extracts, the equilibrium of soluble to total nickel
restores the soluble fraction to almost its original value otherwise it would be
quickly depleted. The procedure obviously cannot reproduce the exact field

conditions but will certainly err on the conservative side since, in the laboratory, we are looking at an equilibrium recovery time of only as long as it takes to replace the extractant with a fresh supply compared with a period of 12 months under field conditions.

The reality is therefore that under field conditions, nearly 6 croppings would keep within the guidelines of 30% removal of available nickel. An undeniable advantage of using a perennial with a life of about 10 years is that resowing would not be needed. After harvesting, the plant would regenerate and would be reinforced by seedlings from the crop itself.

When the pool of available nickel is reduced by 30%, ploughing to bring fresh soil to the surface would be followed by resowing. The latter might even not be necessary if sufficient seed from the current crop were already distributed in the soil.

The Economic Limits of Phytomining

Although there are clearly economic limits in terms of biomass production and metal content in respect of the potential use of any plant for phytomining, the same is not true for the wider subject of phytoremediation. Whereas phytomining is limited by the need to produce a commercially viable metal crop, this is not the case for phytoremediation.

Table 15.3. Metal concentrations (μg/g d.m.) in vegetation required to provide a total $US500/ha return* on hyperaccumulator crops with varying biomass.

Metal	($US/t)	Biomass production (t/ha)					
		1	5	10	15	20	30
Au	13,600,000	36.8	7.4	3.7	2.5	1.8	1.2
Pd	4,464,000	112.0	22.4	11.2	7.5	5.6	3.7
Ag	183,000	2732	546	273	182	137	91
Co	48,000	10,417	2083	1042	694	521	347
Tl	15,000	33,333	6667	3333	2222	1667	1111
Ni	7485	66,800	13,360	6680	4453	3340	2227
Sn	6200	80,650	16,129	8065	5376	4032	2688
Cd	3750	133,333	26,667	13,333	8889	6667	4444
Cu	1961	254,970	50,994	25,497	16,998	12,749	5666
Mn	1700	294,120	58,824	29,412	19,608	14,706	9804
Zn	1007	496,520	99,305	49,652	33,102	24,826	16,551
Pb	817	612,000	122,400	61,200	40,800	30,600	20,400

*Excluding any profit from sale of the energy of biomass incineration.

Table 15.3 shows the elemental content (μg/g in dry matter) that would be required in plants with fertilised biomasses of 1-30 t/ha to give a gross financial return of $500/ha. So far, only cobalt and nickel appear to be suitable candidates for phytomining with plants of a 10 t/ha biomass. Hyperaccumulators with 1% (10,000 μg/g) cobalt are known from Zaïre (Brooks and Malaisse, 1985) and many *Alyssum* species can have nickel contents well in excess of 1% (Brooks *et al.*, 1979). There are no records of high-biomass plants (i.e. around 30 t/ha) whose noble metal contents exceed the values shown in Table 15.3 and neither are there any for cadmium, copper, lead, tin or zinc. It must be emphasized however, that many hyperaccumulators of nickel will also accumulate the much more valuable cobalt. Similarly, cobalt hyperaccumulators often hyperaccumulate copper as well. Though copper is not valuable enough for phytomining in its own right, it can add value to a crop of cobalt. Consider the copper/cobalt hyperaccumulator *Haumaniastrum katangense* from Zaïre. It typically contains 0.2% (dry weight) of both elements. Assuming a fertilised biomass of 7.5 t/ha, a hectare of mineralised soil could provide 0.015 tonne of cobalt worth $720 plus the same weight of copper worth $29. The total of $749/ha would be easily economic.

The above problem can be addressed to some extent by use of hyperaccumulators of higher biomass combined with a sufficiently high metal content. Table 15.4 shows the biomass needed to give a gross return of $500/ha assuming that the plant contains 1% (dry weight) of the target metal. The table shows that values range from 0.0037 t/ha for gold and platinum to 61 t/ha for lead. For annual crops the biomass range is up to about 30 t/ha (maize) with a value of about 5 t/ha for hay. It is not likely that an unfertilised hyperaccumulator annual crop will be found with a biomass exceeding that of maize although there are several large trees that can hyperaccumulate metals. For example, using the data of Jaffré *et al.* (1976), it can be calculated that a mature specimen of the New Caledonian tree *Sebertia acuminata* (see Fig.3.6) contains a total of about 40 kg of nickel. A crop of these trees planted at the rate of 2000/ha would produce 8 tonnes of nickel worth about $60,000 at today's prices. Assuming that this tree would have taken 40 years to mature, the annual yield is then only $7500 after 40 years and much less in a shorter time frame when the tree was initially only a sapling.

The reproductive matter of *Alyssum bertolonii* after fertilising, has a biomass

Table 15.4. Biomass (t/ha) of a hypothetical hyperaccumulator containing 1% (dry weight) of a given metal that would be required to give a crop with a gross metal value of $500/ha.

Metal	Biomass	Metal	Biomass
Gold or platinum	0.0037	Tin	8.06
Palladium	0.011	Cadmium	13.3
Silver	0.27	Copper	25.5
Cobalt	1.04	Manganese	29.3
Thallium	3.33	Zinc	49.6
Nickel	6.68	Lead	61.2

After: Robinson *et al.* (1997)

of about 13.5 t/ha. This value is in the middle to low part of the potential economic range. The South African *Berkheya coddii* (Morrey *et al.*, 1992) has an unfertilised biomass of about 12 t/ha but our field trials in New Zealand have shown that a fertilised biomass of 22 t/ha can be achieved (Robinson *et al.*, 1998). Together with its high nickel content of over 1 %, it is probably one of the best candidates to extend the outermost limits of phytomining for nickel. Experiments with both of these hyperaccumulators are described below.

Examination of the Potential of Phytomining by Pot Trials and Field Tests

Experiments with *Alyssum bertolonii*

Introduction

The work described in this subsection has already been reported by Robinson *et al.* (1997). Therefore, only the salient details will be reported here. The study was based on the principles established by Nicks and Chambers (1995 - see also Chapter 14) and as far as we know, represents only the second report ever published on the potential of phytomining based on field work under natural conditions and using native plant species rather than exotic taxa. The work was carried out in Tuscany, Italy, using populations of *Alyssum bertolonii* growing under natural conditions over the ultramafic (serpentine) soils of the region. The aims of the experiments were to assess: the approximate yield of nickel per hectare; the relation between the nickel content of the plant and the available nickel status of the soil; the effect of plant age and size on the nickel content of the plant; the effect of fertilisers on biomass and nickel content of the plants; the reduction of nickel availability in the soil after successive croppings.

The test areas were located on Monte Pelato (350 m) in Livorno Province and near the village of Murlo (350 m) south of Siena, Italy (Fig.15.6). The rocks are composed of lherzolitic serpentinites emplaced in gabbro and basalt. The soils are often skeletal with a low water-holding capacity (Vergnano Gambi, 1992). The pH ranges from 6.6 to 7.4 on serpentinite and 6.8 to 7.0 on gabbro.

Arrigoni *et al.* (1983) distinguished a specific vegetation community on the screes and debris. It is known as the *Armerio-Alyssetum bertolonii* vegetation type and is spread over all the Tuscan ultramafic outcrops. It encompasses all the serpentine-endemic plants including of course, *Alyssum bertolonii* itself (Chiarucci *et al.*, 1995). This species and its community are absent over the gabbro and basalt.

Climatic data are available for both Monte Pelato and Murlo. Both sites can be classified bioclimatically as Mediterranean pluvio-seasonal oceanic.

For Murlo the mean annual temperature is 13.8°C ranging from 5.8° in January to 22.7° in July. The total annual rainfall is 893 mm ranging from

37 mm in July to 129 mm in November.

For Monte Pelato the mean annual temperature is 12.6°C ranging from 4.3° in January to 21.7° in July/August. The total annual rainfall is 978 mm ranging from 21 mm in July to 137 mm in November.

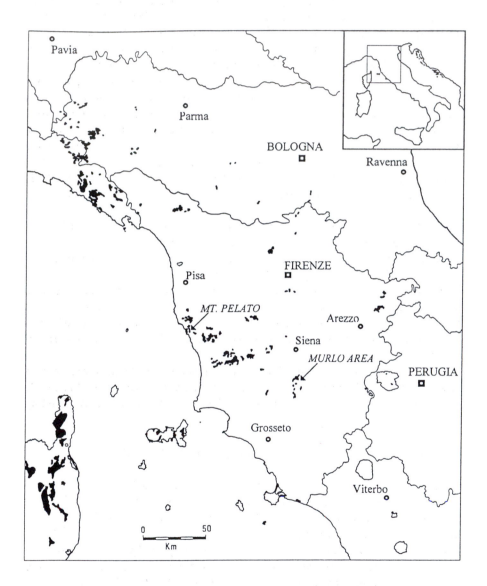

Fig.15.6. Map of north-central Italy showing areas of ultramatic rocks (dark) and sites of the study. Source: Robinson *et al.* (1997).

Brief description of experiments

At Monte Murlo, 35 random quadrats (1 m × 1 m) were selected in Spring 1994 on a gently sloping hillside. In each plot the presence of all vascular plants was recorded and their coverage estimated by the points-quadrat method which estimates the ground cover of a given plant by the relative interceptions of the plant canopy with regularly spaced point observations which, in this case, were made on a square grid with 5 cm spacing of each point. These quadrats encompassed natural populations of *Alyssum bertolonii* that were treated with the following fertiliser regimes in Autumn (October) of the same year: 1 - calcium carbonate at 100 g/m^2, 2 - sodium dihydrogen phosphate at 10 g/m^2, 3 - ammonium nitrate at 10 g/m^2, 4 - sodium dihydrogen phosphate + ammonium nitrate each at 10 g/m^2, 5 - sodium dihydrogen phosphate + ammonium nitrate + potassium chloride each at 10 g/m^2, and 6 - calcium carbonate + sodium dihydrogen phosphate + ammonium nitrate + potassium chloride each at the loadings shown above. There were five replicates of each treatment including five controls. The fertiliser treatment was repeated one year later in October 1995. The increase in biomass of *Alyssum bertolonii* was noted by measuring the increase in cover (the relationship between cover and biomass having been previously established by experiments in which plants were harvested, dried and weighed at the end of each year for a period of two years). The plants were harvested from each quadrat at the end of the two-year period and a soil sample (0-15 cm depth and weighing 100 g) was also taken.

Subsamples (0.5 g) of the vegetative and reproductive parts of the plants were removed and placed in small borosilicate test tubes. These were ashed overnight at 500°C and 10 mL of hot 2M HCl added to each tube to dissolve the ash. The solutions were analysed for trace elements using flame atomic absorption spectrometry.

In statistical tests on the biomass of *Alyssum*, the total biomass was transformed logarithmically and submitted to analysis of variance (ANOVA). Statistically significant differences at the 5% level (P < 0.05) were determined by the LSD test.

Results and discussion of a proposed pilot project for biomining

Earlier in this chapter, it has been shown that fertilisation of the natural populations of *Alyssum bertolonii* was able to increase the biomass of this plant by a factor of three (Table 15.2) without concomitant reduction in the nickel concentration in the fertilised plant.

We have calculated the biomass of a fertilised crop of *Alyssum bertolonii* by multiplying the unfertilised yield (4.5 t/ha) by a factor of two rather than three in order to err on the conservative side. This gives a fertilised biomass of about 9.0 t/ha. It appears that this factor was indeed too conservative because in experimental 1 m^2 serpentine plots here at Massey University, we obtained yields of 3.4 and 13.0 t/ha for unfertilised and fertilised plots respectively. This is in very good agreement with the Italian findings on natural populations of *Alyssum* and suggests that our estimate of 9 t/ha for the biomass is on the conservative side.

From our experience in the field at Murlo, we envisage that a pilot project might involve the following programme:

1 - select a suitable site where the topography would permit harvesting of the crop, ploughing, and fertiliser addition to the serpentine soil;

2 - seed the site directly or plant out seedlings at a rate of approximately 16 plants per m^2 (160,000 per ha);

3 - after a period of 12 months, harvest the reproductive structures with a harvester set to collect all vegetation above 10 cm from the ground;

4 - burn the crop in some type of incinerator and collect the bio-ore which will have a nickel content of about 11%. With application of N+K+P fertiliser, the yield of upper reproductive tissue should be at least 9.0 t/ha. There would be no problem in producing the seedlings as *Alyssum bertolonii* grows very quickly and produces a large quantity of viable seed that germinates in a few days. It must be emphasized that the first crop could not be taken during the first season in order to allow the plants to grow large enough in the second season for sustainable cropping

Nicks and Chambers (1995) have proposed that commercial exploitation of the annual Californian hyperaccumulator *Streptanthus polygaloides* would produce about 100 kg/ha of nickel after moderate application of fertilisers. Our own calculation with *Alyssum bertolonii* have arrived at a conservative value of 72 (more probably 108) kg of nickel/ha containing 0.08% nickel worth $539 at the present world price. The value of the accompanying cobalt from the 80 μg/g of this element in dry plant tissue (0.72 kg/ha) is about $35, making a total of $574 for the entire crop.

However, this Italian plant is a perennial with a life of about 10 years that might be extended with annual removal of the crowns. If only half of this sum represented a net return to the "phytominer", the value of the crop would be $287, a little lower than the net return of a hectare of wheat ($309). This of course presupposes that the costs associated with farming a crop of nickel would be the same as for wheat. It must be remembered however, that native plants growing in their own natural environment should require less fertiliser and irrigation than a crop of wheat. There is also the fact that the *Alyssum bertolonii* crop is perennial and will not require resowing the following year.

The yield of nickel could be increased by removing some of the vegetative tissue of the plant along with the reproductive material. Another approach might be to add a complexing agent to the soil in order to increase the availability of nickel (see above). Such a procedure would not be without risk. In experiments carried out in the United States, a crop of *Brassica juncea* has been grown in lead-contaminated soil and EDTA added to the soil once the plants became well established (Huang and Cunningham, 1996; Blaylock *et al.*, 1997; Huang *et al.*,

1997). At this stage the formerly immobile lead is complexed and taken up by the plant which then starts to die because of the phytotoxic nature of this element. The plants were harvested at this stage to phytoremediate the polluted soil.

Our chelation experiments with *Alyssum bertolonii* are only at an early stage, but there is no evidence of plant death after adding EDTA. It may well be that plants already adapted to high metal uptake (unlike *Brassica*) will tolerate both EDTA and the increase in yield of the target metal to which it is already adapted. We do not believe that EDTA in itself is phytotoxic. It is only the phytotoxic metals that cause plant death. Residual EDTA in the soil might also be a problem. There is nevertheless scope for controlled trials with EDTA or other complexing agents.

From our own experiments with *Alyssum*, a second crop seems to be quite feasible. We obtained 13.0 t/ha of biomass for plants harvested in December 1996 (Southern Hemisphere early summer). Three months later (early autumn) a second crop estimated at 5 t/ha emerged from the stubble.

Another cost-effective strategy that might be adopted would be to recover some of the energy released during incineration of the biomass (17,500 kJ/kg for cellulose material). To quote Nicks and Chambers (1995), if only 25% of this energy were recovered, an additional $219/ha could be recovered making a gross return of $793 t/ha. If half of this sum were recovered by the company after making allowance for capital costs, fertilisers etc, the net return of $396 would be well above the net return of $309 from a crop of wheat obtained by American farmers in the period 1993/1994.

An obvious problem with the use of an incinerator to produce steam for power generation is that the crop harvesting would occur over a fairly short space of time and therefore the power plant should be situated near an urban area where domestic waste might be used as a feedstock to keep the plant going the rest of the year. There is also the possibility of two crops a year as mentioned above. This would not only increase nickel yield but would give more work to a nearby incineration plant.

Although it must be clearly stated that the economics of *A. bertolonii* for phytomining are at the lower range of economic viability, the same is not true if instead this plant is used for phytoremediation of soils polluted with nickel. The costs of conventional methods of remediation such as removal and replacement of polluted soil and storage of the toxic material ($1,000,000/ha according to Salt *et al.*, 1995), are so great, that a "green" method that would also permit recouping some of the costs by sale of an environmentally friendly "bio-ore" will clearly be of economic benefit.

Experiments with *Berkheya coddii*

Introduction

The South African plant *Berkeya coddii* belongs to the Asteraceae and grows to a height of about 2 m. In its natural state it is confined to ultramafic outcrops in the Eastern Transvaal, near Barberton. The nickel content of *B. coddii* has been investigated by various workers (Morrey *et al.*, 1989,1992; Howes 1991) who have shown that it is a hyperaccumulator of nickel containing up to 1.7% of this element in the dry mass.

Berkheya coddii begins to grow at the onset of the rainy season in late winter and finishes its cycle at the end of March. It is a perennial that dies back after

Fig.15.7. Stand of *Berkheya coddii* growing in late summer (March) in an artificial serpentine soil (550 μg/g nickel) in experimental plots in New Zealand.

flowering and re-emerges at the start of the next rainy season.

Experiments were carried out at this university with both pot trials and growth in outside plots (Robinson *et al.*, 1998). The aims of the experiments were to determine the following:

1 - the biomass achievable in outside plots using an artificial "serpentine soil";

2 - effect of fertiliser addition on nickel uptake;

3 - the effect of excision on the nickel content of new growth;

4 - the limiting nickel content that could be achieved under controlled conditions;

5 - calculation of the probable nickel yield of plants grown over serpentine soils in various parts of the world;

6 - the potential of the species for phytoremediation as well as phytomining.

The achievable biomass of Berkheya coddii *under controlled outside conditions*

After one year's growth from seed, the total above-ground dry biomass of plants in two 1 m × 1 m plots were 2.08 and 2.20 kg respectively. This translates to a mean biomass production of 21.4 t/ha/annum. Plants were on average 180 cm tall and are shown in Fig.15.7. This biomass was achieved with moderate addition of Osmocote slow-release fertilisers. However field observations indicate that the plant attains this height in its natural habitat. Poorer soils may however need fertiliser addition for optimal production. The value of 21.4 t/ha is among the highest reported for any natural hyperaccumulator species, and over twice as high as the biomass production of *Alyssum bertolonii* (9 t/ha/yr) that has been shown by Robinson *et al.* (1997) to have the potential of being able to provide an economic crop of nickel.

The effect of fertiliser on nickel uptake by Berkheya coddii

Chemical analysis of leaves collected in from plants subjected to various fertiliser amendments (N and P) showed no statistically significant relationship between the nickel content and phosphorus amendment when the nitrogen content was constant. If however the experimental samples were considered as three separate populations with treatments of 0, 100 and 200 μg/g N added to a 3:1 bark:crushed serpentine mixture, the mean nickel content rose from 2300 μg/g d.w. for zero addition of N, to 3250 for the 100 μg/g N treatment and to 4200 μg/g for the 200 μg/g N amendments.

The effect of leaf excision on the nickel content of fresh Berkheya *shoots*

Plants that were excised at ground level rapidly grew new foliage. This new growth had a much higher nickel concentration than the original plant (Fig.15.8). The difference is on average over three times greater in the optimum range of 600-1000 μg/g in the soil. The same behaviour has been noted by Varennes *et al.*

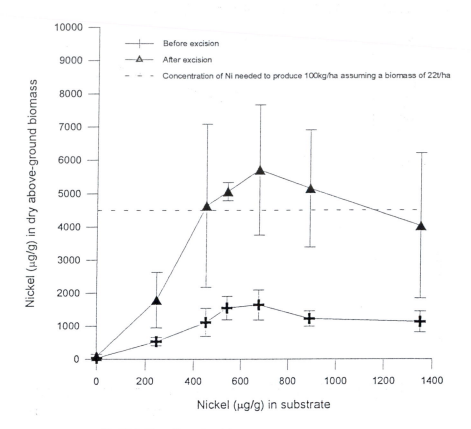

Fig.15.8. The effect of excision on the enhancement of the
nickel content of fresh shoots of *Berkeya coddii*.

(1996) for the nickel hyperaccumulator *Alyssum pintodasilvae*.

Enhanced nickel uptake after excision could be due to two factors. The plant may be removing more nickel from the soil, or it may simply be translocating existing nickel in the plant to the new growth structures. The higher nickel in the new growth would be advantageous to the plant if it inhibited its predation by folivores (Boyd and Martens, 1992). Were the plant to be extracting more nickel from the soil, it may be possible to induce increased nickel uptake by removal of the apical meristem. The new growth could be harvested as another high nickel crop a few months after the original cropping, or the plant may be cropped once a year, removing the need to resow the plants.

Nickel in plants as a function of the total available nickel in the substrate

The results of experiments in which three-month-old whole plants of *B. coddii* were grown in standard seed mix containing incremental concentrations of available nickel (0-10,000 $\mu g/g$) are shown in Fig.15.9.

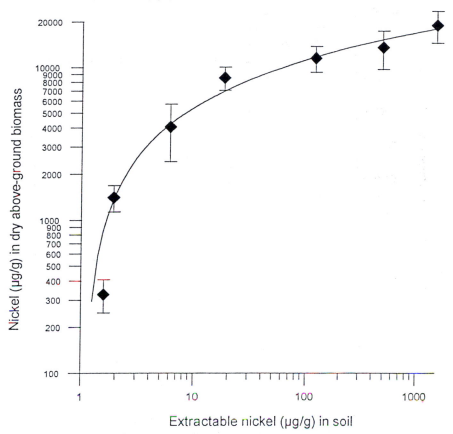

Fig.15.9. Results of pot trials showing the relationship between nickel in *Berkheya coddii* and the available nickel in the substrate.

The plants would not grow in substrates containing more than 3333 $\mu g/g$ available nickel and the highest level of just over 10,000 $\mu g/g$ nickel in the plants grown in pot trials is probably a limiting value. Although under natural conditions the plant can have up to 17,000 $\mu g/g$ (1.7%) of this element in its dry leaves, the whole plant has a nickel content around 0.8% and this is of the same order as our experimental value of 1% in pot trials.

Prediction of the probable nickel yield of Berkheya coddii *grown over serpentine soils throughout the world*

From Fig.15.9 and analogous experiments in which the ammonium acetate extractable nickel was determined in 140 serpentine soils from 11 locations worldwide, it was observed that there was a linear relationship between the nickel content of *Berkheya coddii* and the extractable fraction of this element in the soil. The nickel content of *B. coddii* growing on a nickeliferous soil can therefore be predicted by the extractable nickel as determined by use of ammonium acetate solutions. Obviously there are other factors involved in nickel uptake such as the pH of the soil, nutrient availability, and the concentration of other heavy metals, however the predictions should give a rough guide to a soil's suitability for phytomining or phytoremediation. Predictions of crop yields for nickel are shown in Table 15.5.

Table 15.5. Predicted nickel yields for crops of *Berkheya coddii* grown on nickel-rich soils throughout the world.

Country/State	Location	N	A	B	C
New Caledonia	Plaine des Lacs	1	30.8	2.30	460
California	Red Hills (Chinese Camp)	2	26.3	1.93	386
New Zealand	Coppermine Saddle	6	19.3	1.40	280
Italy	Monte Pelato	40	14.4	1.00	200
New Zealand	Dun Mountain	5	11.7	0.82	164
South Africa	Barberton	2	10.5	0.73	146
Italy	Monte Murlo	76	7.46	0.49	98
Argentina	Vitali Quarry, Cordoba	1	3.40	0.18	39
Morocco	Taafat	1	2.91	0.14	31
New Zealand	Cobb asbestos mine	5	2.46	0.11	24
Portugal	Bragança	1	1.63	0.05	10

A - extractable soil nickel (μg/g), B - estimated nickel content of plant (%), C - estimated nickel yield (kg/ha).
Assumptions: 1 - biomass of 22 t/ha, 2 - nickel yield of the plant, and hence its extractive power, is a function of the extractable nickel content of the soil as determined from Fig. 15.9 and other experiments.

Phytomining by use of Berkheya coddii

It is possible that economic crops of nickel could be phytomined from sites (Table 15.5) with >98 kg Ni/ha projected yields from *B. coddii* provided of course that other factors were favourable, not the least of which would be a sufficiently large area for economic metal farming.

The most obvious regions on earth where metal farming with *Berkheya* might be possible are the ultramafics of California/Oregon, Central Brazil in Goiás State, New Caledonia, Anatolia, and Western Australia. The next step in the development process would be the establishment of a pilot scheme in one of these or other suitable territories.

It is important to mention, however, other considerations that would need to be addressed before *B. coddii* were to be introduced to an area intended for phytomining (or phytoremediation). All the plants in our experiments, as well as plants growing naturally in South Africa were not under water stress. It has yet to be determined how *B.coddii* would tolerate xeric conditions. The plants in our outside plots in New Zealand withstood ground frosts of up to -5°C, though growth will undoubtedly suffer in very cold climates.

The question arises as to whether *B.coddii* could potentially become a weed. The rapid growth rate, and the production of large quantities of wind-borne seeds could in theory make the species invasive of surrounding areas, thus outcompeting native vegetation. Even though this plant is entirely confined to ultramafic environments in South Africa where the limiting factor may indeed be lack of competition from non-serpentine plants, it should not be assumed that there is no risk of its becoming a weed in other environments and this question should therefore be addressed in future field trials in other countries.

Phytoremediation of nickel-contaminated soils by use of Berkheya coddii

The combination of high biomass and high nickel content of *Berkheya coddii*, together with its ease of propagation and culture as well as its tolerance of cool climatic conditions, should render this species a suitable agent for phytoremediation. Sites highly polluted with nickel are less numerous than those contaminated with lead and zinc. There is however a need for some degree of remediation of several sites throughout the world where pollution from nickel is a problem. McGrath and Smith (1990) have reviewed the problem of nickel pollution of the environment. Apart from the obvious local pollution from smelters, a significant problem arises from addition of nickel to pastures via sewage sludges. At Beaumont Leys (UK) for example, the nickel content of surface soils at a sewage farm was found to be as high as 385 μg/g.

European Community guidelines for nickel in pastures receiving sewage sludge have been set at a maximum level of 75 μg/g where background levels are around 25 μg/g for UK (McGrath and Smith, 1990). Assuming a biomass of 22 t/ha for *Berkheya coddii* and a soil depth of 15 cm and density of 1.3, it is possible to calculate the amount of nickel that could potentially be removed annually from contaminated pastures using a crop of this species. Using the data in Fig.15.9, an estimate can be made for the probable nickel content of a *Berkheya* crop growing over polluted soils. It must be remembered, however, that the experiments portrayed in Fig.15.9 were carried out with substrates containing nickel as the totally soluble nitrate, though a high proportion of this element would have been absorbed by complexing with the organic matter of the substrate. In applying these data to a hypothetical situation involving a contaminated soil in which the availability of the nickel might not be known in advance, we have adopted a conservative approach that assumes that only half of the metal burden of the soil would be available to the plants.

The number of annual crops of *Berkheya coddii* that would be required to reduce the nickel burden of soils down to the EU level of 75 μg/g is summarised in Table 15.6. For moderate nickel contamination (100 μg/g) two crops would be sufficient to reduce the metal content to well below the 75 μg/g of the EU guidelines. Even at 250 μg/g (few polluted sites would exceed this value) only 4 crops of *Berkheya coddii* would be needed.

Table 15.6. Number of annual crops of *Berkheya coddii* required to reduce nickel contamination in soils to the EU guideline of 75 μg/g.

Initial Ni in soil (μg/g)	Ni content after one year (μg/g)	Number of crops to decontaminate
10,000	9918	138
5000	4925	74
2000	1932	34
1500	1435	26
1000	939	18
750	691	14
500	445	10
250	200	4
100	59	2

Assumptions: 1 - biomass of 22 t/ha, 2 - only half the nickel is extractable, 3 - nickel content of the plant, and hence its extractive power, is a function of the nickel content of the soil.

Current EU guidelines (CEC, 1986) permit an annual addition of only 3 kg/ha nickel when sewage sludge is used as fertiliser for pastures and cropping. One crop of *Berkheya* would remove the equivalent of 24 years of annual fertiliser additions assuming that only half of the nickel is extractable.

Some Philosophical Observations on the Feasibility or Desirability of Phytomining

In contemplating the possibility of phytomining in the future, a number of questions need to addressed:

1. In the case of phytomining for nickel, would conservationists allow a large area of serpentine soil to be colonised by commercial crops of an exotic or native hyperaccumulator?

To answer this question, it must be appreciated that there is a difference in the acceptability of exotics compared with local hyperaccumulators. To take the example of Italy, it would be technically feasible to carry out phytomining by use of the serpentine-endemic *Alyssum bertolonii*. The species is endemic to serpentine soils and there would be no question of introducing an exotic species. In the worst case scenario the *Alyssum* would merely be replacing other serpentine-tolerant plants.

Nevertheless, it would be unreasonable to expect Italian conservationists to allow unrestricted use of serpentine environments for phytomining unless such use were confined to degraded land such as in the vicinity of former mines. The most likely sites where phytomining could be allowed, either in Italy or elsewhere, would be as a "green" alternative to opencast mining. If *Alyssum bertolonii* were to be used for phytoremediation of soils polluted by nickel as a result of industrial activity, a very different situation would arise, a situation where the blessing of conservationists might be expected.

2. Would phytomining involve using land that might have been used for agriculture?

The answer here is very clear. By their very nature, mineralised soils are extremely hostile for unadapted plant life and are almost never used for food production.

3. What will happen to the price of metals in the future?

This question is hard to answer. Some metals such as nickel have been constantly rising in price for the past few years, but there is no guarantee that their price will not one day collapse. Nickel is a metal whose price is relatively stable unlike metals such as tin. In any case, phytomining will be just another mining technique no more susceptible than others to fluctuating world prices.

4. Where are the most likely sites where phytomining might one day be used?

The technique would be most applicable in areas where there are large areas of subeconomic metal reserves. Unless some way can be found to discover plants that hyperaccumulate the noble metals, we are essentially looking at nickel and cobalt as potential targets for phytomining. Only in central Africa (Zaïre and Zambia) as well as possibly Canada and Russia are there significant areas of low-grade cobalt ores. Nickel is an entirely different matter because of the widespread occurrence of ultramafic soils throughout the world. Table 15.7 lists some areas of the world where major subeconomic deposits of nickel or cobalt are to be found and suggests potential hyperaccumulators that could be used for phytomining. Forest species are not included in the listing as they would not lend themselves to cultivation in unshaded areas.

Table 15.7. Potential regions suitable for phytomining by use of local hyperaccumulator plants.

Element	Region	Suitable species for phytomining
Cobalt/copper	Zaïre/Zambia	*Haumaniastrum katangense*, *H.robertii*
Nickel	Australia	*Hybanthus floribundus*, *Stackhousia tryonii*
	Brazil	Numerous local endemic plants (Brooks *et al.* (1992).
	California/Oregon	*Streptanthus polygaloides*
	Cuba	Numerous plants of the *Phyllanthus* and *Leucocroton* genera
	New Caledonia	*Geissois pruinosa* and several others
	South Africa/Zimbabwe	*Berkheya coddii*
	Turkey/Italy/Greece	*Alyssum bertolonii*, *A.lesbiacum*

5. How hardy are hyperaccumulators and how resistant to insect attack or disease, i.e. could a crop be wiped out in a single year by such agencies?

Hyperaccumulators of nickel and other elements have a strong protection against predator attack because of their high nickel content. Boyd and Martens (1992) have carried out experiments on nickel hyperaccumulators to illustrate this protection. In the course of our own extensive investigations on *Alyssum bertolonii* and *Berkheya coddii* we have not observed any tendency to disease in either plant. *A. bertolonii* is an exceptionally hardy plant that will withstand extremes of temperature. Our tests in New Zealand have shown that it will even grow over asbestos tailings where no other dicotyledonous plant will survive. Although *Berkeya coddii* grows in a warm part of South Africa, it will readily overwinter in New Zealand where ground frosts occur in winter.

6. How environmentally acceptable would phytomining be?

By its very nature phytomining will be seen as "green" by many critics of the mining industry. There will be others however who will decry the suggestion of bulldozing away other native plants to make way for large crops of a phytomining plant. This question is a moral one and hopefully some solution will be found to satisfy all interested parties.

7. What is the potential of obtaining dual crops of two different metals?

There is clearly a possibility of obtaining crops of two different elements whenever a given species of hyperaccumulator has the ability to accumulate two elements. The most obvious possibilities (see also above) are nickel and cobalt from "nickel plants" growing over ultramafic soils that usually contain around 100 μg/g cobalt. Initial experiments with EDTA chelation indicate that there might be a dramatic increase of the cobalt content of some plants after such treatment even if there is no great increase in the nickel content. The other obvious possibility is concomitant uptake of copper when growing a crop of cobalt using some of the Zaïrean dual hyperaccumulators such as *Haumaniastrum katangense*. The copper is far less valuable than the cobalt, but might just tip the scales to make a subeconomic crop viable.

A Model of a Possible Economic Phytomining System

A model of a possible economic phytomining system is shown in Fig.15.10. The system differentiates between annual and perennial crops and encompasses the questions of fertilising and soil exhaustion. Probably the success or failure of a project will depend on whether or not some of the energy of combustion of raw material can be recovered. In tropical regions of the earth, it should be possible to have crops maturing in each month of the year so that the incineration plant could be kept busy yearlong.

If phytomining proceeds beyond the theoretical and pilot plant stages there are two possible scenarios that might be envisaged. The first of these presupposes a commercial project on a very large scale involving a few square kilometres of ultramafic soils or low-grade nickel mineralization.

The second scenario, perhaps the more likely, could involve phytomining being farmed out to smallholders throughout the region in which a peasant farmer might grow a few hectares of plant material and have it collected for processing at a nearby facility preferably close to a large city where industrial waste could also be used as feedstock for the incineration plant which in turn could supply steam for producing local supplies of electricity. A country such as Brazil that has large areas of sub-economic nickel mineralisation and ultramafic soils might be an obvious site of the small-farmer scenario.

I have myself seen in Brazil (Goias State) several fields of failed soyabean

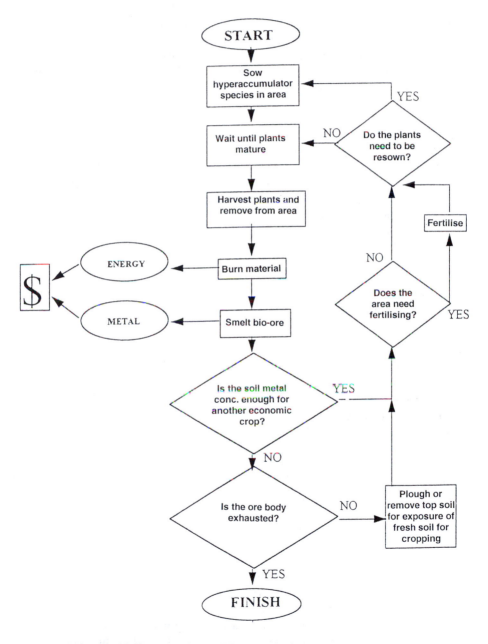

Fig.15.10. Model of a possible economic phytomining system.

crops where peasant farmers had sought to grow these crops over ultramafic soils. The surrounding natural vegetation was in contrast quite luxuriant and included several nickel hyperaccumulators (Brooks *et al.* 1992) that would have grown quite well as an alternative to failed soyabean crops.

The future of phytomining remains unknown at present, but who would have thought 20 years ago that the Brazilians could have grown their own motor fuel (alcohol) from sugar cane. To grow a crop of a metal such as nickel is only an extension of that idea. Only time will tell whether it is merely a flight of fancy or an idea whose time has come.

References

Arrigoni,P.V.,Ricceri,C. and Mazzanti,A.(1983)*La vegetazione serpentincola del Monte Ferrato di Prato in Toscana.* Centro di Scienze Naturale Prato Cat. Quarr., Pistoia, 27 p.

Baker,A.J.M. and Brooks,R.R.(1989) Terrestrial higher plants which hyperaccumulate chemical elements - a review of their distribution, ecology and phytochemistry. *Biorecovery* 1, 81-126.

Blaylock,M.J.,Salt,D.E.,Dushenkov,S.,Zakharova,O.,Gussman,C.,Kapulnik,Y., Ensley,B.D. and Raskin,I.(1997) Enhanced accumulation of Pb in Indian mustard by soil-applied chelating agents. *Environmental Science and Technology* 31, 860-865.

Boyd,R.S. and Martens,S.N.(1992) The *raison d'être* for metal hyperaccumulation by plants. In: Baker,A.J.M,Proctor,J. and Reeves,R.D.(eds), *The Vegetation of Ultramafic (Serpentine) Soils.* Intercept, Andover, 279-289.

Brooks,R.R. and Malaisse,F.(1985) *The Heavy-Metal Tolerant Flora of Southcentral Africa.* Balkema, Rotterdam, 199 pp.

Brooks,R.R.,Morrison,R.S.M.,Reeves,R.D.,Dudley,T.R. and Akman,Y.(1979) Hyperaccumulation of nickel by *Alyssum* Linnaeus (Cruciferae). *Proceedings of the Royal Society (London) Section B* 203, 387-403.

Brooks,R.R.,Reeves,R.D. and Baker,A.J.M.(1992) The serpentine vegetation of Goiás State, Brazil. In: Baker,A.J.M.,Proctor,J. and Reeves,R.D.(eds), *The Vegetation of Ultramafic (Serpentine) Soils.* Intercept, Andover, pp.67-81.

CEC (Commission of the European Community) Council Directive of 12 June 1986 on protection of the environment, and in particular of the soil, when sewage sludge is used in agriculture. *Official Journal of the European Communities* L181 (86/278/EEC), 6-12.

Checkai,R.T.,Corey,R.B. and Helmke,P.A.(1987) Effects of ionic and complexed metal concentrations in plant uptake of cadmium and micronutrient metals. *Plant and Soil* 99, 335-345.

Chiarucci,A.,Foggi,B. and Selvi,F.(1995) Garigue plant communities of ultramafic outcrops of Tuscany, Italy. *Webbia* 49, 179-192.

Cunningham,S.D. and Ow,D.W.(1996) Promises and prospects of phytoremediation. *Plant Physiology* 110, 715-719.

Howes,A.(1991) *Investigations into Nickel Hyperaccumulation in the Plant Berkheya coddii*. MSc Thesis, University of Natal, Pietermaritzburg, South Africa.

Huang,J.W. and Cunningham,S.D.(1996) Lead phytoextraction: species variation in lead uptake and translocation. *New Phytologist* 134, 75-84.

Huang,J.W., Chen,J.,Berti,W.R. and Cunningham,S.D.(1997)Phytoremediation of lead-contaminated soils: role of synthetic chelates in lead phytoextraction. *Environmental Science and Technology* 31, 80-805.

Jaffré,T.,Brooks,R.R.,Lee,J. and Reeves,R.D.(1976) *Sebertia acuminata*: a nickel-accumulating plant from New Caledonia. *Science* 193, 579-580.

Kumar,N.P.B.A.,Dushenkov,V. Motto,H. and Raskin,I.(1995) Phytoextraction: the use of plants to remove heavy metals from soils. *Environmental Science and Technology* 29, 1232-1238.

McGrath,S.P.,Sidoli,C.M.D.,Baker,A.J.M. and Reeves,R.D.(1993) The potential for the use of metal-accumulating plants for the *in situ* decontamination of metal-polluted soils. In: Eijsackers,H.J.P. and Hamers,T.(eds), *Integrated Soil and Sediment Research: a Basis for Proper Protection*. Kluwer Academic Publishers, Dordrecht, pp.673-676.

McGrath,S.P. and Smith,S.(1990) Chromium and nickel. In: Alloway,B.J.(ed.). *Heavy Metals in Soils*. Blackie, Glasgow, 125-150.

Marten,G.C. and Hammond,P.B.(1966) Lead uptake by bromegrass from contaminated soils. *Agronomy Journal* 58, 553-555.

Minguzzi,C. and Vergnano,O.(1948) Il contenuto di nichel nelle ceneri di *Alyssum bertolonii*. *Atti della Società Toscana di Scienze Naturale* 55: 49-74.

Morrey,D.R.,Balkwill,K. and Balkwill,M.-J.(1989) Studies on serpentine flora: preliminary analyses of soils and vegetation associated with serpentinite rock formations in the south-eastern Transvaal. *South African Journal of Botany* 55, 171-177.

Morrey,D.R.,Balkwill,K.,Balkwill,M.-J. and Williamson,S.(1992) A review of some studies of the serpentine flora of South Africa. In: Baker,A.J.M.,Proctor,J. and Reeves,R.D.(eds), *The Vegetation of Ultramafic (Serpentine) Soils*. Intercept, Andover, pp.147-157.

Nicks,L. and Chambers,M.F.(1995) Farming for metals. *Mining Environmental Management*. Mgt. September 15-18.

Norvell,W.A.(1972) Equilibria of metal chelates in soil solution. In: Mortvedt,J.J.,Cox,F.R.,Shuman,L.M. and Welch,R.M.(eds), *Micronutrients in Agriculture*. American Society of Agronomy, Madison, pp.115-138.

Ortiz,D.F.,Ruscitti,T.,McCue,K. and Ow,D.W.(1995) Transport of metal-binding peptides by HMT1, a fission yeast ABC-type vacuolar membrane protein. *Journal of Biological Chemistry* 270, 4721-4728.

Ow,D.W.(1993) Phytochelatin-mediated cadmium tolerance in *Schizosaccharomyces pombe*. *In Vitro Cell Development Biology* 29P, 213-219.

Robinson,B.H.,Brooks,R.R.,Howes,A.W.,Kirkman,J.H. and Gregg,P.E.H. (1998) The potential of the high-biomass nickel hyperaccumulator *Berkheya coddii* for phytoremediation and phytomining. *Journal of Geochemical Exploration* (in press).

Robinson,B.H.,Brooks,R.R.,Kirkman,J.H.,Gregg,P.E.H.   and   Gremigni,P. (1996) Plant-available elements in soils and their influence on the vegetation over ultramafic (serpentine) rocks in New Zealand. *Journal of the Royal Society of New Zealand* 26, 455-466.

Robinson,B.H.,Chiarucci,A.,Brooks,R.R.,Petit,D.,Kirkman,J.H.,Gregg,P.E.H., and De Dominicis,V.(1997) The nickel hyperaccumulator plant *Alyssum bertolonii* as a potential agent for phytoremediation and phytomining of nickel. *Journal of Geochemical Exploration 59, 75-86.*

Sadiq,M.   and   Hussain,G.(1993) Effect of chelate fertilizers on metal concentrations and growth of corn in a pot experiment. *Journal of Plant Nutrition* 16, 699-711.

Salt,D.E.,Blaylock,M.,Kumar, N.P.B.A.,Dushenkov,V.,Ensley,B., Chet,I. and Raskin,I.(1995) Phytoremediation: a novel strategy for the removal of toxic metals from the environment using plants. *Bio/Technology*, 13, 468-474.

Varennes,A.de,Torres,M.O.,Coutinho,J.F.,Rocha,M.M.F.S.   and   Neto, M.M.P.M.(1996) Effects of heavy metals on growth and mineral composition of a nickel hyperaccumulator. *Journal of Plant Nutrition* 19, 669-678.

Vergnano Gambi,O.(1992) The distribution and ecology of the vegetation of ultramafic soils in Italy. In: Roberts,B.A. and Proctor,J.(eds), *The Ecology of Areas with Serpentinized Rocks; a World View.* Kluwer Academic Publishers, Dordrecht, pp.217-247.

Wallace,A.,Rommey,E.M.,Alexander,G.V.,Soufi,S.M. and Patel,P.M.(1977) Some interactions in plants among cadmium and other heavy metals and chelating agents. *Agronomy Journal* 69, 18-20.

Index